P9-COP-014

HEAVEN
ON
EARTH

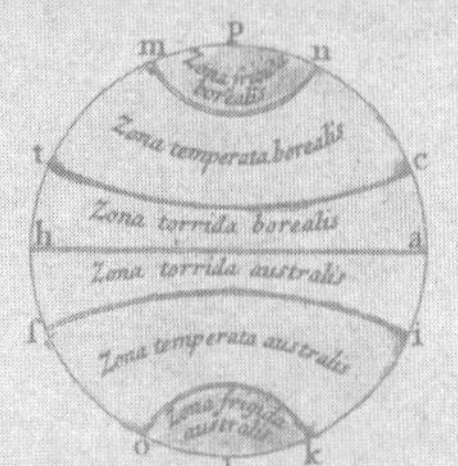

HEAVEN ON EARTH

How Copernicus, Brahe, Kepler, and Galileo Discovered the Modern World

L. S. FAUBER

PEGASUS BOOKS
NEW YORK LONDON

HEAVEN ON EARTH

Pegasus Books Ltd.
148 W. 37th Street, 13th Floor
New York, NY 10018

First Pegasus Books cloth edition December 2019

Interior design by Maria Fernandez

Library of Congress Cataloging-in-Publication Data is available.

ISBN: 978-1-64313-204-4

10 9 8 7 6 5 4 3 2 1

Printed in the United States of America
Distributed by W. W. Norton & Company

Accurate scholarship can
Unearth the whole offense
From Luther until now
That has driven a culture mad.

—W. H. AUDEN,
Another Time (1939)

"No mathematician is believed to be worth anything unless they have first been in trouble with the police."

—JUVENAL,
Satire VI (ca. 100)

Contents

Introducing the Stars

In Europe, all throughout the sixteenth century, that dreadful era of civil strife and bloody rebellion, there lived four men who loved to watch the sky. Though they differed in nation, age, religion, and class, they were all united by a single discovery, a wondrous discovery, a truly unbelievable discovery that became the herald of every other social change within their violent, paranoid era. This discovery pushed the spirits of all four men together with such force that, even when separated by the chasms of time and space, they became, to one another, "fathers," "brothers," and "sons." When they made contact, in books, letters, or person, it was with all the possible intimacy of long-lost relatives. They offered one another the same much-needed community they learned from their siblings, parents, children, and wives. Like most families, there was much love between them, but also much error. Like all families, their drama was too immense for a single life to contain.

In the following pages, I tell the story of this drama, from its genesis as a passing daydream in the head of a little Polish boy all the way up to the grand world stage, in the trial of the century, awaiting judgment before the greatest political power of its day. The characters of this drama are now so famous they may each be introduced with only one name: Copernicus,

Tycho, Kepler, and Galileo discovered the modern Earth, a moving Earth, and in so doing, discovered the uneasy conditions of modern life itself. Theirs is not only a story of individual genius. To tell such a story would not be a lie, but a withholding of other, more essential truths. Theirs is an intergenerational epic, a family saga, of the most unusual sort.

HEAVEN ON EARTH

NICOLAUS COPERNICUS

A life, in which a boy stumbles into manhood

DRAMATIS PERSONAE

Lucas Watzenrode	his uncle & Prince-Bishop
George Joachim Rheticus	his acolyte & friend
Tiedemann Giese	his dear friend & fellow canon
Andreas Osiander	Lutheran minster & dilettante stargazer
Domenico Maria da Novara	his favorite teacher
Erasmus Reinhold	his second acolyte
Martin Luther	reformer of legend
The Teutonic Knights	a war-making crew
The Popes	whose passings keep the time as the second hand of a clock

MAJOR WORKS

Little Commentary • *First Account* • *On The Revolutions*

YEARS ENCOMPASSED

1491–1551

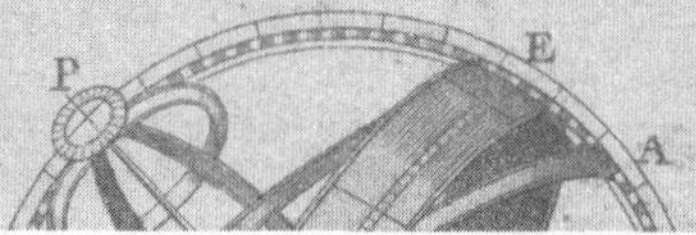

Nicolaus in the Old World

When Nicolaus Kopernigk was a gangling, unattractive teenager, with no lofty ambitions, he did not mean to trouble anybody. He liked math. He was quiet. He did not like to cut his hair.

In the fall of 1491, he was packing his bags, preparing for a sobering departure from his childhood home in Torun. Within his mind lay the seed of a troubling idea concerning his favorite subject, astronomy, and the possible movement of the Earth. Such a seed had not yet had any time to grow, but he was, at that moment, readying himself for college, one of the rare places in medieval Europe where an intellect could flourish. He did not look impressive. He looked rather silly, if his portraits are any guide, with a tiny mouth inherited from his mother's side of the family and big white rings around his eyes, which made the whole rest of his face seem dirty. So unremarkable was he that no one made any note of his leaving for a two days' ride south through the Polish countryside to begin his adult years at the University of Krakow.

Time makes strangers of us more than distance ever could. The world Nicolaus journeyed through was an old one: small, wild, and weird. There were no narrow buildings. There were no giant cities. There were no factories coughing smoke into the air. There was no place named "America," no light bulbs, no vaccines, no nationalism, no cheap steel, no secular state, no accurate clocks, no feminism, barely any guns, virtually no coffee, absolutely no democracy, and almost no books. A healthy town would, at least, have a new mechanized corn mill to serve its citizens, outside which urchins, lepers, and women of ill repute would loiter in the shade; here was the first hint of modernity rising.

Traveling near the fertile Vistula River, the boy saw enough sprung-up shanty villages to form a sorry picture of his old world. Most people were peasants, and most peasants were dirt poor. They had almost nothing, he observed: a single cow, a lonely pig, one "she-goat," a sack of grain, all providing a meager diet of homemade cheese and black bread. From this pittance, food-rent, worth one day of work a week, went straight into the thankless bellies of the noble elite. Sometimes a farming man, usually one inspirited by a lusty new bride, got it into his head to run away in quest of a better life; such couples always returned dejected, within six months, punished by reassignment onto untilled, hardscrabble land. They had nothing better.

The poverty of their condition brought many to the point of rebellion, especially in countries more distant from Rome, seat of the Catholic Church, the organizing principle of medieval Europe. Excepting a handful of Jews demanded by authorities for their necessary sin of banking, every European was Catholic by default; every poor little town had built a poor little church of stone, standing room only, with no stained glass in the windows. There the common folk crowded in. For them, this church was not only a religious institution but the means by which they understood themselves as social beings. God was in the audience of their every public ceremony, from the sacrament of marriage to the coronation of kings to the baptism of children to the divine appointment of clergymen, who comforted them every Sunday with words from the wonderful Bible. The Church soothed the peasant spirit and led many through an honest, peaceful, happy life of religious devotion.

In the poor country, a parish was lucky to get a priest who could read and write, but ideally, a clergyman was the locus of wisdom in the Catholic community, entrusted with knowledge, which confused and complicated the mores of simple faith. Nicolaus was already skilled in Latin, and those close to him were expecting his college venture to result in a successful career with the Church. At the age of eighteen, he was already full of knowledge denied to common people.

Pythagoras, he knew, was the first ancient to propose that mathematics was the key to understanding nature. This idea had so excited the Greek mystic that he had even started a pagan cult about it. This cult told all its members to never eat beans, and supposedly drowned a man for proving that the square root of two was not a fraction, so people naturally started to assume that they were a bit crazy. But their doctrine that math could be used to understand nature turned out to be pretty sane. After them came Plato and Plato's student Aristotle, whose surviving works span every genre and were read by every serious student of Latin and Greek.

Aristotle's philosophies had become so hopelessly interwoven with medieval biblical interpretation that only the sort of extensive study Nicolaus made could distinguish the two. Aristotle had divided the world into physics, the study of change, and metaphysics, the study of the unchanging, just as the Church served as the holy intermediary between the material world and their transcendent deity. Aristotle had even referred to metaphysics as "theology." Physics concerned life here on Earth, he said, which was dirty and smelly and rotten, while metaphysics involved things above the Moon, which were perfect. While these things above the Moon may move, their movements were not subject to change; they moved in ways predetermined by a "first mover." The Greeks called this first mover Logos, Reason, the Divine Word, but all good Catholics called it God.

Aristotle also argued that the Sun orbited around the Earth, but he did not spend long on it, for this fact was plain to the merest child. Just look at it. The Sun is obviously moving.

The Sun moved in its great arch above Nicolaus, as he continued his journey southward to Krakow. The light from auburn maple smeared out in reflection on the placid blue river Vistula. Farmers paced along their furrows, planting hops and barley for the fall. They, more than anyone, knew how the Sun apparently moved. Their working lives were limited by its daily motion, the giant circle they believed traversed around the Earth every twenty-four hours, causing day and night. They also believed that the Sun had an annual motion, around a second circle, making daylight in some days of the year shorter than in others. This annual orbit caused the seasons, their frigid snows, lively greens, humid nights, and ruby leaves. Astronomers, who often loved fancy names, called the path of this annual solar orbit the *ecliptic*, because it was only on this path that eclipses could occur. The ecliptic was also named the *zodiac*, or "circular zoo," because it traveled in a circle through twelve animal constellations. A farmer who was not alert to the motion of the Sun and its changing seasons risked a weak crop, and therefore a dead family.

All the signs of the city presented themselves to Nicolaus as he made his approach to Krakow; the churches grew fatter, the people grew richer, the roads became worn and the air ran thick with the noise of business. After a passage through the blooming suburbs, the red-capped fortifications of the north gate roared up before him, opposing his old home in Torun, swiping at the hallowed sky. The city of old Krakow was less than a mile across, but the idea of it, the idea of the city, was the limitless future. Nicolaus entered in and went to college, in search of a new way to live.

In 1491, the University of Krakow ranked among the greatest universities in the whole of Europe; that is to say, it was nothing special, a few cramped buildings stuffed into the northwest corner of the city which taught a couple hundred students at most. The medieval college was born of the monastery, giving first place to the study of God and the preservation of culture. In second place was language and rhetoric, a prerequisite to biblical scholarship and the translation of ancient texts. The bottom rung in this ladder of respectability was science, then called "the philosophy of nature," and mathematics, which had little cultural value. Regardless of this hierarchy, every discipline was respected, mixed together under the liberal arts, so called because they were taught only to free men. The aesthetic wonder of a church hymnal or religious icon was to them also contained within a shapely polygon; the forceful logic of analytic philosophy was also found in the thumping meter of poetic verse. Knowledge of the world had been fiercely divided into distinct categories, but experience of the world had not.

Nicolaus embraced this interdisciplinary education as any early humanist would. "Variety gives greater pleasure than all else," he liked to write. Can there be any doubt, in such an atmosphere, that he attended lectures on astronomy? The professor at Krakow was a scholar typical of his kind, a mouse of a man with a beard as big as his head. He taught an academic astronomy which affirmed the common belief in an unmoving Earth, explaining all the more formidable details that such a worldview implied. And the most formidable detail of them all, he explained, was the bewildering orbit of the planets—or, as romantic Christians would have it, "the divine revolutions."

To observe these divine revolutions, sky-watchers in the fifteenth century had not yet discovered the benefits of glass; the naked eye was all they had. By looking up, they could see, alongside the Sun, only six more wandering objects: planets Mercury, Venus, Mars, Jupiter, Saturn, and, of course, the Moon. But rather than orbiting in a perfect circle around the Earth each year like the Sun, these six planets made all sorts of nasty motions: crisscross, turn around, stop, get bigger, go farther away. These nasty motions are called retrogression, Latin for "a backwards step," and they have plagued astronomers since, without exaggeration, the dawn of history. The ancient Egyptians must have thought the planets drunk.

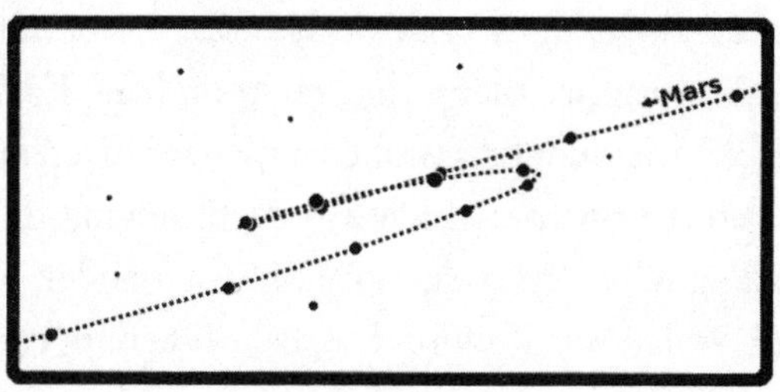

Observed retrogression of Mars in Virgo. Each point is exactly ten days' time apart.

To predict these nasty retrogressions was, Nicolaus learned, nearly the entire purpose of astronomy, since its inception up to his present day. Aristotle had provided the philosophy, but it was for a man named Claudius Ptolemy, born in Alexandria five hundred years later, to provide the predictions. For those who thought like Nicolaus, this was when the real fun began.

Just as the Roman Empire entered its decline, Ptolemy wrote *Almagest*, the culmination of his civilization's knowledge of astronomy. The impact of *Almagest* upon Nicolaus's culture cannot be overstated; the Arabs provided its reverential name, which simply means "The Greatest," and the Europeans agreed. So epic was its argumentation and mathematical skill that it planted in almost every scholar's mind the truth of an unmoving Earth. Despite being a fiction, *Almagest* is brilliant science; it is a book full of beauty, pragmatism, and craft, containing the first plausible method of astronomical forecasting, using nothing but geometry and a clever, creative explanation of planetary motion.

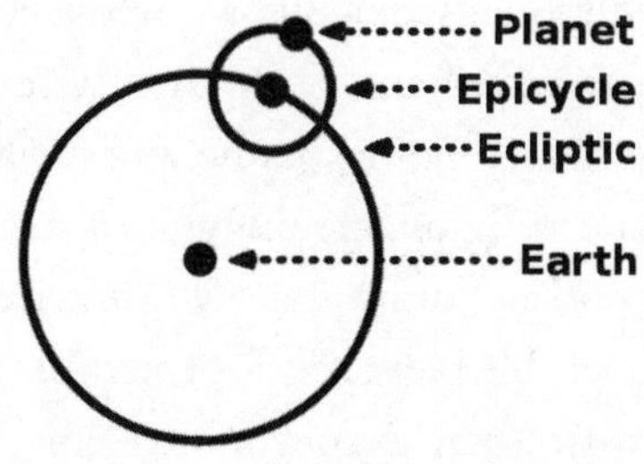

A diagram for the Ptolemaic system of a single planet orbiting the Earth.

Ptolemy explained the ugly, backwards retrogression of a planet with one simple addition. Rather than orbiting around the Earth on a perfect circle each year like the Sun, he assumed a planet moved around on a different circle, called the *epicycle*, which was itself moving in a circle around the Earth. A planet on an epicycle moved like a wooden horse on a little merry-go-round, which was itself on the seat of second, enormous merry-go-round. This combination of circular motions produced a planetary orbit with balletic dips and loops, which often looked like petals of a flower. That no one pointed out that an orchid-shaped orbit was physically absurd is hardly a surprise, for no one thought orbits had a whit to do with physics.

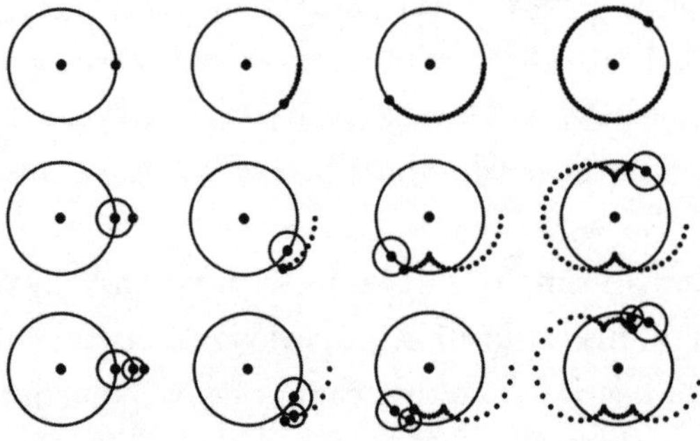

The Ptolemaic model, seen here with zero, one, and two epicycles. The trails of dots represent the orbit of a planet around the Earth.

Nicolaus was delighted by the grace of this Ptolemaic system. It was fluid and flexible; "perfection," he called it, "almost perfection." Generations of astronomers would modify it, adding and removing epicycles, although just one per planet was already enough to offer consistent and respectable predictions.

As Nicolaus learned more astronomy, a queer feeling overtook him. A successful member of the Catholic clergy was wise to spend some time growing acquainted with astronomy, but he was spending more time than did benefit to his career or religion. He was enjoying mathematics too much for its own sake. "All good arts draw us away from vice," he wrote, "but this art also provides unbelievable pleasures for the mind."

Nicolaus loved his astronomy lessons more than any other subject, and he yearned to drink it straight from the source. Fluency in ancient Greek, he was sad to discover, was not acquired overnight. Copies of Ptolemy's *Almagest* in Latin translation were rare, but he decided to drop by a bookstore downtown and see what he could find. There, he purchased the first Latin edition of the classic of Greek geometry, Euclid's *Elements* and a copy of the *Table of Directions* by a man named Regiomontanus, with the baby Jesus impressed upon its cover.

For Nicolaus, reading this work by Regiomontanus was not like reading a work by Euclid or Ptolemy or Aristotle. Regiomontanus was not some stuffy Greek; he was a pleasant German who had died less than twenty years back. His life, alongside that of his teacher Georg Peurbach, forged a fresh European myth, at once both allegorical and attainable, of the modern astronomers who dared strive amongst the ancients.

When young Georg became Professor Peurbach at the University of Vienna in 1453, student interest in astronomy had been so faint that he had to lecture on Latin poetry instead. But he was a hungry reader, devouring Ptolemy and the Arabs, and he managed to seat a classroom soon enough. By the end of August that year, the thirty-year-old gave the first major public lecture on planetary theory in Europe.

In the half-awake crowd sat Regiomontanus, alert, carefully taking notes. He was an unheard-of specimen, the educated country boy, whose father's lucrative milling business had sent him to college. His nickname, meaning "Mountainside," was a send-up to this rural origin, but his birth name was John Muller, and he was Georg's brightest student. Once he graduated and obtained parity with Georg as a teacher in Vienna, the two could hardly be separated. The bachelors lived side by side for the next four years. Their final collaboration was titled *Summary of Almagest*, in an effort to make Ptolemy's difficult classic more accessible to Europeans. This *Summary* had many tricks up its spine, not only condensing Ptolemy but changing him, throwing out wrongful observations and tossing in elaborate mathematical proofs. Timid repetition was spiced by daredevil revision.

At work on the sixth chapter, all this had so obsessed Professor Peurbach that he barely noticed he was dying. Regiomontanus recalled the scene in the preface to their *Summary*, which Nicolaus would soon read. "The memory is doleful and bitter," he wrote.

> Shortly before his life had fled, squeezing my hand, with his head in my lap, he said, "Farewell, my sweet John. Farewell, and if the memory of your teacher might live beside you, the work of Ptolemy, which I leave incomplete, you must finish."

That all these math equations and abstract philosophies had blossomed into a friendship so rich and meaningful would surprise everyone except the mathematicians. Such seemingly austere things for them always carried a great emotional weight.

Regiomontanus fulfilled Georg's dying wish as best he could, completing their *Summary of Almagest* on his own. Then, facing life without his dearest friend, he traveled. He taught and wrote and studied, and though he never did find another friend to match, he left behind a host of students turned teachers. These teachers were German, Hungarian, and Italian; a cross-cultural community of scholars emerged through his effort.

The wanderings of Regiomontanus served as an exemplar to Nicolaus, who dropped out from the University of Krakow to travel across Europe in 1496. This was supposedly for the sake of his career in the Church, as he reenrolled at the University of Bologna, in the Catholic heartland of Italy, to study for a doctorate in Canon Law. But working on his doctorate did not refocus his studies onto religion. On the contrary, it was in Bologna where he made his first friendship with an astronomer.

Domenico Maria da Novara was himself a student of the legendary Regiomontanus, and this attracted Nicolaus to him immediately. One collaborator recalled Nicolaus as "not so much the pupil as the assistant and witness of observations," but this hardly scratched the surface. Nicolaus rented a room in Novara's house. The two had sleepovers, staying up at night in March, staring at the Moon. Decades later, Nicolaus would still source observations taken in Novara's company. He listened to Novara intently, and though nothing remains of their conversations, Novara was known to be an adamant critic of Ptolemy.

By the turn of the century, Nicolaus was reading professional astronomers, living with professional astronomers, and lecturing on professional astronomy, but as a profession, it was not for him. Astronomy was an uncomfortable science. The night sky did not contain, as far as any yet knew, money. Should the stargazing hopeful seek to earn their dinner, they had to convince patrons that the firmament was a matter of literal life and death. This was astrology, and all but the most flamboyant astronomers believed it the way admen believe their commercials: not a total fiction and great at moving product. The production of astrological horoscopes was so profitable that astronomers were often forced into it;

Novara was legally obligated by his university post to create them. Nicolaus must have learned it, his own friends would adore it, yet nowhere in his writings does it figure. Astrology was a foundational part of the world in which he lived, but it would not be foundational to the world he orchestrated.

So began Nicolaus Kopernigk's seven years in Italy. He became more international than national, spoke more Latin than Polish, and even vacationed in Rome to celebrate the jubilee year of 1500. There, he Latinized his name such that he became, in title if not yet in spirit, Copernicus, Coppernicus, or Copernic. One might also name him canon; after his first year in Bologna, he accepted, by proxy, his first official employment in the Catholic Church, as an administrator back in the Polish province of Warmia. This post was a sinecure; it provided a tidy salary, did not require his presence, and demanded no work from him for the next fifteen years.

Instead, Nicolaus put in a request with his Catholic chapter to allow him two further years of education at the University of Padua, which they granted, according to their letter of approval, "because Nicolaus promised to study medicine, and as a helpful doctor, he might one day advise our other members." This promise was not a lie so much as a misdirection. Nicolaus continued lessons in Greek and even took up oil painting. As for astronomy, the university held a notable sect of Aristotelian philosophers, who spoke openly about the perceived failures of the Ptolemaic system.

In Aristotle's philosophy, uniform motion around a circle was said to belong to the perfect heavens, because this was the most perfect sort of motion. Perfection meant simplicity. But Ptolemy, having sullied his philosophy with applied mathematics, knew that planetary movements were anything but simple. To conveniently model the backwards retrogression of the planets, he was obliged to subvert Aristotle, with the final key part of his astronomical system, the equant.

In Ptolemy's astronomy, the equant was an imaginary point within a planet's orbit. A planet ordinarily orbited the Earth with uniform motion, but with an equant, it moved in such a way that motion would *appear* uniform if you were standing at the equant point. The closer the epicycle got to the equant, the slower it actually moved.

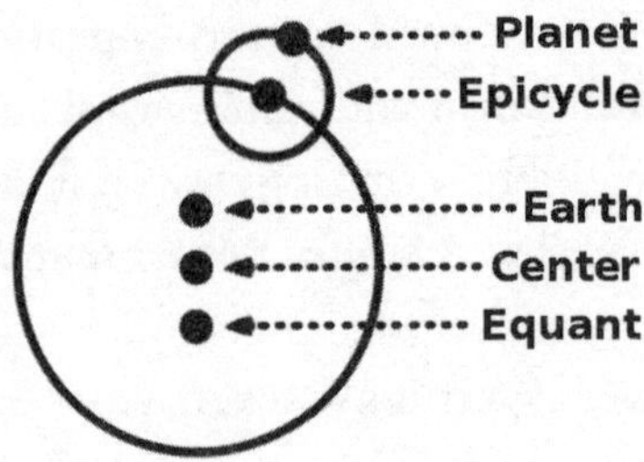

A diagram for the Ptolemaic system of a single planet orbiting the Earth, with its movement modified by an equant.

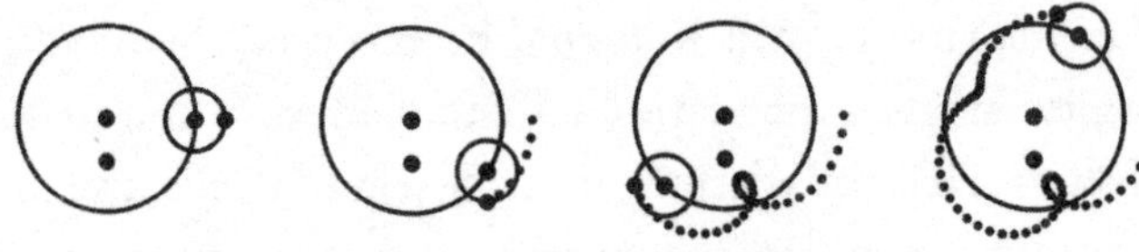

Ptolemaic model of a planetary orbit with one equant and one epicycle. The epicycle slows as it approaches the equant, but its planet continues to rotate with the same speed.

To every true follower of Aristotle, Ptolemy's equant was an outrageous scandal. Aristotle's heavens were aesthetic bliss, meant to contain perfect, uniform, circular motion, but motion with an equant was not perfect. It was nonuniform and ugly and had been catcalled as such for well over a thousand years by other Greeks, Arabs, and now, Europeans. Yet Ptolemy's equant survived all its criticisms and was taught to Nicolaus, because it was so useful.

Young Nicolaus had not been studying astronomy to be useful. He listened silently to the equant-based criticisms of Claudius Ptolemy and imagined different astronomies, for his own pleasure, unsuspecting of where such thoughts would lead.

In 1503, Nicolaus concluded his additional studies in medicine and at last obtained his doctorate in canon law. He had just turned thirty, held good claim to be the most educated man in Europe, and had contributed absolutely nothing to society at large. Such ludicrous financial security and intellectual development had been possible only because of his family.

Family was an unexpected dependence for Nicolaus to have, as he was an orphan. In 1483, when he was a mere ten years old, his parents had vanished: his father from a mysterious illness, his mother's fate unknown. By a strange trick of life, the loss of both parents, which would have doomed most children to illiteracy, had steered him into the land of intellectual opportunity. He had been adopted by an uncle, Lucas Watzenrode, and Uncle Lucas was rich.

A holy man, the Prince-Bishop of Warmia, Lucas's nepotism had afforded Nicolaus his easy career in the Church. For all this guiding affection, though, Lucas was motivated by an old school of thought, which held that no relation was more binding than blood and no subject more appealing than Church doctrine. "Where there is justice, there is God," he believed, "and our justice system forms the very foundation of a friendship." When he finally beckoned Nicolaus to return to Poland with an offer of work as his private secretary, Nicolaus could hardly refuse; his uncle had supported him for twenty years. He set off back to his old home to return to the man who was father to him, to whom he owed everything.

Before he left, he finished an oil painting of his face. Like so much of his life story, that painting would be burnt to a crisp in the fast-approaching fires of the seventeenth century, but if we are to trust a reproduction, it was a fitting testament to his education. Its style is refined, brushstrokes invisible, each colored object clearly delimited by its border, indicating an unhurried pace. Its subject, the painter Nicolaus himself, is smartly observed, yet his many features appear a little too disjointed, too inhibited to form a unified whole. His pursed lips are a scab on the flesh rock. His black hair is a mop on the head. His right eye wanders outside the skeleton. A better painting would have to be entirely different, possessed by the imaginative frenzy of medieval illustration, or the realist concerns of Renaissance formalism, but this quality of in-betweenness is its charm. A young man was crafting an art that was not yet his own. Much of the painting is obscured by a large off-yellow slab, on which is inscribed the words, "a true portrayal of Nicolaus Copernicus, made out of the self-portrait."

The Fall of the House of Watzenrode

Uncle Lucas Watzenrode lived in a monumental castle, and he was a lion. He was a funny-looking lion, to be sure, with a tonsure and no mane. He had stalked authority, lain his traps, and captured it, lashing at any who scavenged the prey. The Teutonic Knights, a rotting leftover of the Crusades, had repeatedly made incursions onto his land. Each night, their junta prayed for his death; their grand master called him "the devil in human shape." One of the burghers Lucas ruled over proclaimed that he was "a learned man, a pious man, skilled in many languages, leading an exemplary life, and yet . . . no one has ever seen him laugh."

He was the protector of the old guard, stern and sullen, with a crucifix round the neck. As an ordained, chaste bishop, he had a bastard, to whom he gave money freely and made sure received almost as fine an education as his adopted child Nicolaus. Family was, for him, if not an object of love, at least a badge of honor. To the millions he excluded from such a label, this was but a standard example of nepotism, corruption in the Church, a staple of medieval life somehow even more common than the Church itself. It drove the common people mad, but Nicolaus, as its foremost beneficiary, kept his thoughts apolitical and his mouth shut.

Now that Nicolaus was officially a doctor, his Uncle Lucas requested him for a personal secretary, physician, and jack-of-all-trades; Lucas even convinced the Church to give his nephew a small boost in paygrade for the service. When the old tough fell miserably ill in 1507, his nephew loyally nursed him back to health. Nicolaus's placebos were obsolete, saccharine stuff. He would take a few teaspoons of some light floral essence, black cardamom, violet, or rose, add a pinch of cinnamon and ginger for flavor, and then calmly fetch a bag of sugar and pile on half a pound. "Mix with distilled water," he wrote with confidence, "Make pills in the shape of a pea." A doctor's genius he had not.

Hidden by this quiet of daily life, there was a great violence coming. People could sense it. Nicolaus sat in on parliamentary proceedings, and witnessed his veteran uncle in peace negotiations with the Teutonic Knights, which were a spectacular failure. He was inducted into war planning and began to harden, if only a bit, in response to this new labor. In time, the unspoken purpose revealed itself. Nicolaus, the only male Watzenrode descendant with a good head on his shoulders, was being groomed for his uncle's legacy, for the protection of their old province and old way of life. Prince-Bishop Lucas had offered Nicolaus everything he could, up to this final gift, his livelihood.

His nephew looked away. Now over thirty, Nicolaus took the first useful independent observation of the skies in his life. It was an eclipse. He was unable to quit astronomy, unable to halt his artistic pursuits.

While his uncle wanted him invested in church politics, he was dawdling with the ancient Greek he had been learning since college. This insensible hobby came to a head with the publication of his translation of an obscure poet, Theophylactus Simocatta, which was distributed by none other than his favorite old bookseller back in Krakow, and dedicated, with a most sincere and retrospective love, to Uncle Lucas.

> To you, Right Reverend Bishop, do I dedicate this modest gift which, however, can by no means be compared to your generosity. For everything of this sort which my meager talent attempts or produces may be rightly considered yours, as that which if true (as it surely is) Ovid once said to Germanicus Caesar:
>
> "In rapport with your mien,
> my inspiration stands or falls."

Often, academics of the age published collections like this, dispassionately, as proof of learning and a small contribution to the new culture of European scholarship. The translation likely began as a simple exercise in Greek. But perhaps some of the stanzas really did sneak their way into his heart. At least the beginning: "The cricket is a musical being," Copernicus chirped, "At the break of dawn it starts to sing. But much louder and more vociferous, according to its nature, is it heard at the noon hour, intoxicated by the Sun's rays."

In 1510, he declined to follow in his uncle's footsteps. Sixty kilometers away, in the blanched coral brickwork of Frauenburg Cathedral, Nicolaus took up his post as canon of Warmia.

On March 29, 1512, Uncle Lucas died. Three months prior, as morning light was breaking through the clouds, Copernicus observed Mars leaving occultation.

From out behind the planet, he wrote, was revealed "in the Claws of the Scorpion . . . its first bright star."

He would always refer to his uncle as a "blessed memory," but when pressed for comment on the passing of this man of last generation, a more vindictive young bishop spat out, "The noose has fallen. We are free."

*

In Opposition

A rare privilege had been extended to Canon Copernicus by Pope Julius II just before he left for Frauenburg, allowing him to acquire multiple positions in the church hierarchy, but he never exercised it. The obvious next step in his career, his ordination, consecration, and acceptance of Holy Orders, he never trod. He already had an income. He had a horse and a page. At home there was a maid's room and a working toilet. Most people had none of these things. He was comfortable enough and had other business to attend to.

Frauenburg Cathedral became home to his starry passions. An entire room with a view was dedicated to his astronomical contraptions, all of ancient origin. His *triquetrum*, or Ptolemaic ruler, was a shafted triangle ten feet high, with appendages curling out like the fingers of a wooden giant; one joint could be directed out the window to take altitudes of a star. His quadrant, a trim square of wood a head or two above a grown man, was carved with a quarter circle and fixed with a small perpendicular block, whose shadow measured the appearance of the Sun throughout the year. Copernicus regretted not making it with stone, because it was warped by the cold. Old standards were present like the sundial and an armillary sphere, a handheld model of the solar system, its unlabeled orb hanging wanly in the center. All these instruments were glued together with Copernicus's own sweat and blood, for no one could find him a worker more skilled in the art.

Serious study demanded that he dirty his hands, but Copernicus was much more invested in beautiful theory. For a few inquiring friends, he wrote out a quick five-thousand-word treatise of his thoughts on astronomy. "The theories of Ptolemy," he wrote, "seem very dubious, as they imagine certain equant circles, on account of which a planet never moves with uniform motion respective to its proper center."

A boy who learns to paint in childhood will grow into a man first motivated by aesthetic principles. Copernicus had been such a boy. It was his aesthetic principles, his quest for beauty, that turned the key in his mind, unlocking the shackles of tired philosophy. Sometimes these principles were archaic, even reactionary, and led his more obvious foundations into radical collapse.

Copernicus agreed with, of all people, old man Aristotle. Uniform motion in a circle was, to them, more gorgeous than the nonuniform motion of Ptolemy's equant. But in both Aristotle's metaphysics and Ptolemy's astronomy, planets were also held to orbit along the outside of spheres, all of which were perfectly centered about the unmoving Earth. In the manner befitting a mathematician, without any fanfare or further explanation, Copernicus wrote out a list of theses which he thought to be "more reasonable."

1. There is no center of all the celestial spheres.
2. The Earth is not the center of the universe. It is the center towards which heavy things tend.
3. All the spheres revolve about the Sun, as if the Sun were the center of the universe.

This was heliocentrism: the Earth moves around the Sun. This idea, which would slowly infect the world like a righteous virus, began its ascent here, in this unremarkable, handwritten pamphlet, which Copernicus never thought to title or publish. Instead, he wrote out a few copies for fellow scholars, who wrote copies of their own, circulating it around like scientific contraband. Future generations would name it *Little Commentary*, as if to mock their humble origins.

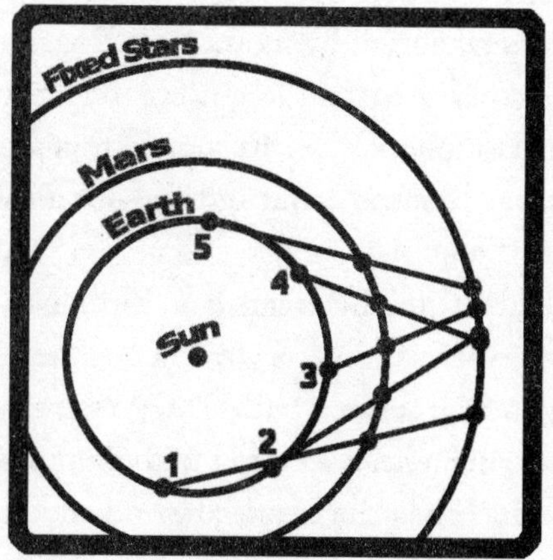

In this heliocentric system, the retrogression of Mars occurs between points 3 and 4, as can be seen by the intersection of Mars's projections onto the night sky.

Copernicus wrote down further assumptions until he arrived, at last, to the problem of retrogression, the backwards movement of the planets. Here, he knew, was the true elegance of his theory, for "the retrogression of the planets belongs not to them, but to the motion of the Earth." The backwards motion of a planet was only an appearance, *Little Commentary* claimed, actually caused by the moving Earth, overtaking the position of a planet in orbit as it moved through the orbit of its own. The moment of opposition, when Earth split the line between another planet and the Sun, was the exact moment at which retrogression began. In theory, this was the most stirring and simple solution to the tricky orbit of the planets. In practice, it was less wonderful.

"Thirty-four circles suffice to explain the entire ballet of the planets," Copernicus expressed in his urbane conclusion. Even with the motion of the Earth, he could not totally remove the use of Ptolemy's loopy epicycles, or the basic principles of Aristotle's metaphysics. Modern sun-centered astronomy did not spring out from Copernicus like Athena from the brow of Zeus; it took the knowledge of countless generations and would take generations more. But *Little Commentary* represents a specific turning point, a new world of possibility. It reads like a manifesto, disarming in its brevity; Copernicus promised to provide the requisite mathematical proofs in a "larger volume."

Copernicus had moved the Earth, but only out of a desire to return the planets to the beautiful uniform motion of Aristotle. He was a staunch conservative in this way, dotted by moments of sharp radicality. Church dogma was simply too pervasive for the spirit of total revolution to consume an educated mind. Its monopoly on European thought was a thousand-years strong, and it was about to come undone—by an accident, of course, a murderous accident.

As Copernicus had written his treatise on astronomy, a few years later, another mostly conservative man was writing up a treatise of his own. He was a stocky German friar, brown haired, aggressively shaven, with rolls of scholarly fat obscuring a head teeming with opinions, more than a few of questionable worth. "The princes of the world are Gods," he would one day remark, "the common people are Satan." The Jews had "uncircumcised hearts," he said, "We are at fault for not slaying them." On Copernicus, he was less sanguinary and half right: "the fool will overturn the whole of astronomy."

This man was Martin Luther, who had written out ninety-five theses, not entirely distinct in scholarly intent from those found in Copernicus's *Little Commentary*. Both were supposed only to kindle debate, but both possessed an alarming radical subtext. Luther was decrying the corruption of his modern Church, encouraging Catholics to unlearn the power of money and return to the humble words of Christ. He sent his theses off to a friend, handwritten, but had them printed as well.

That made all the difference. Copernicus had printed nothing. He too held conservative beliefs about money, but without religious grandeur, offering a timely warning in a letter to the Catholic elite of Prussia from that same woebegone year, 1517.

> Of the countless scourges which debilitate kingdoms, the principal four (in my view) are dissent, mortality, inarable land, and debased coinage. The first three are so obvious that no one is unaware of them . . .

This was not so obvious to the sluggish Church, who took four years to excommunicate Luther for his dissent. The Holy Roman Emperor then declared him an outlaw, but it was too late. His public support was overwhelming. Copernicus's worries about debased coinage would be a meaningless footnote to the ensuing struggle. He cried out, "Woe to you, O Prussia, you who pay the penalty for a maladminstered state by your ruin, alas!"

Revolt was the name of the hour. In Germany, a vast army of peasants rose up against their aristocratic overlords, inspired by Luther's example. But they did not have Luther's support. This was the start of the violent Protestant Reformation, something he could never have foreseen. Luther composed a pamphlet, *Against the Murderous Thieving Hordes of Peasants*, whose title speaks for itself. "Stab, smite, slay whomever you can," he advised the nobility, and they did.

Never has an attempted revolution been so utterly futile as this German peasant uprising. With a few broadswords, pitchforks, a rare musket, and no cavalry, their only successful attacks were upon monasteries and the regular clergy. The powerlessness of the peasants made even their violence pathetic; in one small village, a group of housewives smashed in the heads of two Catholic priests and, having nothing better to do, continued to beat them with shillelaghs for thirty long minutes, in the middle of the town square, until both men were a crimson mash. Peasant deaths were hardly better: they were so viciously decollated, disemboweled, and dismembered that a traveling bard could walk two kilometers outside of nearly any German town and still see body parts in the fields. When women and children refused to leave their homes for execution, they were simply burned alive inside. Mercenaries earned a florin a kill, half a florin for spare fingertips, keeping foul grocery lists of murder to ensure payment: "Eighty beheaded, sixty-nine eyes put out, plus, two months' back salary . . ."

Luther approved. He wrote, "The peasants have brought down the wrath of God and man upon them."

The Protestant bloodbath spilled over Europe like a glass of red wine tipped over by a thoughtless nobleman. In Switzerland, a reformist leader was slain and ripped into quarters, upon which a Catholic mob swirled his innards with pig entrails and feces. In France, so many Protestant corpses polluted the river Seine that a polite royal edict was issued to ask that citizens please stop killing each other. In England, chopped up Catholics were mailed to all four corners of the kingdom and shoved up on pikes by the coastline as a celebratory display. If Copernicus's Poland fared better, it was mostly because it could not have fared worse. The Teutonic Knights, whose grand master was about to convert to Lutheranism, made regular pillages across the country's northern rim. When one farmer dared fight back, they took his hands for safekeeping. "The robbers have intensified their activities," the Church canons begged the king for protection. "The population has endured bloody assaults and acts of robbery. The chapter has begun to resist."

Whether Copernicus would have it or not, war was coming for him. In 1516, the Warmian Chapter had ruined his leisure by assigning him administration of its largest town, Allenstein, forcing upon him the annoying gift of a castle. One frozen morning in January 1521, he awoke, strolled out between its crenelated walls, and saw the Teutonic Knights preparing a siege. He was the unfortunate city administrator, whose job description, he now remembered, included military commandant.

In his astronomy, Copernicus had kept his contentions civil, bound to the willing friends for whom his *Little Commentary* had been written, but in his life he could not prevent the war. Guided by the spirit of his belligerent uncle, against his better judgment, he readied for the oncoming siege. As it began, he looked to the skies, observing Jupiter and Saturn in opposition.

General Copernicus was remarkably composed. Right before the siege he had sent off a formal request for reinforcements and obtained twenty extra cannons for the castle defenses. News came to him of the latest skirmish, where the knights had broken through the town gates before being repelled. It seemed to Copernicus that he would end his life on the battlefield. The reluctant grimace he wore to this fate is still detectable in his words. "Canons want to act nobly and honestly as faithful subjects," he wrote. "They are even ready to die . . ."

Such even-readiness was not called upon. A temporary ceasefire was brokered at the start of next month. There was a quiet church back home, waiting for Copernicus and his trepidations.

After the Knights were pacified, Allenstein resurrected, and peasants reintegrated, he settled back into his silent way of life. He continued work on his astronomy, steadily and surely, painting a projection of the Sun onto the walls of Allenstein castle. He was uncertain about whether to publish. The idea of a moving Earth felt somehow more dangerous, more treacherous than it had when he was writing *Little Commentary*, before Luther.

Work on this larger volume of astronomy would be constantly disrupted by Church bureaucrats, who relished boredom above all else. Copernicus was tasked with accountancy, taking inventory; at one point, he was ordered to write an essay about bread, a tall step below the excitement of watching paint dry. To make his asinine day job tolerable, he often set to work in the company of a fellow canon, Tiedemann Giese.

Tiedemann was a wealthy yet unassuming man, with a passing interest in astronomy, whose scientific knowledge did not preclude a selective naivety; thus, he thought doctor Copernicus's workaday remedies were prescribed by a "second Aesclepius," god of medicine. His boldest writing always concerned peace and shared compromise; "Love endureth all things" was his mantra. In the unfolding political landscape, this made him a rebel.

"I completely reject the war," Tiedemann would grumble. He longed to shout it from the rooftops. He was a far more deliberate Christian than Copernicus, and spoke with him often about mending the grim fracture in their Church. "Nothing forbids the diversity of human feeling," he wrote, demanding "justice between brothers." When Tiedemann felt concerned about expressing such peaceful lunacy in a book, Copernicus urged him on. Their opinions were not widely shared. The two men became best friends.

Certain topical gossip always ran through their circle. Villagers entered and left, aging visitors to Copernicus's castle slowed to a trickle, certain canons ascended and certain canons passed. In 1537, Tiedemann was promoted to bishop, in the province of Kulm, over sixty miles (a hundred kilometers) away. Though Copernicus had letters to write and books to read, he was, for all intents and purposes, quite alone. He had had almost twenty slow, solitary years.

Once achieved, such independence is bittersweet. His promised "larger volume" was finished but forgotten, a private affair, with no plans for publication. This epic hobby was ready to die with its owner.

Copernicus was nearly seventy. He had managed to forget the war, to no small pleasure, until a Lutheran showed up on his doorstep, begging to discuss astronomy. "The whole world is dragged into the fight!" cried Tiedemann in sorrow.

The First Copernican

In the middle of May 1539, a lonesome mathematician by the name of George Joachim was about to hop the loose border cordoning his German University of Wittenberg from the Polish Frauenburg Cathedral. He was stopped at a rustic inn, crafting a letter to the first scholar he had gone to see, one John Schoner in Nuremberg. After that it had been a Stoffler in Tubingen, then an Apianus in Ingolstadt, but it is likely from Schoner he had heard of Copernicus and gotten it into his head that this was the man worth visiting. Having just made professor at university, he nonetheless took leave to continue this most curious journey. Copernicus had received no known warning.

Georg shared with his target a childhood trauma. When he was a boy, his father had died, beheaded for petty theft. The son was first feminized with his mother's maiden name, de Porris, then dehumanized by his mother country's name, Rhaetia. So became the man called Rheticus. He and Copernicus had little else in common, excepting their mutual love of astronomy, but this was all that was required. Rheticus was over forty years younger, a Protestant, and an irrepressible homosexual, a fact which had filled his life with misery ever since it first arose within his dolorous soul. To be an actively gay man in sixteenth-century Europe was to risk exile, castration, and death by fire; such was the influence of "natural law" in medieval Christianity, which Rheticus embraced with all his heart. His public Lutheranism and his private sexuality were two halves of his whole, locked in a spiritual war to which he could find no resolution. His confessions, made in fear and trembling, left priests in a state of abject sorrow, which they could only ward off through a quiet prayer for his soul. "Satan tempts him still," one whispered in horror. The academy offered Rheticus his only respite from this battle within. He was an amazing, excitable scholar and could lose himself completely to the work. "We young men desperately need the counsel of older and wiser men," he wrote around this time. "The opinions of older men are better."

"He is," recalled one of these older men, "above all an astrologer."

Rheticus reported not one bit of conflict between him and his unlike mentor. Copernicus even received him in Warmia right after the ascension of a partisan bishop, who had declared all Lutherans to be damnable heretics, demanding they leave the country. Thankfully, this orthodox astronomer had a long history of changing, or even ignoring, the rules.

Nor has Rheticus ever written of his meeting Copernicus. Equipped with formal letters of recommendation and several rare books as gifts, he must have been cordially received that day. Copernicus was always reluctant but never withholding; elderly, decrepit, and technically an amateur astronomer, living in what he called "the most remote corner of the earth," he had no cause to expect any visitors, and no notion of deserving praise. But it could not have been a total shock; most astronomers worth their salt knew of *Little Commentary* and could name its author. The heliocentric idea even made its rounds all the way to the Pope Clement VII, where his secretary had been well-received for lecturing on the Copernican worldview. Amusing for its novelty, it angered some and delighted some, no more.

Such ideas will travel, but the body is taxed. Suffering from fits of exhaustion on arrival, Rheticus spent the following weeks in convalescence, with Copernicus at the palace of Tiedemann Giese. His first real introduction to a sun-centered cosmos was through this fog of illness; truly, there are endless ways to receive the universe.

As the flush of his cheeks returned, young Rheticus entreated Copernicus to publish his astronomical work. At first, the old man refused. "As much as I can," he had written the new bishop just last year, "I want to avoid offending good people." "I fear I may arouse anger," went an older, private letter on astronomy, "I wished to leave these matters, just as they are, to the attention of others." Tiedemann once reported that Copernicus liked retirement; let the youthful have their poems and manifestos, and let the old live peacefully according to their will.

Rheticus reported that it was Tiedemann who revived the astronomer's ailing ambition, who gently pressured his friend with quotes from the ancients, and "won from the teacher a promise to permit scholars and posterity to pass judgment on his work." Tiedemann was only repaying the favor. Copernicus had, after all, encouraged his work too, which preached a unity between Roman and Lutheran creeds. Tiedemann had written, "When everywhere there is warfare and rebellion, when everything is carried away by this sudden violent storm, who is creating? Who is seeking to improve? . . . For we are estranging ourselves entirely from love." "Nicolaus Copernicus, a man of otherwise sound mind, persuaded me to commit these trifling comments to print."

Copernicus's agreement to publish had one caveat for his student. Rheticus would first study the manuscript back in Frauenburg, the reason for which he had come. He would make his reputation by publishing a synopsis, and secure an audience for the hesitant canon. If this appetizer sold well, the main dish would follow.

For this task, Rheticus received the work proper, fetched from Copernicus's study.

It was something of a mess, according to Copernicus, with the work "buried among my papers," tucked between sketches, letters, and his other writings, order and quantity known only to him. His gifts from Rheticus would now be on the bookshelf, a fresh Greek edition of Euclid near the Latin copy he had bought in Krakow, decades ago, when he was his visitor's age. The floor below held the tired medieval machines he had built in midlife to observe the stars. They were ineffective and barely touched, never so necessary to him as his private capacity for invention. Walking over these obstacles, he collated each chapter from around the room, hundreds of pages slowly piling on his desk. Perhaps he then sighed deeply, looking up at the earnest young scholar waiting in his doorway—a child, really—and then down at the sixth chapter in his hands, the last new tool of the old astronomy.

At that point, no printer had made a title page. Rheticus learned its name.

De Revolutionibus.

On the Revolutions.

The First Account

Rheticus spent the next ten weeks with his nose buried in a dense collection of words, the only copy of *On the Revolutions.* This pleasure was tempered by a torture: he had to write as well! Through his window at Frauenburg Cathedral, the groves of linden and birch began to ocher and crimson. By the start of fall, his *First Account* was completed. He was quite frank about its failure. "I have mastered the first three chapters, grasped the general idea of the fourth, and begun to conceive the hypotheses of the rest," Rheticus confessed. It was a rush job.

First Account is, like most book reports, a promotion for its subject, a quick bit of theater, and Rheticus was a natural thespian. There was his layer of indirection; the text is phrased as a letter to Rheticus's previous "older and wiser man," John Schoner. There was his undeniable tact, the way that Copernicus is only mentioned in the title; elsewhere he is always "my teacher." And there was a sort of rising action. In *On the Revolutions*, as in *Little Commentary*, Copernicus stated forthwith that the Earth orbits the Sun, as a matter of course. Rheticus, however, drenched him in honeyed praise, as a new Ptolemy, "in every field of knowledge equal to Regiomontanus," "worthy of the highest admiration," protecting him until the slight reveal.

> Linear, stationary, retrograde, near and far from the Earth, etc: these phenomena may all be demonstrated, as my teacher shows, by the regular motion of the Earth. The Sun occupies the center of the universe, while the Earth revolves in place of the Sun on what it has pleased him to name the Great Circle. Indeed, there is something divine bound up in these regular and uniform motions of the terrestrial globe.

The cunning effect all this engenders is that Copernicus had hardly changed anything at all! Why, he fit in snugly with the familiar, for according to Rheticus, "there is nothing better or more important than to walk in the footsteps of Ptolemy and to follow, as Ptolemy himself did, the Ancients and those who came before him." As Europe fell apart to religious hatred, he begged that this new science work in the spirit of unity. "I well believe that Aristotle would support my teacher," he beamed.

To make the Copernican system sweet on the palate, Rheticus drew upon every inner resource he had, not least of which included the arguments of beautiful astrology. He devoted a long paragraph to how "the kingdoms of the world change with the motion of the Great Circle," relating the moving Earth to the fall of the Roman Empire, the rise of the Islam, and finally, the second coming of Christ. A page is given over rapturously to the holiness of the new number of planets.

> What could be more commodious and dignified than the number six? By what number could anyone more easily have persuaded us mortals that the whole universe was divided into spheres by God the Creator, worker of the world? For of all the divine prophecies of God, left behind by the Pythagoreans and other philosophers, none are more celebrated. What could be better than that God's works, timeless and perfect, should converge upon this timeless and perfect number?

The boy spoke like a prophet. As a work of science, *First Account* is useless; as a work of science propaganda, it is unparalleled. Many more inaccurate judgments of Copernicus's theory were to come, but none would be so tender, earnest, and exuberant as this, its first.

The First Dissent

Yet another one of Rheticus's older teachers idled in Nuremberg. He was Andreas Osiander, leader of the Lutheran vanguard, with deeply sunken eyes offset by a quirky bowl cut and squared beard. He had a sharper knowledge of mathematics than most theologians, but only as a consequence of a freaky fascination with the numbers found in Jewish Kabbalah. This sort of independent study made him a real oddball. He thought and acted always on his own volition.

"But enough about these topics," Osiander wrote to Rheticus, after reading of his stay with Copernicus, "What remains is that I ask over and over, just as you offered me your friendship, obtain the friendship of this man for me too. I have not been able to write to him, and did not even want to, being certain you would not conceal my peccadilloes with him. I heartily honor him for his intellectual talents and way of life."

After completing his own book, Rheticus eagerly passed around advance copies to all his past mentors. Osiander chimed back, "I have received several copies of your *Account*. They have pleased me very much." This response was typical; scholarly opinion appeared so engaged and accepting that Copernicus, much surprised, happily readied his *On the Revolutions* for print.

Soon after the affable Osiander entered communications with Copernicus himself. "I have always been of the opinion that hypotheses are not articles of faith, but bases for calculation," he wrote, "It would be a good idea for you to say something on the matter in a preface."

Osiander's idea had serious merit. Europe was teeming with thousands of powerful Aristotelians, theologians, and scholars, many of whom had naught but wind whistling inside their heads. On the opposing side stood precisely two Copernicans. Against such odds offense seemed inevitable and defeat probable. To ease the tension, Osiander proposed an entente by proclaiming that the motion of the Earth was merely useful, rather than true.

Copernicus did not answer Osiander. He was aging quickly and knew he was about to die. He had written his book, and a young follower had come for it, dreaming to continue the work. The big questions of war, peace, truth, and fiction would have to fall, as they always do, to a generation with time left for answers. Rheticus the believer, working by his teacher's side, wrote out a fair copy of *On the Revolutions*, cutting and revising with Copernicus's approval. In September 1541, he bid adieu to this house of hidden knowledge, autograph in hand, and returned to the University of Wittenberg.

The route to publication was not clear. Wittenberg was a Lutheran capital, but at this time Protestants might just as likely condemn Copernicus as his own Catholic Church. One could never be sure with Martin Luther. Adding to this confusion were lurid rumors of Rheticus's homosexuality, which roiled the air around his college campus. For his own safety, he quit the city within the year, opting for an obscure press in Nuremberg.

This placed him in a bind. He had to move to a new university post before the start of the academic year. Book printing was then an infant process demanding great labor, especially for densely illustrated mathematical texts. Almost one hundred and fifty woodblocks would have to be cut. Most printshop workers were ironically illiterate and could hardly be expected to format mathematical calculations and tables on their own. The movable type for each page had to be arranged and justified by hand, which would take weeks. Rheticus did not have the time.

Searching for a solution, he entrusted the rest of publication to that recent friend and skeptic Andreas Osiander, whose criticisms had appeared written by unassuming hands. They were false impressions. Without permission, Osiander inserted a lengthy unsigned preface of his own. The printer, if he even noticed the addition, did not much care. "These hypotheses need not be true nor even probable," Osiander warned the reader. "On the contrary, if they provide a calculus consistent with the observations, that alone is sufficient. So far as hypotheses are concerned, let no one expect anything certain from astronomy."

Osiander's preface was certainly one valid reaction to the heliocentric hypothesis: good for computing, bad for believing. But unlike most interpreters, he forced his beliefs into the very structure of the work. It was a stain upon Copernicus's intentions, and the belief that he was insincere, that his heliocentrism was never intended to be taken seriously, would be idly parroted by Church conservatives for generations to come.

To sensitive readers, Osiander's trick was not even successful. "Indeed," wrote one university professor, "the words and thinness of style reveal them not to be those of Copernicus." The professor's rambunctious student was more crass: "A little ass wrote this for a big ass!"

Osiander was, like Copernicus, an intelligent and amiable man, with all the right and reason to believe what he wished. He may have even inserted the preface with an eye toward helping his friend's worldview gain support, but this meant little to those whose trust he betrayed. The peace-loving Tiedemann Giese called it a "crime of deceit," and Rheticus, upon receiving his copy, drew a big red X in crayon upon the page like an angry toddler.

As fast as Osiander had entered their friendly circle, did they now excise him from it.

Yet it was done. Hundreds of copies were out, preface and all, floating around Europe for astronomers, lay scholars, and anyone hungry for mathematics to consume.

The Second Account

Certain excerpts of *On the Revolutions* are haunted by the younger spirit of *Little Commentary*, with its high poetry and aesthetic sense. "The Sun is appropriately called the lantern of the universe," Copernicus declared, "As though seated on a royal throne, the Sun governs the family of planets revolving around it."

After this delightful introduction, however, the book grows thick with highly technical calculations. Rheticus often alluded in his text to a *narratio secunda*, a second account, but his flighty temperament drove him elsewhere. It was never written. *On the Revolutions* contains many concepts other than heliocentrism, only some of which are mentioned in *First Account*, others entirely implicit. Another summary would have done them good.

In *On the Revolutions*, Copernicus stands up to Ptolemy like his clever, rebellious little brother. Both begin with philosophical claims that clash against one another like knights in a joust. Ptolemy's argument for the Earth's stillness was Aristotle's, roughly the same as the common Catholic peasant's. If the Earth did orbit, Ptolemy wrote, it must be "the most violent of all motions, seeing that it makes one revolution in such a short time." The Earth must move so fast, he argued, that any time a bird lifted off its perch, any time a cat leapt up a stair or father tossed his son in the air, these poor creatures should go howling off into the deepest regions of space, their world ripped out from under them. Careful experimental testing showed this not to be the case, hence, the Earth was immobile.

"This problem has not yet been solved," Copernicus replied, as though Ptolemy were still alive, sitting across from him in friendly expectation. "We merely say that not only does the Earth have this motion, but also no small part of the air, and whatever is linked to Earth in the same way."

Copernicus did not argue for the correctness of a moving Earth, but its probability. He did not replace, but added an opinion, one deliciously possible proposition, which need not but *could still be* said. He did not yet claim any ancients were necessarily wrong, and the fact that he personally accepted heliocentrism was not, he knew, deductive proof.

Deduction, however, did help the cause. The rest of the first chapters in both *Almagest* and *On the Revolutions* are reserved for basic theorems of geometry and a discussion of spherical triangles. The two books share several proofs, though *On the Revolutions* is a better teacher; the first chapter is punchy, orderly, and unobscured. From there, many readers of the time lost resolve, though the dedicated remained, leaving behind a labyrinth of annotations. "Anyone can rightly wonder," wrote one anonymous commentator, "how from such absurd hypotheses of Copernicus, which conflict with universal agreement and reason, such accurate calculations can be produced."

They were right. These first skeptics against early Copernican theory had no end of protestations, many of which were more than justified. Professional mathematicians could have questioned more than the nature of hypotheses. The calculations in *On the Revolutions* were curious, if masterful, some rather confused, most quite conventional, yet tremendously odd in purpose; they were, in short, very much like Copernicus himself. The sweetest sample of this, which Rheticus talked about at some length, was the little-known third motion of the Earth.

In addition to the daily rotation and annual orbit of the Earth, so too moves its axis, pushing the North and South Poles of the Earth around in an almost imperceptible circle.

This motion of the Earth's axis, due to gravitational pull, is slow enough to kill a man and melt his skin, revolving once about every 25,772 years. Copernicus estimated it to surprising accuracy at 25,816 years, but unable to explain its motion with physics, resorted to some clumsy explanatory guesswork.

While Earth's poles were traversing the 25,816-year circle, he proposed, they were changing speed, slowing down then speeding back up, in accordance with a separate, ethereal circle, which was itself traversed every 1,717 years.

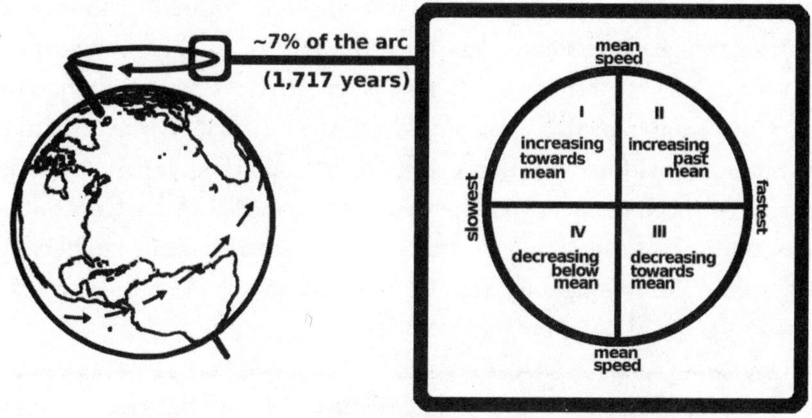

The axis of Earth orbits in a circle about once every 26,000 years. Copernicus's proposed model is to the right.

If the modern reader feels that this is nonsensical, they are right, up to the bounds of modern nonsense. Aristotle had identified astronomy as a "branch of mathematics," and Copernicus, like most astronomers of his era, agreed; any calculation which respected the observations was a valid option. He was only providing a heliocentric version of a nonuniform motion in the stars suggested by his teacher's teachers, "Regiomontanus and Georg Peurbach, from whom I differ very little," he wrote with pride. "They were my immediate predecessors."

The reason that such an absurd motion was invented is as heartbreaking as scientific theories can get. The observations Copernicus had used, taken from astronomers past, were mistaken, but in his persistent respect for his elders, he could not identify when one was an unreliable narrator. Ptolemy had given observations that implied the motion of the Earth's axis was much too slow, while a famous Islamic astronomer, Al-Battani, said it moved too quickly. Copernicus was stuck trying to satisfy two opposing parties, both of whom were wrong.

When the next generation of astronomers read about such baffling motions, the possibilities of newer, better observations became obvious. Many of Copernicus's nitty-gritty solutions acquired a similar sort of ugliness, and as such, the book lacks much of the aesthetic beauty he had hoped heliocentrism would produce. This was not the end, but the start of a new way of thinking about the world.

Copernicus's resolution of the problem of parallax is another strange tale of revolutionary orthodoxy. Parallax is the shift in an object's appearance when it is viewed from two different perspectives; it is to view the same narrative from another's eyes. The conceptual difficulty is that, without any background for context, it is impossible to tell whether parallax occurs from an object moving or you, the viewer, being moved. All of Copernicus's silly, boring, ridiculous, wonderful calculations were designed to illustrate this subtle and foundational transition between basic ideas.

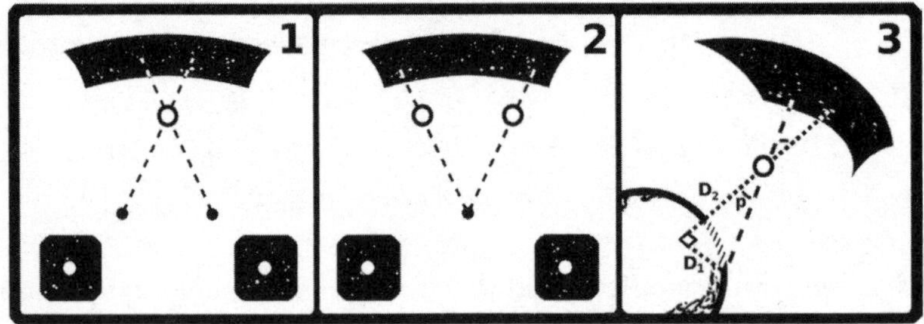

This image demonstrates three different kinds of parallax in Scorpio. Image 1 occurs when the observer is moving, as with the Earth's orbit. Image 2 occurs when an object is moving around the observer. Image 3 is parallax using only the size of the Earth, rather than the size of the Earth's orbit. The parallaxes in images 1 and 2 can only be differentiated by context.

The parallax, which occurs when the Earth moves, rather than the observed planet, was new to Copernican theory. Though it featured in some of his most confused calculations, it also helped him to estimate planetary distances and orbits with admirable accuracy. But even in the age of Ptolemy, astronomers would measure the parallax of a planet in a similar way, by taking measurements of the same planet from two different parts of the Earth. To start, they would take one measurement from somewhere close to "the center of the earth." Then they would take another observation, which could be anywhere, as long as the distance between the two observations, D_1, was known. The parallax was given as p, the angle representing the shift in the planet's position. With these two measurements and some trigonometry, astronomers could figure out facts no one had ever expected of them, like the distance of the planet away from Earth, D_2, using the formula for tangent, $D_2 = D_1/tan(p)$.

Parallax offered far more than distance measurements; it could even be used to prove whether or not the Earth moved. These proofs leaned upon a central dogma of Aristotelian astronomy: that the universe was closed and finite. Because everything rotated around the Earth, the reasoning went, everything must be some set distance away from it, otherwise there could be no center about which to rotate. Such a universe was not far different from a giant snow globe. The stars were believed to be hung up, fixed upon this boundary of the universe like gargantuan fairy lights, with all the planets glittering inside.

If the Earth truly did move about the Sun, and the universe was as tightly enclosed as the ancients proposed, a parallax should be seen in the fixed stars. In Copernicus's time, no such parallax had been measured, and for many decades this was a powerful argument in favor of the Ptolemaic system. To resolve this, Copernicus needed only wave his hands, and explode the size of the universe.

> Near the Sun is the center of the universe. Moreover, since the Sun remains stationary, whatever appears as a motion of the Sun is really due to the motion of the Earth. This motion is truly enormous compared to the size of the Earth. *But the size of the universe is so great that the distance from Earth to Sun must be imperceptible in relation to it.* This should be admitted, I believe, in preference to perplexing the mind with an almost infinite multitude of spheres . . .

Such an admission can, in fact, be calculated. If the smallest angle of detectable parallax is given as p, the minimum size of the cosmos could be found with the previous formula, $D_2 = D_1/tan(p)$, using the radius of Earth's *entire orbit* for D_1, rather than the radius of Earth, as was done with the unmoving Earth in the Ptolemaic model. It was, suffice to say, aggressively larger than previous determinations, by over a million times. Copernicus could not even imagine such a monstrous number; he just called it "immeasurable."

These calculations do feel depressed by a certain lack of imagination. To expand the size of the universe is a tremendous magic act, yet compared to the unknowably infinite, the mere shifting around of finitudes can seem meaningless. Copernicus had no previous examples of infinity in nature, and no impetus to produce one; to have then suggested an infinite universe would have transcended beyond calm reason into mad poetry; beautiful, but far out beyond the limits of necessity.

On the Revolutions contains untold failures. Copernicus's admission, revised from *Little Commentary*, that the center of the universe is only "near the Sun" is the worst betrayal of his lacking physical intuition; the first Copernican system is more accurately described as heliostatic, rather than heliocentric. The Moon is quietly subordinated to the only earthly satellite, and while the lunar theory remained confused, it was more elegant than Ptolemy's, which had a mobile equant. The correct order of the six planets is given, with many tables to estimate eclipses, planetary positions, and other helpful data, but they are strewn about, higgledy-piggledy, in an extraordinarily unhelpful fashion. The motion of each planet is considered in-depth, which consumes most of the last half of the book.

In short, the first robust, novel theory of the heavens in over a thousand years is laid out in full—a little patchy perhaps, often wrong he knew, but altogether present. Before any science of celestial physics, with the aid of countless computations, several surprising friendships, and forty more years, a middle-aged hobbyist had ushered the physical solution out into the confused old world. The fact that this truth preceded its foundations has its reasons and contingencies, scientific, historical, biographical, but none have ever fully explained the inverted discovery, for none can. The triumph of the mind resides within. It is his last great mystery.

Postmortem

A thick torrent of blood was leaking out over the brain, snuffing each neuronal ember. The canon had been trapped in bed for days, senile, waiting for the hemorrhage to take him. His right arm dropped, and Copernicus was gone.

At his side was his book. Tiedemann gave a heartfelt eulogy to Rheticus that Copernicus had "seen the treatise, only on his last breath on his last day." That day was May 24, 1543. He was "our brother," Tiedemann wrote. "By reading his book, he seems to come back to life for me."

Of those last sane years, Copernicus had spent all with Rheticus, together preparing his lifework.

After a modest run of a couple hundred copies, it saw a second printing in Basel, Switzerland, in 1566, along with Rheticus's *First Account*, which has undeniably always been the more widely read of the two.

There were better signifiers of Copernicus's importance than sales. The masses loved *ephemerides*, or star almanacs, calculations of planetary locations for each day over the coming years, which court astrologers used to craft public horoscopes and prophecies. Rheticus, having developed into a wizard mathematician, realized his first heliocentric ephemeris in 1551. Copernicus had also attracted a new supporter in a college professor named Erasmus Reinhold, who published his heliocentric *Prutenic Tables* that same year, though Rheticus's calculations were more purely Copernican, having trained with the man. To those two fairy-tale years in Warmia the rest of his life was indelibly tied, "for nobody," wrote one of his future students, "has understood the mind of Copernicus better than he."

The simultaneous release of and significant differences between the two tables were unfortunate consequences of a failure to communicate. This had always been the outstanding problem of science. Mathematical monks had Aristotle in every library, but no means or motivation to broadcast discoveries and ideas to the world. Now there was the printing press, cheap and developed enough for scholarly work. Translation ran rampant. Literature was turning into conversation, rather than commentary, a tool for the individual voice, which might shun and accept others as it saw fit.

Martin Luther, resting on his deathbed, had understood this. He was the first journalist of theology, publishing constantly, contradicting himself daily, never afraid to revise his babble. The greatest schism in the history of modern Christianity could be better represented by no man. But Copernicus had only ever wanted to be a builder. Pope Paul III, to whom he dedicated *On the Revolutions*, set the schism in stone with a vengeful program of Catholic renewal known as the Counter-Reformation.

"I hesitated for a long time and even resisted," Copernicus wrote in that dedication, "but my friends helped me on." One of these friends, he continued, was the irenic Tiedemann Giese, who would one day be buried at his side. Of Rheticus the Protestant, Rheticus the heretic, Rheticus the sinner, Copernicus wisely, but cruelly, avoided all mention.

The Lord, which the Catholic and Lutheran sects share, would have known the good for which the boy longed. In 1551, Rheticus was subpoenaed to defend himself in court against charges of "sodomitical misdeeds." Without a moment's hesitation, he fled Germany forever. As an old man in Prague, distraught from exile and homosexual shame, but still a flourishing scholar, he was visited by a disciple of his very own. The acolyte had barely arrived before Rheticus burst out, "You come to see me at the same age I myself was when I visited Copernicus! Had I not visited him, none of his works would have seen the light."

And when, in 1557, orphan Rheticus had occasion to look back on his childless teacher, he was sure to correct, "not only my teacher . . . my father."

TYCHO BRAHE

A life, in which there is a very rich light

DRAMATIS PERSONAE

Sophia Brahe	his sister & gardener & genealogist & a very bright mind
Frederick II of Denmark	his king & kindly patron
Christian IV of Denmark	Frederick's son
Nicolaus Reymers Ursus ("The Bear")	an associate of Lange & noted opportunist
Erik Lange	his fourth cousin & an obsessive alchemist
Michael Maestlin	his academical friend
Kirsten Jorgensdatter	his quiet wife
Rudolf II, Holy Roman Emperor	King of Bohemia
Queen Sophie of Mecklenburg-Gustrow	wife of Frederick & mother of Christian
Johannes Kepler	
Galileo Galilei	two young children

MAJOR WORKS

On the New Star • *On the Aetherial World* • *Prelude to a Restored Astronomy*
Instruments for a Restored Astronomy • *Theater of Astronomy*

YEARS ENCOMPASSED

1546–1597

New Stars

For years there was darkness. No comet raged in the sky. No further fire swallowed up Europe. As many as could be hoped, which is not many at all, read Copernicus's *On the Revolutions*. Copies bled across the continent and into England. Nonastronomers heard about it. They laughed. They disagreed. They quit thinking about it and went to bed.

And then there was light.

In November 1572, stargazers all the world over looked up and saw something new. A supernova. To them it looked a star, a new star, a day star, a space dragon, blazing in the heavens alone with the Sun, brighter than Venus in the night, usurping the throne of the princess, constellation Cassiopeia. Four hundred years later, it had changed everything.

None understood what this new star meant, especially once it began to fade. The Chinese grasped the matter best; they had previously recorded the appearance of "guest-stars," though still held them as portents, which caused their emperor a great deal of worry. "It was as large as a lamp," their astronomers wrote, "seen before sunset, pointed rays of light streaming out in all directions." For a short while longer, this civilization remained almost entirely foreign.

European astronomy was in greater disrepair. Its forefather Aristotle had proposed the heavens were changeless and eternal, and thus caused by some changeless and eternal being. But, here, now: if there was no parallax in this new star, then it must lay in the heavens; if it lay in the heavens, then the heavens admitted change; if the heavens admitted change, then Aristotle's physics, and metaphysics, must be revised. Revision, in its turn, entices more active minds to wholesale replacement. The philosophical bedrock of all known astronomy became a quivering line of dominoes, waiting for a reckless child.

The new star did not have a visible parallax. A talented astronomer needed little more than his wits to prove this, and two young Europeans took this incredible event as a chance to announce their wits to the world.

Having just graduated, at twenty-two years of age, with thoughts split between being a teacher or a priest, the German Michael Maestlin proved the new star had no parallax using a simple piece of string. By stretching the thread between two fixed stars that sat on a line with this new one, he could see over time that none of the stars moved off the line. "The truth is brought out by accurate observations," he wrote, "this prodigious star is no meteor, no planet, but numbers among the fixed stars"; "As for what this bodes, I leave that to others." Thrown into a state of philosophical anxiety, Maestlin reviewed his copy of *On the Revolutions*, acquired two years earlier, where he uncovered a trail to solace. "I agree with Copernicus," he scribbled in the margin, unaware of what a rare triumph this comment was.

In college, a vibrant Dane by the name of Tycho Brahe had begun his lifetime of observations using the same method. Still, both young men acknowledged that a piece of string was inadequate for their ultimate desire: nothing less than a complete restoration of astronomy.

"Restoration" was a word Tycho threw around a lot after this new star, but he could never really define what it meant. To him astronomy needed to be "restored," "reformed," "renovated," "reparated," "rejuvenated," "redintegrated," and whatever else sounded right from the "re-" section of his dictionary. If these words had any practical meaning, which can be doubted, it was to make astronomy "without assumptions," as Tycho wrote, with new assertions resting on better observations than Aristotle's philosophies had used. But what such a new astronomy would be was yet unclear. More than anything these words were emotional, aspirational, a new language for talking about an old scientific pursuit. They were poetry. Tycho loved poetry. He wrote it constantly. His favorite poet was Ovid, who sang of metamorphoses and shocking transformations. For obvious reasons, these themes were most popular with Lutherans. To them, the past might not only grow but burn, and from the ashes would arise, they believed, a "Phoenix of Astronomers." "This almost collapsed science will be greatly amplified in its new splendor," wrote Lutheran Maestlin in his ephemeris, but if the task could belong to one man, Lutheran Tycho would have claimed it. His thousand-page monster, *Prelude to a Restored Astronomy*, would not finish publication in his lifetime.

Whatever restoration meant, it was obstructed by as many mental blockades as practical ones. Inherited beliefs can play tricks on the mind, as Tycho knew. Of the new star, he wrote:

> It happened just before supper . . . in the middle of my walk back home, I was contemplating different bits of the sky. It seemed to me then most clear, as from a wish to guarantee continued observations after supper; when lo and behold! Just overhead, a strange kind of star from an unexpected place, blinding the eyes and shining brilliant, radiant, fulgent light. I was amazed, practically stupefied, thrown into such perplexity by the impossibility of it all that I began to doubt my own eyes.

Tycho gaped at his traveling companions, astonished, until his dry mouth could form the obvious question. They all replied in a faithful chorus, "It is glorious!" Still he could not believe. Like a madman, he began to screech at the nearest carriage and poke at the sky. The peasants inside looked up, then screeched back: *Real, you noising fool, real, it was real!* "Finally," he concluded, after a chance to calm, "certain my vision had not deceived me, I marveled that the heavens had summoned up some new phenomenon. It demanded to be compared. Immediately readying my tool, I attacked."

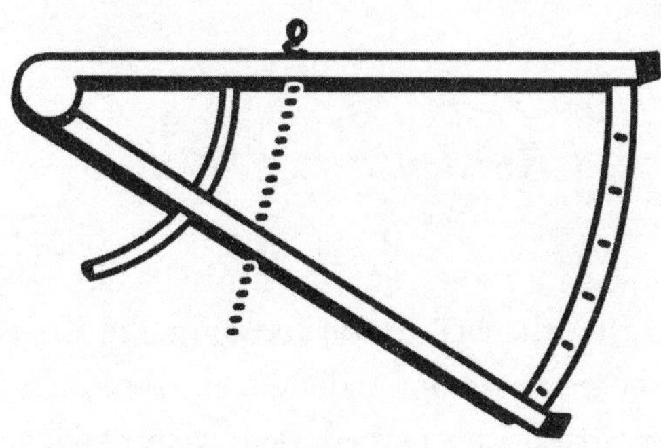

Tycho's immature sextant.

Aged twenty-five, all young Tycho had yet made was a T-shaped stick about five feet long, called a *cross-staff,* which could be swung up to the eye to measure the angular distance between stars, and a handheld joint of wood, which he had named the *sextant*, for the same purpose. With this simple sextant, he took unusually detailed observations of the new star, as he mentioned with pride, "noting very carefully its size, form, color, and other visible aspects."

There were far more commentaries on the star, even from Tycho alone. The Lutheran offered up an unsubtle prophecy, that certain unnamed "religions which have a jovial splendor and pomp will fade, if they do not die completely, just as this fake star did." There shall then come "a great light, which will slowly swallow the darkness, just as does the Sun over the course of spring." Some theologians proved even less reserved. It is as the Star of Bethlehem, they said, let us ready for the second coming of Christ.

Christ did not arise, but another man did. Today, this new star is known as "Tycho Brahe's Nova." It has long since vanished from the heavens, but its impact upon the world remains. "God will come shortly," predicted one eminent Oxford scholar, "and consume all with fire."

Of course, to a great deal of the young, old, and poor, whose stories are not yet told, this star meant very little. The nine-year-old Galileo Galilei, though perceptive beyond his years, gave it not a thought. The baby Johannes Kepler, not even a year old, could not have cared for this Christmas star. Of a comet five years later, he reported only that he heard of it often and that, despite his physical frailty and his father's abandonment, his tired mother carried him to some hilltop one night to watch.

A Burdensome Privilege

A few months after the fact, Tycho was dining in Copenhagen with a wealthy French ambassador and a doctor of medicine when he was alarmed to discover that neither of them were aware of the new star; indeed, hardly a Dane in the country was talking about it. When confronted, the host ambassador thought it was a bad joke. Tycho smiled and said he hoped for a clear night.

To their credit, when the two realized their mistake, they urged Tycho to publish his writings, which would become his first book, *On the New Star*. But they must have found his astronomical approach—one of always looking, validating, measuring—very curious in a culture that had for centuries celebrated inner contemplation. It was more than curious, in fact, for it must always be remembered how discordant it is to live, see, and think unlike others; it is terribly, irrefutably, irrepressibly *strange*.

Tycho was branded with strangeness before he could have chosen it. Firstborn son of a high-ranking noble lineage, birthed in 1546 with a stillborn twin, he was stolen, as he said all too coolly, "without the knowledge of my parents," by an uncle and aunt who could not bear children. Out of love for their brother, Tycho's parents acquiesced to this impromptu adoption after the birth of their second son.

This different set of parents did nothing to change Tycho's class. He was not only a noble but an aristocrat, the kind of rare noble who rubbed elbows with royalty. Aristocratic children were raised, in word and action, under the assumption that they were better than everyone else. Denmark was the first declared Lutheran kingdom; every Danish noblechild was reared in the Lutheran way, nationalistic, independent of a universal church, all the way up to King Frederick II. This novel combination of aristocratic pretension and Protestant individualism provided wealthy Danish youths with an easy path to self-doubt, and they were always demanding constant reaffirmation of their own magnificence. There was no ethnicity of noble more industrious, extravagant, vainglorious, and uncertain of itself than the unhappy Dane.

Tycho was, in this regard, most definitely a typical aristocrat. But rather than situating himself within the extensive courtly rituals and social faux pas of a large and noble family, to share his inheritance with his siblings, he was raised in splendid isolation by a boorish uncle and his coy wife and would receive a similarly large inheritance all his own. His adopters loved him all the more, being their only child, and tended specially to the needs of their one very weird son, but this opening act of the family romance was already on the wane. By the age of twelve, Tycho became the first Brahe to attend university.

At Copenhagen, the usual student was eighteen, but it was all a matter of upbringing. Some children as young as eight enjoyed a college education, and occasionally grown men of peasant stock had the gall to try to improve themselves. Tycho was of the highest class; he ate up knowledge until he grew sick of the local flavor three years later. Shifting to Leipzig, his foster parents hired a student a few years his superior to carefully engineer his studies. This warden had a knack for putting things bluntly. He diagnosed the Danish condition with such clinical force that it was painful.

> When a Dane comes up in the world, people in many places, even Europe, are inclined to think that they are hearing talk about the New World. There are even people who have heard or read a little about Denmark, but who are, nonetheless, of the opinion that we are a stupid, unlettered, barbarian nation, unacquainted with art and good order.

Tycho knew good order, but he would have to learn stridency and willfulness, if he was to get respect. In boyhood, he had spent his days roughing up an almanac, enthralled by the stars. Now at university, he planned to audit astronomy classes, except that, he wrote, "my warden, pleading the wishes of my parents, did not like it and opposed it." They wanted him to study law or another career befitting a privileged noble. "I had to buy and read astronomical books in secret," Tycho remembered. "In a month I learned all the constellations in the sky. For this purpose I made use of a small celestial globe, no greater than a fist, which I used to take with me in the evening without mentioning it to anybody."

Soon the sneak got his hands on a cross-staff and initiated what would become a thirty-five-year streak of consistent observation. These first attempts, he had to concede, were "childish and of doubtful value," but they were still enough for him to recognize inaccuracies in astronomical observation he felt to be grievous. He designed improvements. A millennia and a half of learned instrument making and theorizing was not, apparently, anything a headstrong teenager thought beyond his own measure.

That teenager was not so for long, abandoning Leipzig at twenty-one. "The M.A. degree is immaterial to me," he would reflect years later, "I would prefer that one really be a master of arts." Rather than be passively educated, he decided to travel the world.

When Tycho's dear foster father died saving King Frederick from drowning in a lake, Tycho felt obliged to return to Denmark for a spell to bond with his birth family. This did not go well; hardly any nobles approved of a career in astronomy, but the hardhead was set. Few of his brothers and sisters were even present, but the youngest, little six-year-old Sophie, was. Despite his ignoble quirks, to the tiny girl he soon became "the good brother." Tycho drew closer to her as he realized that no grown member of his family liked him. They gave him no reason to stay in Denmark. Dismal meditations on expatriation raced around his mind.

Such embittering thoughts kept with him through the frigid winter when, on tour of Germany, he got viciously drunk at a pub. He squawked, louder and louder, in a language almost no one could understand. One of those who could, a Danish drinking buddy, squawked back—they were, people guessed, arguing about math—and the two traded indignities until, worked into a silly rage, both foreigners stomped outside with their swords at the ready. All Tycho's life he was acting in a duel, but only this once did it become literal. A local girl by the bar, being the only person in the room with the slightest bit of sense, instructed the men around her to go halt a likely death, but they were too late. There was a flash of steel in the night; a patch of Tycho hit the ground, and a red river ran through his face. He collapsed, defeated, but survived, for no reason other than that fortune can smile even on an idiot.

Portraits of the blond mustachioed Dane are as tactful as honesty permitted, with most only alluding to a thin line carved around his face, but his disfigurement was ghastly. The rest of his life, Tycho wore a regal prosthetic of gold and silver atop his nose stump, fixed in place by an adhesive gel from a snuffbox he carried with him everywhere. His manly honor was preserved, but his manly face was ruined, and this, a friend of his suggested, "may conceivably have had something to do with the selection of the humble life-companion." For returning to Denmark in 1571, he ruined himself further by taking up with a common preacher's daughter, Kirsten. Then he took up with her again. And again. It was an amour polluted by caste, chivalry, and social expectation, none of which could overcome the cleansing powers of mutual lust and affection.

Such a sincere relationship between classes was considered, at the time, to be like a physical deformity, rather rare, and quite as unbecoming. The two could not even legally marry. Though King Frederick was sympathetic, most of the Brahe line frowned further upon their black sheep boy, and could barely stomach his choice of partner. "Tycho's harlot," they would call her, partaking in "an evil life." "I do not like this society," Tycho shot back, "Its customs are rubbish. They always want more from me." The appeal of peasant girls was that they could not ask for more.

Of these two major life events, Tycho, by whatever exact mix of etiquette, egotism, and embarrassment he was burdened, did not much speak. His letters reveal a simple rule of verbal conduct: he would mention his girlfriend to no one, and no one would mention his noselessness to him. Such were the differing priorities between Tycho and the more normal nobility. Even neutral onlookers described Kirsten through a veil of dehumanizing contempt, "of an admirable and, to her husband, satisfying fecundity." History would remember her better were she truly such an instrument; then, at least, Tycho could have dedicated to her a description as loving and detailed as he gave his own astronomical tools.

A year before he met Kirsten, he told, he had found himself discussing his instrument designs outside of a local shoppe in Augsborg with two especially cultured friends, while scouring Germany for cities he might immigrate to. One of them, a wealthy alderman who "seemed addicted" to astronomy, leapt to underwrite the production of Tycho's first serious invention, and offered up his backyard to host it. "With this we immediately switched from play to work," Tycho wrote. He sketched out a picture of a quarter circle, the quadrant, an extreme revision of his toy sextant for measuring angular distance. It would be capable of unprecedented accuracy, down to a single arc minute, or sixtieth of a degree.

In the days of naked-eye astronomy, there were a couple clever tricks that used free space on measuring instruments to increase accuracy, but such cleverness had limited returns. The most obvious, expensive, and aggravating solution was (and still is) to scale up.

The Great Quadrant (Tycho Brahe given for scale).

Tycho's instrument was big. Very big. He called it *The Great Quadrant*, or *quadrans maximus, permagnus*. When he had calculated the proportional enlargement required by his obscene desire for precision, the result was over five meters in radius, supported by a fat, carved oak tree, held hovering above the ground by an even weightier frame. From all reports, it was a theoretical success, but a practical failure—a youthful blunder Tycho would never repeat. Forty grown housemen were pulled off duty to heave it across the soil to its final resting place, and more than a few were required in its daily operation. These first laborers serving Tycho's scientific life could not have been pleased, pushing around this foolhardy piece of equipment on a frosty March day, stomping up a perfectly good garden.

From here on out, Tycho's intellectual fever would only quicken. Of his populous family, he managed to uncover a lone older relative, another uncle, who was vaguely supportive of his career in the sciences. He lived out two hermetic yet luxurious interim years at that uncle's estate, an abandoned monastery in the southern tip of Sweden, occasionally circling back to Copenhagen to spend a night with his girlfriend Kirsten. That uncle had a happy hobby of alchemy, which Tycho found he enjoyed so much that he promptly quit astronomy and prepared to become a chemist.

Then, at almost the exact moment that he impregnated Kirsten, the light of the supernova that would one day bear his name touched down upon Earth. He looked up, and his aspirations took a shuddering jump back into their old domain. "It was the greatest wonder that has ever shown itself in the whole of creation," he wrote in ecstasy; though chemistry would always remain his secret second study, "the new star which flared up in 1572 made me give up my chemical labors and turn towards the study of celestial phenomena." He left his uncle's home to start a family with the mother of his child.

During his investigations into the new star, he revisited the house of his parents, where he observed an eclipse with a most precocious assistant, "my lovely sister Sophia Brahe, then a maiden of about fourteen. She's a charmer." He made this note of her help publicly, in his first book, *On the New Star*, commenting how "many think it a travesty, that some descendants of a noble species should venture into this sublime science." So discouraged was the outset of his career that he almost released the book under a pseudonym, but his friends in academia talked him out of it. Tycho affixed his name to his many intraclass quarrels as a matter of public record. *On the New Star* concluded with a poem decrying snobbish joys, the "raging cups of Bacchus," "insane romance," and "noble grandeur" of his fellows. Though, he conceded, he enjoyed these too, he believed the purest pleasures were found outside of the self.

Happy on earth, happy beyond the heavens
Is he who delights in the heavens before the earth.

More than one scholar doted over this image of their "prince astronomer." His presence glorified their profession, and all astronomers rose in social status by association. Most warmly supported the restoration of astronomy he was dreaming of, but few had similar resources or freedom to pursue it.

Michael Maestlin, as example, would become a pen pal Tycho much enjoyed hearing from. Though a Copernican at heart, he had found his calling as a devote schoolteacher, whose frequently reprinted textbook was entirely Ptolemaic. As he justified it, "the familiar ancient hypotheses are retained for the young and, being easier to understand, are taught as correct, but all specialists as a body agree with Copernicus's demonstrations." Despite this concession, Maestlin was an earnest, hardworking man, who would later add a subtle passage to the appendix of his textbook explaining Copernican ideas. He argued vigorously over issues like calendar reform and, his more radical students professed, walked hand in hand in with pioneering science. But in the eternal coupling between education and discovery, he always graced the former.

Tycho was, in contrast, not so certain about the Copernican system, though he gave its inventor an abnormal amount of praise. Nor was he a schoolteacher, though he did deign to guest lecture at the University of Copenhagen, a behavior other nobles judged to be unthinkably base. Maestlin wrote extensively, but his crowning work was his textbook. Tycho was more ambitious, to the point of hubris; he believed his crowning work would be a *Theater of Astronomy,* a massive ten-volume compilation of his modern thinking and practice, of which his planned, already massive *Prelude to a Restored Astronomy* would be only the first third. But he claimed that his desire to restore astronomy arose "not out of arrogance or in contempt of the ancients in any way, but because I am in harmony with the truth."

Tycho was living out his truth, and he had a severed nose from a senseless duel, a common-law wife despised by his relatives, and a first book in the presses to show for it. Lying awake one night on his bed in his old home, contemplating how to best properly emigrate from Denmark, he received a curious summons from Frederick, the king his foster father had given his life to save.

"Because you have not even asked for the things that others covet and fight to get, I do not know what you are thinking," the king confessed, "I suspect that you do not want to accept a great castle as a sign of royal favor because the studies you enjoy so much would be disturbed by the external affairs . . . I looked out of the windows, and I saw the little island of Hven . . ."

Hven

> It occurred to me that Hven would be very well suited to your investigations of astronomy, as well as chemistry, because it is high and has an isolated location. Of course, there is no suitable residence, and the necessary incomes are lacking, but I can provide those things. So if you want to settle down on the island, I would be glad to give it to you as a fief. There you can live peacefully and carry out the studies that interest you, without anyone disturbing you. I gladly support your investigations, not because I have any understanding of astronomical matters or know what is involved in all of it, but because I am your king and you my subject, you, who belong to a family which has always been dear to me.

Tycho could not abandon this family, not now. He was not so cold as to refuse such an empathetic gift.

Clustered at the joint of the elbow-shaped islet, he and his closest friends anointed the first laid stone of what would become his home, spilling several bottles of wine out upon the rock. "Liquid and light," Galileo would call it, and for the rest of Tycho's life, that bit of porphyry would lie, stained with light, tucked into the corner of the east dining hall.

The countrymen of Hven, about two hundred in number, could only look on hopelessly. Their lives as free subjects were at an end, and all now owed two full days of work a week to this obdurate, impertinent sovereign. A lovely bit of apocrypha goes that, as the peasants began to anger, Tycho turned to his court jester for what to do. "Fill them with ale," he chortled, "all the ale they can hold."

The serfs built him a mansion. King Frederick had proven true beyond his word; he gifted Tycho fief after fief to provide him with natural resources, and soon enough a private boat for his free use. The lowborns took to the sea, felling wood on the mainland for return to the isle. After a day lost in bondage, the mass shuffled back north, through a small copse of hazelnuts and alders to their little subjugated village. The air was tinged with brine.

For the sake of beauty and astronomy, the walls of Tycho's mansion were built in the cardinal directions. The central building was cube-like, high ceilinged, three floors and a cellar tall, with two semicircular appurtenances on either side. A curved roof fed into a cupola mounted by a gilded weathervane of Pegasus, the symbol of everlasting glory, on its hocks, aflutter. The rooms on each floor were simple subdivisions of that squarish space. The mansion was anomalous this way, lavish but not beyond reason: proportion and symmetry were the emphasis. Most of the rooms had fireplaces, and there was—wonder of wonders!—running water. A marmoreal nymphet was carved out in the center of the anteroom, splashing water about in a fitful display of decadence. The whole project, Tycho estimated, cost the royal family a literal ton of gold.

When his manor was finally completed, four more years had passed, and he was approaching his late thirties. The mansion he christened Uraniborg, after Urania, the muse of astronomy, but to Tycho the name meant something more. His little sister Sophie had grown nubile, soft-skinned and huge-hipped, always rouged for a family outing. She was, all the men agreed, that subtle goddess.

Urania Through the Years

A noblewoman, if it be her desire, could lead a life of elegant waste. Sophie Brahe was not twenty before she became Sophie Thott, married off to a man nearly twice her age. Mr. Thott was tremendously rich, she learned; he lived in a castle that was five stories tall. As for Mrs. Thott, the world was not her oyster, but she was its pearl: baroque, expensive, cultured, with a center obscured by a hardy "mother-of" exterior. Having produced a male heir for Mr. Thott straightaway, she had earned the right to roam about, money new-minted in her fat purse.

The mother was still such a young thing. It was not yet out of line for her to play with a ten-year-old cousiness, both as guardian and as equal. Around 1582, the two girls took a boat trip out to Tycho's island to see what wonders of science he was building. Here at Hven, Sophie was forever welcome, always a Brahe, and sometimes a simple "Soph."

Tycho would even call her Urania. She well knew her older brother had named his mansion in her honor, or at least, had named her in honor of his mansion. He was converting Uraniborg into a serious institution right before her eyes. Tycho had found maintenance staff, was searching for fellow researchers, and would soon be consumed by his astronomical endeavors.

In his travels, Tycho had acquired a giant wooden globe, almost as tall as his sister, and certainly wider, now mounted in the center of his library. It had been warped by the cruel German weather, but he spent two years smoothing the asperities until it became, to his eye, a perfect sphere. It was to be a catalog of all the fixed stars, he decided, representing what he thought were stars stuck up on the spherical border of the finite universe, shrunk down to human comprehension. Layered in brass, decorated with the animistic creatures of the zodiac, the star globe's primary function was to be spectacular: to convince potential patrons that Tycho was awesome and intimidate voyeurs through its sheer celestial majesty. Over the years, it would acquire nearly a thousand stellar markers, but this project was just beginning. All above and around the globe, hundreds of books towered over it.

Prize among these many shelves was an inconspicuous copy of an untitled manuscript, all but lost until its current loving home. Tycho had realized its true author and shared knowledge of its existence with his learned friends. He had also invented its modern title: *Little Commentary*, by Nicolaus Copernicus. Copernicus's disciple Rheticus, who had died eight years earlier in exile, bestowed his library to a friend of Tycho's, who spread its secret wisdom about.

In his later years, Tycho would be swamped by the growing Copernicus mythos, but he was presently more concerned with clocks. He had been purchasing clocks at an alarming rate and expense. Tycho's focus, from the beginning, was precise observation; even in astronomy, clocks were essential to this. They were needed to time lunar eclipses to check against predictions. The stars, because they moved in a perfect circle each day, could have their horizontal coordinates represented with clock time, by measuring the second they crossed a specific line in the sky—the meridian. Doing so would speed up work on Tycho's star globe tremendously. The most important use of clocks, however, was not astronomical. The workers of Uraniborg and peasants of Hven required discipline, fixed work schedules, and timed breaks for rest, meals, and sleep. Tycho had spent years seeking out the best clocks, yet no timepiece was self-consistent enough. In truth, no one back then even really knew what time was. After many wasted hours he was finally starting to accept this initial setback of not knowing what time it really was, a consequence of not having a good enough timepiece available. Uraniborg would have to operate without a second hand, and workers would have to reset clocks each day. It is a difficult business, this timekeeping, and nothing more quickly disorders it than contact with any of the arts.

After arriving to these first wonders at Uraniborg, Sophie and her kid cousin planned a sleepover at Hven. During the evening, the two were up late outside, having forgotten the time, observing the stars, when Tycho's large and imposing bailiff, ignorant of the noble blood before him, ordered them off the premises. Sophie's good brother apologized to her with his quill, dedicating to her a parody in verse.

Little girls begin to wonder:
"What path is made by the Moon?"
But faithful slaves shove them away
Thinking the planets do them harm.

This was the start of Tycho's habit of writing poems for every occasion, sometimes out of genuine emotion, but mostly as an affectation, to appear civilized and smart. For Sophie, he would produce all of his most considered verse.

Tycho did not cater only to his sister, although she was his favorite. By the next year he had an entire research staff to oversee. Many a young postgraduate was at his service. He had even hired a poet or two.

Uraniborg, bustling with clever people, beautiful nature, and mysterious science, quickly became a tourist trap, and King Frederick's promised isolation fell by the wayside. Tycho's tangled family tree was already people enough; in 1584, his fourth cousin Erik Lange, heading a retinue of noblemen and servants, visited Uraniborg for two weeks after his sister married Tycho's brother. (Nobles inbred themselves like pureblood hounds; they were all related twenty different ways.) The two supped together and had a cheery drunken conversation. Eight years the junior, Erik seemed to Tycho a "bright young man," despite having a penchant for unwise monetary investments. He was "in love with love," Tycho laughed, "in Cupid's castle," chasing after any girl with hips to hug and a mouth to kiss. When Erik asked for advice on such affairs of the immature heart, Tycho was so pleased that he gave it in moralizing poem. "You chase sex if you are weak," he wrote. "Let the stars above be your nighttime love."

Traveling with Erik was Reymers Ursus, a startling autodidact who had risen up from penniless childhood in German Dithmarschen to serve as a scientific consultant in his middle age. Associates called him "the Bear" because of his last name, which he had picked for himself as part of his rise to wealth. He was a tenacious, bawdy misfit, but none took special note of him at the time. He later sent Tycho a poem, where he thanked the "very great astronomer" for the food, drink, and money. He signed it as a "dear friend." All seemed well, and there was no time to reflect; some very famous guests were coming.

Sophie was back at Hven on August 12, 1586, for two whole weeks, to prepare for the arrival of the inconveniently named Queen Sophie, wife of Frederick, along with the queen's parents and several members of the Brahe family from court. Sister Sophie dutifully took the place of Tycho's wife Kirsten, who, being of humble origin, was made to hide back in Uraniborg with her six children, so as not to taint the royal proceedings. The queen could never fathom such unthinkable depravities as a happily married commoner at the front of her welcoming party.

Both Brahe siblings were esteemed enough to have sat for portraits. Their clothing was status they could wear; Sophie was the poorer, in a black satin gown embroidered with silk, a single ring on her squat finger. Tycho wore an haute midwaisted doublet, a velvet cape over the shoulder, lined with fur. A stately plumed beret sat on his head. Both added to their raiment many fine jewels, and a ruff, out of which their heads burst like the carpel of a flower. When the ship docked, Queen Sophie emerged dressed in greater regalia.

In a carriage to the center of the island, Tycho was surprised to discover that the duke and duchess, like their queen daughter, delighted in the natural arts. The duke was particularly amused by fire and chemistry. They had come, Tycho rightly assumed, "to gaze directly upon the astronomical machines, which they knew crammed the house," and he had a host of elaborate set pieces to show them.

Two years ago, Tycho had bid an assistant visit Poland, to verify Copernicus's geographic coordinates, in order to ensure the accuracy of observations Copernicus gave from that location in his *On the Revolutions*. The apprentice returned not only with a corrected latitude but several surprise gifts from an old canon. Out back the opposite side of the Uraniborg, up the staircase, in the northern observatory, a copy of Copernicus's self-portrait stared peaceably out. Copernicus's old sighting instrument, the triquetrum, rested by its side.

The device had become so anachronistic that Tycho never used it, but he simply had to put it out on display. It had made him "so delighted, because it reminded me of the great master who was said to have constructed it, that I could not help writing, immediately, a heroic poem."

That, I say, battle-ready Copernicus once attacked so much
with this slight cudgel, was easily able to ready the art . . .
What can genius not overcome?

Last decade Tycho had personally visited the son of Erasmus Reinhold, author of the Copernican *Prutenic Tables* and, he complimented "the illustrious astronomer of our age." His life was being overrun by Copernicus, the very man he sought to surpass.

There was a huge quadrant built into the corner of the north-south wall for the royal family to observe on their way to the dining hall. Tycho's first attempt at a great quadrant, back when he was a fresh college drop-out tripping through Germany, had long ago perished in a storm, but he had learned from its mistakes. This was no facile reproduction. Two meters in radius and five millimeters thick, the new quadrant was smarter, more restrained, and more sublime for it.

Spiced meats and meads ran from the downstairs kitchen to the banquet hall, blessed by Bacchus, god of wine. The queen may have told her parents the superstitious tales she learned on her last visit, Ovidian fables of hubris and incest. When the late lunch was over, there would be the cellar furnaces and a hidden laboratory for Tycho and the duke. There they might discuss fire and chemistry, in hushed tones, until the day's humidity gave for a brief tour of the outside grounds.

Left out the east door, through an exotic herbary of elecampane, angelica, sweet flag, and apricot, down past the printing press, the man-made, fish-filled waterholes, short of the aviary and right of the grange, there lay a small constellation of rooftops poking out the soil like furtive groundhogs. This was Stellaeburg, an underground observatory Tycho was having workers excavate to house his best instruments. Its rooms were small like a cell, three meters in height, with plans for servants to dig out two more crypts on the far north side. As it stood now, incomplete, cold and dark in the evening, Stellaeburg was of most use in demonstrating to the royal family what further glories were to come.

When the royal family boarded the ship once more, an hour before dinner, sister Sophie departed with them. Both Sophies were young, their husbands old, while brother Tycho lived in between, struggling to create a scientific legacy that would outlast him.

In 1588, Mr. Thott died, and it was, by all lack of accounts, not a particularly great loss. Sophie Thott inherited a brilliant estate, with her eight-year-old boy, a great admirer of his lordly uncle. To many noblewomen, this was their footling forever, a life of long corridors and prepared meals, hoarfrost leaping worse upon the trees each winter.

"She has," wrote Tycho, "as can easily happen to a widow, had many sorrows and worries, so she has sought different means of softening them." Sister Sophie, almost thirty, began to visit Hven compulsively, more than once a month. Her brother was a man of such scholarship, with such a vast library. She decided to start training in Latin, the scholar's tongue.

By now, Tycho had finished virtually all of the preparatory work on Uraniborg. His massive quadrant had been decorated with a celebratory mural on its inner surface, a painting Tycho had commissioned of his own image, made giant, slanting imperiously in a chair, minding only the stars. It was a visual love poem addressed to himself. Perhaps it was best the royal family never saw it.

The underground observatory Stellaeborg was also completed. It was accessed through an exquisite portal, with three coronate lions sculpted atop, the largest with an imperial scepter in paw. Yet another poem was inscribed on the stonework.

Urania discerning from heaven this earthly cavity,
What strange trick has been readied under the earth? says she
I sink under: that I hide the stars of the heavens . . .

The foyer was warm and cozy. Heating had been added, a necessity in the unpleasant winter months. The roof itself was closed, but Tycho had rigged up a clever architecture that a servant could open out, as a sliding window, onto any part of the sky he desired. Through this window one could measure interstellar distances, or heights from horizon, with a suitable instrument. That was precisely what Tycho's mature sextant was for.

Of the many instruments Tycho designed, his sextant was the oldest, least imposing, and most deserving of recommendation. Only a sixth of a circle (hence the name) it was more portable than a quadrant, and so was carried to one of the freshly interred crypts.

The sextant took two to operate. Luckily, Tycho had an ambitious little sister, who was now offering up her many free hours to him, and had been his aide since she was a teenager.

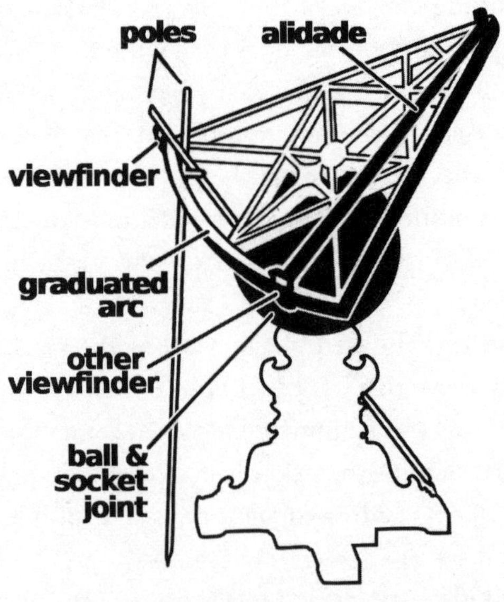

The mature sextant.

Rotating the sextant on a ball-and-socket joint until it was on equal plane with two stars, the first would appear visible to Sophie through a viewfinder in the corner of the sextant, which could then be fixed handily with two poles, sharpened like the endpin of a cello. Tycho would then move the alidade, a sliding post, until the second star shone through another viewfinder attached to it. If the plane was off, the trial would have to begin again, but this rendition was flawless. The angular distance between the two stars was now the distance between the two viewfinders, and could be read out off of the graduated arc between them.

Such an observation would then be recorded into an in-process astronomical journal, which lay open, numbers for all to see, on a large table to the east side of the foyer. These diaries were now gargantuan in extent, outranking all previous astronomers in quantity and rigor: an uncharted mountain of richly observed experiences. Fully completed tomes were studied in the library.

On the west side were six plus two portraits of famous astronomers, all lined in a row: Timocharis, Hipparchus, Ptolemy, Al-Battani, Alfonso, and Copernicus. The penultimate was Tycho himself, never one for humility, and to its right was his "Son of Tycho," a mysterious portrait reserved for his great imagined student yet to come. "May you venerate your father," the inscription below read. Tycho was a man most terribly provoked by his place in history.

His sister was also such a woman. Sophie had begun prolonged investigations into the genealogy of the Brahe line, and Tycho was sincerely impressed. The family chronicle and maintenance of memory were women's work of the highest order. These studies would have made clear her blood relation to Erik Lange, a man who had, since his visit to Hven, remained close to the Brahe side of his family. He was spiraling into a mad passion for alchemy, for those who practice it, and above all, for the legend of the philosopher's stone, said to enable the alchemical production of gold. In search of the stone, he had been tossing away his inheritance as though it were so many trifles. Sophie espied him, a handsome nobleman burning with desire. He espied her, a newly single mother with plenty of love, and money, hoping for an honest taker.

She took up medical alchemy, Tycho saw, incredulous, "and has been so successful that she has not merely distributed potions to her friends and well-off people, as needed, but has also given them to the poor, free of charge, such that both parties benefited to no small degree." One day in

April, she reappeared at Hven on a date with Erik Lange. Upon the start of that season, a birdwatcher added an extraneous note to Tycho's meteorological journals: "The song of the skylark has returned."

The last addition to Uraniborg that Tycho desired was a remodeling of his parterres. He visited the garden at Sophie's castle for inspiration, where a change had come over. The hoarfrost was gone. Everything was green and full of renewed life. "Tremendously beautiful," he recalled, "such can hardly exist in these northern climes; it has cost her no end of strife and diligence and hard work," but it was not enough, though nearly.

Sophie, as Uraniborg, required one final addition. "She zealously threw herself into the making of horoscopes, either due to her sharp wit which continually demands harder and harder tasks, or because those of her sex have a tendency for contemplation of things to come, and not without a little superstition." Tycho bore further witness, in the way of a highborn gentleman and protective older brother.

> I gave her help with directions and guidance, as she wanted it, but seriously warned her to desist from astrological speculations, because she ought not strive after subjects too abstract and complicated for a woman's wit. But she, who has an unbendable will and such great self-confidence that she will never yield to men in intellectual matters, cast herself all the more energetically into her studies. When I saw the clear signs of this, I quit opposing it and simply advised her to moderation. I figured she would grow weary when she realized how difficult these investigations were. When she guessed my thoughts—clever, she is!—she sent me a rather long letter, in which she clearly details her progress, showing she could acquire this science.

In 1596, Tycho was preparing a vanity publication of his many letters with members of the academic community, when he rediscovered Sophie's screed. "I know it would be with her applause that I include it," he thought to himself, but the taboo against a female scholar was too strong. The problem was not only that it would give offense to readers but that they would not believe it was real. "I cannot believe this actually arouses disgust," Tycho seethed. "This should surprise no one." "Disregard common prejudice. I have seen the truth of my sister. I know her in an especial way."

Treasures on the Broken Road

As Uraniborg approached the height of her glory, Tycho slid into the most triumphant phase of his scientific life. It would be full of new discoveries, but just as full of mistakes.

The first hints appeared at the tail end of 1577, as his mansion was still being built. During holidays, he would sail in to Hven to check up on construction. Around the tenth of November he was out with servants by a freshly dug pond, dragging a net to haul in captive fish for dinner. The sun was setting, covering the island with a golden veil. Between sweaty heaves of the net, he wrote, "I diligently looked up, that, perhaps, the customary serenity of the night should yield an observation: out of this improv I caught a sudden star." Tycho looked closer. This was not as unfamiliar as the star of 1572; it had a "curly lock of hair on its head, as they say. It was a comet." He had been waiting five years for this, another new phenomenon, which demanded to be compared, measured, and discussed.

The question about the comet on every astronomer's lips was that of parallax. If the parallax of the comet placed it above the Moon, then Aristotle's philosophy of an unchanging heaven would be dealt yet another critical blow, but determining parallax of a moving comet was much harder than the static star of 1572. "Uncivilized and ignorant proclamations have leaked through," Tycho wrote. Many reports were wrong, but he knew that the comet was in the heavens and grew obsessed with setting the affair aright. Even after the comet disappeared the next year, when everyone else had forgotten the matter, he continued to collect literature on the subject. It was to be over a full decade before his portly three-volume treatment, *Of the Aetherial World,* was published. In that interim, he had been quiet, but now like a comet he came roaring into view.

The scholarship in *Of the Aetherial World* was magisterial. The entire second half of the text, a massy two hundred pages, gives a robust account of every other account of the comet Tycho could get his hands on (the essay by dear friend Michael Maestlin was his favorite); this type of peer review was completely unheard of in the sciences. No other astronomer had a library and collection of instruments so fine as Tycho's, so none could threaten his position as judge, jury, and executioner. He became the dean of astronomers, not by virtue of brilliance, but by hard work, constant reading, independent wealth, and the forced enslavement of a couple hundred peasants.

Sandwiched awkwardly between his critique and a lengthy discussion of the comet was the most important part of the book. "Chapter Eight," Tycho wrote: "The Discovery of a Place where the Comet can Sit Comfortably, and a Hypothetical System which Excuses its Apparent Motions."

The chapter was a quick fifteen pages, but it had been awaiting publication for nearly as many years. In 1574, when Tycho still gave guest lectures at the University of Copenhagen, he would introduce the Copernican system and then tease that it could be "adapted to the immobility of the Earth." Now, in 1588, he had an engraving made and titled it "The New System of the World! Made Recently by Yours Truly, which Prevents the Excesses and Uglinesses of Veteran Ptolemy, and also the Recent Physical Absurdity of the Earth's Motion by Copernicus, and also is Most Suitable to the Apparent Motions." Tycho had no desire to follow Copernicus or Ptolemy; he longed to be famous as his own man, with his own system.

The Tychonic system, also known as the geoheliocentric model, is a chimerical beast. Tycho held that the Earth was unmoving, as did Ptolemy. But he also agreed with Copernicus, that every other planet orbits the Sun—except that for Tycho, as these planets orbit the Sun, the Sun was at the same time orbiting his unmoving Earth.

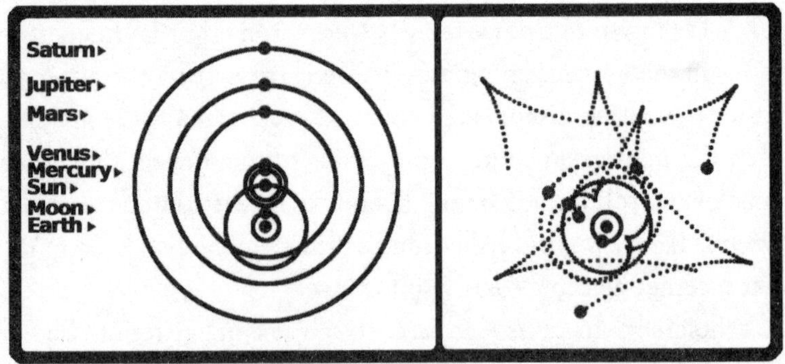

The Tychonic model. Note the overlapping orbits between the Sun and Mars, and how the Earth does not move, the Sun orbits the Earth, and all the other planets orbit the Sun.

Mathematically, this Tychonic system offers nothing that the Copernican system cannot provide with equal complexity. In practice, the two systems could be distinguished by a parallax in the fixed stars, but excluding such details, the two are equivalent. If there were a bird's eye film of the Copernican system in action, centered on the Sun, the Tychonic analogue could appear by simply re-centering the camera on the moving Earth, with none the wiser.

Tycho's system of the world was very much false, but technically, the Copernican system, in assuming planets moved around in perfect circles, was also false. By falsehoods, one may be brought to truths and falsehoods alike. As Tycho pondered his false system of the world, he realized something true: that Aristotle's metaphysics, which had been so damaged by the new star and comet of the past five years, must be dismissed.

Aristotle and Ptolemy had held that the orbits of planets took place on the surfaces of giant, transparent, hollow, rotating spheres, which could not intersect. This was a wild fantasy of the universe, but a widely accepted one, and no matter how Tycho worked out his system, he could not stop the orbits of Mars, Venus, and Mercury from intersecting the orbit of the Sun. To defy Aristotle at first broke his mind. "I could not bring myself to allow this ridiculous penetration," he wrote, "Thus it happened that for some time this, my own discovery, was suspect to me." Over the years he became more confident; the spheres became "unreal," "imaginary." "They are not really in the heavens," he wrote. With these metaphysical spheres dismissed, Tycho wrote, it was even possible for an orbit to be an oval! Tycho's desire for greatness pulled him like a magnet, away from thoughtless obedience, towards corrective narcissism.

So nervous and uncertain was Tycho in his own invention that he spent years of effort to strengthen his conviction. For the sake of his confidence, two principle rejections were required, of the Copernican and Ptolemaic theories both.

Some of Tycho's rejections of Copernicus were commonplace. In the act of challenging Aristotelian metaphysics, he still clung tightly to Aristotelian physics. Though he approved of Copernicus as a "second Ptolemy," he disregarded the Copernican system as "contrary to physical principles." Tycho was not a physicist and never analyzed these principles closely. His respect for Copernicus, at least, led him into less rote criticisms.

While Copernicus had mistakenly proposed that the third motion of the Earth, the change in its rotational axis, was nonuniform, Tycho noted that "his erroneous ideas on the matter are a consequence of the incorrect observations of the ancients." Here Tycho was right; no one had more authority than him when it came to observation, not even the ancients. But in his complete rejection of the Copernican system, he was utterly wrong.

Surprisingly, the strongest reasons for Tycho's fracture with Copernicus were observational. He searched repeatedly for a parallax in the fixed stars, but of course, found none; combined with a medieval sense of heavenly scale and proportion, this formed an extremely coherent argument against the Copernican system. No observational evidence, however, would exceed the paradox of Mars.

Tycho realized the distinction between Copernican and Ptolemaic systems could be put to an empirical test. In the Copernican system, Mars could approach the Earth more closely than the Sun. This meant that, if the Copernican system was true, when Mars appeared during sunset of the winter months, it should exhibit a larger parallax than the Sun.

This basic practical test was imbued with so many blunders that it descended into a bizarre comedy of errors. Tycho blindly assumed the ancient measurement of solar parallax as three arc minutes, which is well over twenty times beyond its true value. Meanwhile, expecting the parallax in Mars to be at least that large, it was actually less than half an arc minute—far too small for Tycho to ever possibly detect. This was like trying to read a book from fifty yards away, while nearsighted, without glasses.

With these blunders, the integrity of his project was entirely compromised, but the truth of the matter seemed deadly real to Tycho. He wrote, "I determined at the end of 1582 and the beginning of 1583 from the parallaxes of Mars, taken at sunset, that there is no evidence for the annual motion of the Earth."

Having rejected Copernicanism, the second act of this farce was Tycho's rejection of the Ptolemaic theory. Much of his justification was the same as Copernicus's (the nonuniform motion of Ptolemy's equant was a "sin against first principles"), but the primary experimental evidence was, unbelievably, also the motion of Mars! For after Tycho developed his system of the world, he realized that just as in the Copernican model, in his model Mars may draw closer to the Earth than the Sun. For Tycho to accept his own theory, he would have to contradict himself. He did exactly this, in writings years later.

> I had considered by means of subtle and accurate observations, taken especially in 1582, that at opposition Mars is brought closer to the Earth than the Sun itself, and on this account the long-accepted Ptolemaic hypothesis cannot stand.

The passage of time has conflated the very order of things. Tycho spoke so routinely with this ambivalent, contradictory language that he came to believe the dual rejections it implied. Such is possible in the human mind, but history accords its solutions.

A drinking straw shows refraction because air, glass, and water are all different media.

It was only later, after 1582, that Tycho reinterpreted his results from that fateful year by way of refraction—the bending of light. Starlight, as it greets the eye of the astronomer, propagates differently across different media, and the atmosphere varies a great deal as a medium. When the light from a star or planet penetrates the atmosphere at an angle, as during sunset, its appearance is correspondingly deformed. Tycho came to this effect while contemplating a mistake by Copernicus and was brought to make the earliest correctional tables of refraction in Europe. Being the first, they were filled with errors, errors upon errors, which when applied to Mars, combined with his mistake in solar parallax, could variously correct for themselves, or compound. In the old results of 1582, they created a parallax for Mars that had not been there previously. How often has he done wrong for the right result and right for the wrong one!

Tycho's greatest contribution to the new astronomy would not be as a theorist, though he was a fierce one at that, who wished it were so. It would be his many observations, dusty upon library shelves, cushioned in the arms of students, open upon the table in Stellaeburg. Awareness of this final misreading of his life, as of his many others, may have dwelt somewhere within him. For down below the Earth, hidden from Urania, on the ceiling of the Stellaeburg lobby was a painting of his Tychonic system. Along the row of portraits, to the left of the "Son of Tycho," his own portrait pointed up to it, with its tentative inquiry inscribed beneath:

Is this it?

The Parvenu

It brought Tycho great joy to see *Of the Aetherial World* leaving his printshop in 1588. It was the first note of his *Prelude*, his *Theater of Astronomy*, but it was marred by a double tragedy.

First was the passing of Tycho's metaphorical father, friend, and financier, King Frederick. He had died, according to his funeral oration, from "injurious drink." Black velvet was draped over his fat coffin, the height of opulence. His eleven-year-old son Christian climbed the throne, a boy king, for the time coddled by a council stacked with Tycho's relatives and supporters. When the boy grew older and attained the throne, Tycho's horoscope predicted he would be just like his father, cheerful and generous, with the chance of "heavy drinking, extravagance, and immoderation."

The second tragedy was far more grievous. By way of shadowy remark, from a friend of a friend, Tycho was informed that his planetary model had already been proposed by another.

Had his brainchild been stolen? Was it never his at all? The implication was unacceptable!

A ferocious search named the competing author, or villainous thief: it was Reymers Ursus (the Bear!), who had stayed with Tycho and Erik Lange four years back. Reymers had published his own system of the world, uncannily similar to Tycho's 1584 prototype.

Suddenly, bilious stories burst from the gut of all Tycho's companions and servants, of the little Bear whom none had written of before. No recollection is so telling as Tycho's own.

> That Dithmarschian Bear was here many years ago, accompanying my dearest friend Erik Lange, whom he served at the time. During those fourteen days, the same Erik lived with me amiably, with certain other servants and noblemen he had brought. Dismissing them, he came to me to settle philosophy (for he is not ignorant of the liberal arts, and loves them, especially our Urania—as my sister Sophia, now a young widow, is also referred to under this name, and he is accustomed to calling her by it, so long as she approves, you should know that this means he loves her too, and has his hands full pulling her into wifery). So I say, Erik now tired of Bacchian liquors, and we desired to return to Urania. He began to seek, among other things, whether some reason could be given he could use to guard against the absurdity of the motion of Earth introduced by Copernicus, which would still abolish Ptolemaic superfluities and incongruities, so that the heavenly appearances would be cleanly and completely satisfied.
>
> For he had deeply sensed me to strive for something so great, had heard what sometimes fell out of my mouth, that neither this nor that reason had been in agreement with the truth, but were more probable and convenient. However, that Erik desired this displeased me, every bit of me, as I am wont to robbed, so looking back I would have paid attention to that Dithmarschen standing by the tables between the other servants. For the rest of our conversation he was shameless, dumb mouth hung open.

Drunkenly, Tycho had whipped out a piece of chalk and, in lieu of a blackboard, ruined his fine green tablecloth with pictures of his Tychonic system of the world. After each new diagram for Erik Lange, he smeared out the chalk of the old one, so that no untrustworthy soul could filch this impossibly valuable and definitely true hypothesis. He was not careful enough, he now wrote, leaving open his "verse" to Reymers, "the inferior poet." A later account from Lange's old secretary found that Reymers had stowed away in the library, where he scavenged garbaged blueprints, behaving "like a raving maniac."

Soon enough, Tycho would refuse to even speak the name of "that Bear" in polite company. Reymers became a comic villain, of impossible variety: a fool, a visionary, a jackass, the master of evil. It was libel. Reymers was a skilled linguist and astronomer who had translated Copernicus's *On the Revolutions* into German. Having used his enormous talents to escape poverty, no spoiled brat was about to shove him back down. In boyhood, he had survived as a swineherder, living with pigs. Tycho was not the only beast who could sling filth.

Reymers ran out with both arms flailing, publishing that Tycho used his detachable nose as a viewfinder on starry nights, and that his "kitchen-maid" Kirsten, mother of his eight children, had been found relishing the naked company of a "merry band of men." Then, even worse, he feigned an admission of guilt. "Let it be theft," he wrote, "but it was intellectual. Learn to safeguard your possessions better."

When Tycho began to allude he had proven his system by 1582, this was his provocation. He desperately needed the provenance, to show that he had invented it first, independently, that it belonged only to him. Not only Reymers but tens, even hundreds of dreadfully self-serious astronomers all across Europe had been playing at this new game of centrist astronomy, searching for ways to celebrate Copernicus's invention while preserving Ptolemy's "common sense" about the Earth's immobility. They shared and borrowed their reservations and revisions across an elite and highly privatized whisper network, one which Reymers, having grown up in poverty, had little access to. To call his own discovery "theft" was, in this context, nothing less than a performative utterance of absolute class warfare: steal from the rich, give to the poor!

Though he would not wish to admit it, Tycho did not arrive at his system without the creations and aid of various assistants, but he borrowed rather more than he stole, and had the pretensions of nobility besides. Reymers, who had been destitute, acquired a position in Prague as imperial mathematician on the back of his supposed plagiarism. Tycho grew even more effusive. He became master of the furious run-on sentence.

> The issue would not be so grave had he not with the most outrageous and criminal lies shamelessly attacked with his scurrilous and misshapen pen in that disreputable book teeming with falsehoods and insults my dignity and reputation, nay, my country, ancestry, and most honorable house, defaming them as much as he can, shamelessly diffusing such stuff in print to the public, while at the same time admitting in the preface that . . .

What is the crueler irony is that none today would speak the name of Reymers Ursus were it not for Tycho's inconsolable worries he was being eclipsed. It was a fear so great as to realize itself.

Goodbye to All That

King Christian IV was nineteen years old when in August 1596, twenty Danish councilors, who had been ruling in his name, surrounded him to place a crown upon his delicate hair. The ensuing revelry and celebration lasted an entire week, for which Tycho was present. The court was a glamorous tavern, with hooting, hollering, and free-flowing wine.

Tycho returned to Hven in a state of unusual contentment, completely unaware. He had been carrying on with astronomy and observations at Hven for nearly twenty years. A month later, he learned that King Christian had rescinded his most valuable fief, blocking the main tributary of cash into Uraniborg. Before his mind could process the loss, his new king also terminated his annual pension. Tycho's excessive funding was dammed to an ebb.

He was in shock. Never in his life had he been at the whim of another, and now this boy who barely knew him had rolled back his life's work without the slightest ado. "Astronomy has always existed for the glory of kings," he begged Christian's economic advisor. "I ask of you, O magnificent secretary, give counsel to our king. Restore the fief taken away from me." A cold response informed him that the king had "absolutely no interest."

Christian had chosen to be an absolutist monarch. He believed in austerity, except when it came to his personal expenses. He deliberately estranged the royal nobility in order to shore up his private power. There was no better victim than Tycho Brahe, the noble so few loved, practitioner of everything useless to the crown. Not only was his money stripped away but his honor.

There was gossip in court that Tycho acted "against ritual," that is, in an un-Christian way. His name was smeared with the mud of his own making: of his constant misprision, of his despicable treatment of peasants, of that awful concubine he lived with. Within that short month, the alienation that had lain dormant in him for thirty years of adulthood awoke from its teenaged slumber.

At the start of next year, a fire broke out in the Uraniborg laboratory; another, the following month, in the winter room. Tycho hastily completed the last 223 stellar markers to complete his giant star globe. On March 3, 1597, he met his sister Sophie on his island for the last time. She must have known his plan; she felt physically ill. Following on March 15, Tycho's final observation at Hven was recorded. The next week, his meteorological journal, which had kept daily observations for all of fifteen years, sat silent. The gilded weathervane of Pegasus, symbol of everlasting glory, spun meekly above Uraniborg, an empty house.

The anxious possibility of Tycho's adolescence had come to pass. He was an expatriate.

King Christian did not think too hard about what he had done. Hven would be a pleasing addition to his private domain; as an island, it was the kind of property he enjoyed. Christian liked to sail. He had grown up sailing daily, on a lake outside his castle. It was the best way to fend off a hangover from a long night with a whore. His slick pompadour lolled in the breeze. He was the king. He was free.

The Outside World

The Sun still rose over Hven the next morning. Come the spring, hoopoes still alit upon its cliffs. The air still smacked of sea salt and the faint tut-tuttering of the tide still rang out along the coast. The common folk of Hven peered out of their bucolic glade, whole years of their free lives returned to them, for their lord was vanished.

No sooner had Tycho left Uraniborg then did he mourn his loss. He importuned King Christian a final long letter, courteous in tone yet crude in meaning, to which Christian retaliated in kind. The two formed a dissonant counterpoint.

> TYCHO: I have been deprived of what I should have had for the maintenance of my studies . . . in addition to which much else has happened to me (as I think) through no fault or error of mine.
>
> CHRISTIAN: Things have unexpectedly occurred and happened to you through no fault or error of yours, so you think.
>
> TYCHO: If I should have a chance of continuing my work in Denmark, I would not refuse to do so.
>
> CHRISTIAN: If you wish to serve as mathematician and do as you are told, then you should begin by first offering your services and by asking about them as a servant should.
>
> TYCHO: It is by no means out of fickleness that I now leave my native land and relations and friends, particularly at my age, being more than fifty years old, and burdened with a considerable household.
>
> CHRISTIAN: You do not blush to make your excuses . . . as if you were our equal.

All this vitriol was thinly disguised by endless feudal courtesies. The pride of both parties prevented any chance of understanding. Christian held to divine right and believed himself beyond the judgment of man. Tycho thought the inherent worth of science extended to free money, a mansion, and a private island. As he wandered about, lost in Germany, he turned for solace, as ever, to poetry.

> Denmark, what is my offense? How
> Have I offended you, my fatherland?

You may think what I did was wrong,
But was it wrong to spread my fame abroad?
Only a few Danes honored my work,
Herculean though it was, as they say.
Many have been led by me to life's greatest secrets.
Many have I given the goods and foods of my home.
Ye gods above! For this they drive me out!
But rightly seen, though driven out, my wings
Are free at last. And Denmark was my exile . . .

So fare thee well!

This ran for over a hundred lines, smack in the middle of his astronomical journal. His elegy was an outpouring of emotion, but as therapy, it was incomplete. Tycho still felt the need to send the poem to his old Danish friends, where it was soon seen by Christian, who could no longer bring himself to care. It was enough, however, for Tycho to begin looking forward.

Out past his own nose, he took stock of his remaining family. He had, always, his wife and children, some now of marriageable age, a few most faithful assistants, a set of retainers bound to follow, and the various objects they carried out from Hven: his journals, deconstructed instruments, parts of his printing press.

Tycho learned, of necessity, how to properly appeal to the sensibilities of nobles. Part of this was from his search for patronage; while he had found guaranteed residence with an old German friend in Wandesbeck, it was not of the epic scope to which he was accustomed. Through his many contacts, his vision came to rest upon the imperial court of the Holy Roman Emperor, Rudolf II, a man with inestimable wealth and a collector's sensibility. Like Tycho, he was a strange bird: an occultist of royal blood, who had even tried his hand at alchemy. The king showed promise. Better yet, the position of court astronomer was presently held by Tycho's nemesis, Reymers Ursus, whose hillbilly mannerisms were not to aristocratic taste. Nothing would please Tycho more than to knock him off the payroll.

He dedicated to Rudolf a new book, *Instruments for a Restored Astronomy*, illustrated by numerous copper plate etchings made in years past. The book is a technical report, with densely labeled diagrams and detailed expositions of his decades of research, but such wonkery is tinged with the bitterness of a home lost. It includes an obsessively detailed diagram of Uraniborg, right down to its tables and statues, even though Tycho swore he would never return. "With a firm and steadfast mind," he ended the description of his giant armillary sphere, "one should hold that everywhere the earth is below, and the sky above, and to the energetic man any region is his fatherland."

Tycho soon received confirmation to set out to see the King Rudolf in Prague. Offered an autographed copy of *Instruments*, His Majesty was most pleased, but the book most desired was not present. The next volume of Tycho's *Prelude*, first of his *Theater of Astronomy*, remained incomplete. The lunar theory, his latest project and last puzzle piece, was unsolved.

Tycho's new thoughtfulness was not just for kings and queens. It applied also to his family; he knew he was nearer his end than his start. For the future of his children, who no longer had any Danish inheritance, there was a growing concern. So too did it apply to his assistants. Having approached poverty for the first time in his life, he had discovered how hard it is to hold friends without home and hearth.

"You will always find a home in your sister," Sophie comforted him. This was miserably awkward. She had never had to write her brother before. He had always just been there, a short cruise away. "After a long long time without a message from you, I hear you have settled in Prague. So I compose, with all the love in my heart, a prayer to God: that you be well, and find some tranquility and people who might appreciate you, that you might be allowed your work and enjoy some good." She spoke of their family, her garden, her son, and her suitor, Erik Lange (or "Titan," as she had come to affectionately call him), who was roping her into his money troubles. She was the only person to ever bother with the name of Tycho's wife, Kirsten, giving best wishes to all, until, "I can see nothing else to write you, my dearest brother." And then once more, Urania vanished.

It was a symptom of his misfortune that Tycho sent several personal letters to his old helpers, asking them to join anew, and further, that several refused. All the more tragic, then, that a certain letter had arrived at Hven to an empty house, whereupon it wandered through Germany for three months, until it finally met Tycho, just before the road to Prague.

A letter from a stranger. It was these uncommon moments of complete novelty that delighted the medieval man of letters, to which Tycho was now most open: to fully displace one's own narrative—to contemplate another—first the letter, then the man, then the work.

A Letter Received

Since you, greatest of men, are king of mathematicians, not of just this age, but of eternity entire, as established by incomparable teachings, and are of most excellent judgment: I would create a disadvantage, if my fresh little work about heavenly proportion (under the title: *The Secret of the Universe*) is to bring to light any praise, were I not to let such flights of fancy seek your judgment and approval. By which reason moves me, as an unknown stranger out in this dark corner of Germany, to send off this letter, begging out of my immense love of Truth, which your reputation suggests is in you too: that you might see to this labor with the integrity and humanity for which you are renowned, unknot my entanglings, and show the results in a tiny letter. O happy me, if that which Maestlin, the same does Tycho see! With these two great defenders I will not waver to keep a strong spirit at the criticism already arising to circle about. But if however whatever is pursued here is weak, all silliness and puerility, natures which have trickled out of me through the many years, I will wish to make note of such assessments: or would I not again prefer this type of refutation to the approval, if it is that, of the whole world? I do not waver at models you have considered before: by these means I submit with ease, so the difficulties of communion through such a space ought not discourage you to message me. However, of your works nothing reaches us. What I knew of you I had learned from Maestlin. You will therefore forgive me, if in spots where your name is mentioned, I do injustice to you and your hypothesis. My reverence for you restrains me from more talk. Open the way to my struggling voice, by writing me the littlest something: for you will bring me to show you how much more I love learning than praise. Farewell, O most noble Sir, and my study awaits your considered recommendation! The Ides of December. Year 1597.

To your Noble and Great Reverence:
Magister Johannes Kepler
In Lofty and Bright Styria
Mathematician to the School in Graz

To Most Noble and Great Sir,
Danish Lord Tycho Brahe,
Foremost of Mathematicians,
by hand.

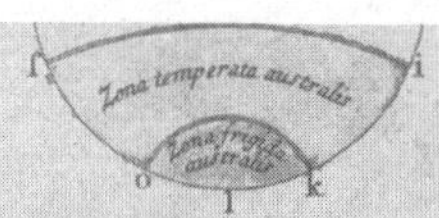

JOHANNES KEPLER

The first part, in which there is a great meeting and a great work

DRAMATIS PERSONAE

Tycho Brahe	an astronomer & his associate
Michael Maestlin	his teacher & pen pal & dear friend
Franz Tengnagel	a bureaucrat & employee of Tycho
Barbara Muller	his wife
Nicolaus Reymers Ursus ("The Bear")	a famous man & imperial astronomer
Rudolf II, Holy Roman Emperor	King of Bohemia
Christian Longomontanus	his hardworking friend & employee of Tycho
Giordano Bruno	Italian mystic & Dominican friar
Galileo Galilei	his infrequent pen pal
Elisabeth Brahe	Tycho's daughter

MAJOR WORKS

The Secret of the Universe • *The Optical Part of Astronomy*
New Astronomy • *The Harmony of the World* • *Rudolphine Tables*

YEARS ENCOMPASSED

1571–1611

Fathers, Sons, Ghosts

Johannes Kepler was astigmatic. He could see things, but not rightly. He was, of course, fully aware of this; "More than once, I tormented myself in order that I might find the cause," he wrote, "in vain." After years of labor, he discovered how his condition could be improved. His theory concerned eyeglasses, and he wore a pair of convex bifocals any time he looked up to the heavens.

It is difficult to discern the truth, especially a truth so all-embracing as in the ambitions of Kepler. He was not after mere equations, but their conceptual unity; not music but harmony; not art, but the nature of light itself. The ghosts of philosophers past had rightly and wrongly divined many such things through their reasoning faculties alone. Kepler's way of thinking was so akin he may well have joined their number, yet he did not. What is it that tempers the spirit of a theorist?

Protestant reformers roved through the country, smashing the idealized altarpieces of Catholic worship in the churches where Kepler would spend so much of his youth. The long history of purely symbolic religious art was diverging into a macabre, peculiarly German strain. Jesus on the cross, an architecture of torture, with gangrenous feet, collapsed ribs, fingers gnarled, rasping at the dead, black sky: such horrors were borne first witness in the sculpture of nearby monasteries. The contemplation of God was amplified by that grim reality of his creation, its most unpleasant aspects.

Four days before the new year of 1572, Johannes Kepler tumbled out of his gravid mother, an ugly, bloody, sickly mess.

Of these traits, it was sickliness that would stay; disease followed all living during the sixteenth century. The first child to two young parents, Kepler used their wedding date to deduce he was two months premature, a weakness further justifying his poor health. Considering his parentage, however, it seems that his conception was the cause of the marriage rather than the other way around.

"She is tiny, witty, skimpy, swarthy, scrappy," Kepler wrote of his mother, "a luckless spirit." Whatever she was, he admitted, was a vital part of him: "I take from my mother my bodily constitution. It is more suited for study than other kinds of life . . . the imagination of the mother imparts much upon the fetus." During her pregnancy, she had engaged in herbalism and what might today sound like a light magic, and Kepler always maintained a certain sensitivity, if not sympathy, for what it meant to be a woman.

His father had no such sense. "Treated shabbily," Kepler continued, "mother could not overcome the inhumanity of her husband." While the family name technically designated the Keplers as minor Lutheran nobility, Dad had been doing his honest best to run this into the ground. He gambled away the family fortune, danced around in the military but could not keep a job. In and out for years, one day he vanished for good, a mere apparition. "Criminal, inflexible, and obstinate," Kepler remembered, "he ruined everything."

Little wonder, then, that a brainy, effete youth, forced by his father to work in the fields, nearly blinded by the pox, had begun to look inward. The wonder is that he would not stop looking. He alluded playfully to gonads and masturbation. He wrote candidly of his various diseases. He judged himself squarely, and would, on rare occasions, write in the third person.

> In boyhood he began his reasoning early, by way of poetry. He tried to write plays, pulled out the longest Psalms, which he committed to memory. From the Latin grammars of Martin Crusius he let each example touch his heart. In verse he gave labor to acrostics, sketches, anagrams, after which out of well-deserved scorn he recovered his judgment, and undertook those diverse and difficult sorts of lyric. He wrote Pindaric odes, he wrote dithyrambs. The subjects embraced are unusual, the tranquil Sun, the origin of running waters, the sight of the Atlas Mountains through the clouds. He delighted of riddles, sought the most salty satires, amused himself with parables, which he pursued to the tiniest detail, yanking them out by the hair. In imitation he strove to retain almost every single word, which he brought to his own stuff. Of puzzles, crafting para-

> doxes would please: that the Gallic tongue ought presage the Greek, that literary studies is to be a sign of the destruction of Germany. When he disagreed, he could not bear to lie. Concerning his descriptions and discoveries, something else can always be found in this exemplary world. Before all else he loved mathematics. In philosophy he read the work of Aristotle, but questioned his *Physics*.

Teachers under Lutheranism, modern and quasipopulist, tried their best to promote talent, whatever its rank. Kepler received a prestigious scholarship; his hands and feet broke out in putrid sores. He advanced to baccalaureate; an old playmate got drunk and beat him up. The year of his father's final desertion, he was admitted to the Tübinger Stift, a promising city seminary some hundred kilometers away. He was ready; to his present seminary in Maulbronn he would soon be but a memory.

Other memories haunted the halls of Maulbronn. Less than a century earlier, the occultist Dr. Faust, whose first name was also Johannes, was said to have lived in the south tower. Legend had it he sold his soul to the devil for knowledge. All he left was myth, for artists to tame and refit to their own design. Shakespeare's compatriot Christopher Marlowe, in *The Tragical History of Doctor Faustus*, reinvented Faust in a moralizing fable about the hubris of man. "Faustus is gone," wrote the Catholic Marlowe in 1589, the year of Kepler's departure to his new Lutheran boarding school, "regard his hellish fall, whose fiendful fortune may exhort the wise only to wonder at unlawful things."

Such words educed a split between the religious and scientific character, but to Kepler they would have been foreign thrice over. In him, both the wondering and its investigation were products of belief. As a kid, if he was too sleepy to remember his evening prayer, he forced himself up early to repair the crime. When confronted with a worse sort of sin, he recited his favorite sermons from memory. His whole youth was made of misfortune but for religion.

So the child decided he was to be a theologian.

In Tubingen, most days were spent in the harsh discipline of scholarship, yet boys must be allowed some play. When he was nineteen, for both his personality and build, Kepler was picked to perform in a bit of winter theater as the comely, penitent Mary Magdalene. This aggravated a case of scabies, naturally; the boy never could quite escape his physical ailments. In spite, perhaps even because of, his illness, Kepler would forever recall his role as performance in a "*Comediae*."

There is no such opposition in Marlowe's tragedy; it ends with a literal descent. Faustus, in bargaining with the devil, contracted twenty-four years' service from the daemon Mephistophilis, whose name out of Greek is "Hater-Of-Light." When Faustus asks Mephistophilis to "argue of divine astrology," the imp tells, in accordance with the age, the Ptolemaic story.

Forced to break from biblical study, Kepler arrived awkwardly to his first astronomy class freshman year, to learn a textbook that was quickly becoming a modern classic in its field. He knew nothing. Before him walked the author of *Epitome of Astronomy*, a bright acquaintance of that Danish legend, Tycho Brahe. He glowed, with a style lucid and direct, a beacon gleaned from the dark seas of childhood unreason. So Kepler recalled,

> At Tubingen I was disturbed by the many difficulties of the universe. Thus, so delighted was I by Copernicus, whom Mr. Maestlin often mentioned in his lectures, that I not only frequently defended his opinions at debates with candidates in physics but even wrote out a thorough disputation on the first motion, arguing it comes about by the Earth's revolution.

Michael Maestlin had once desired the priesthood, serving as deacon to Württemberg for several years, yet now he was a math teacher. He was "my first leader and guide in the art of philosophical argumentation," Kepler wrote, "and much else. He ought to receive the highest mention."

From Maestlin, Kepler received his worst marks, an A- in astronomy. This would never do. The mere absence of disdain was enough for Kepler, who had no preference for his fellow students, to begin pestering teacher Maestlin after hours without end.

Kepler's immense desire to please was constantly betrayed by a complete lack of social grace, which is, in some strange sense, the most admirable aspect of medieval German character. Professor Maestlin was, to state the matter most generously, round, and Kepler, like a fantastic elf tailing fat Saint Nicholas, would speak in jest almost on accident: "The lighter the weight on his delicate bones / The more swiftly he flies to his heavenly homes!" Maestlin was, most luckily for his inept student, not without a sense of humor. When Kepler's kooky and endearing single mom rode out for a college visit, he insisted on getting her drunk.

Through his favorite professor's guidance, Kepler began to study Euclid's *Elements* carefully and publicly. He rediscovered proofs in advance of the text. His contemplations on astrology and theology grew at least as idiosyncratic. He graduated at twenty, with a broad education, and reenrolled to specialize in theology.

As the Copernican delved further into Christendom, he was hounded with the same gigantic contradiction. The Bible apparently claimed the Earth stood still; Copernicus claimed opposite. From the vantage of stony literalism, the two were irreconcilable; yet Kepler reconciled them, and much else, with such an easy confidence as to embarrass his opponents.

Theologians had noted centuries ago that all biblical contradictions (and there were many) could be removed by introducing a distinction of kind. Kepler made endless use of this throughout his life, insisting over and over that "astronomical context does not carry over into common usage." This obvious if unsatisfying resolution between science and religion had become so routine that the choice of kind was often used for comedic effect; when one professor suggested Maestlin thought God had hung up the Sun, still like a lantern, he told Kepler that "I am accustomed to opposing these jokes with jokes . . . while they are jokes." The bitter fight was not over meaning, but intent. Scholarly quibbling was all fun and games until somebody lost an eye.

Young Johannes had still no idea just how fraught this problem would be. He was just going to be a priest, albeit, a priest who loved astronomy, with some very curious opinions on both. He listened to a wide variety of religious sects and found different parts of each agreeable, yet professed to remain a devote Lutheran. "Christ is God and man," Luther had taught, "The two natures are one person. The person cannot be divided in two."

"Difficulty is born out of a union of these two kinds of life," Kepler wrote, in a well-considered formal letter, which could not help but reveal his surprise. The Tubingen council had recommended him not for the priesthood, but for a lowly position as mathematics teacher! Out in Graz, no less, in faraway Styria. It would have been rather rude to refuse. Besides, as he thought on it, the position made more and more sense. Thus he reasoned in his acceptance letter: "I do not declare that I am only to study sacred texts, because so total to me is the delight made by divine grace, that, whatever I eventually become, presently the Lord Our Protector preserves my freedom and soundness of mind."

A quarter-century later, Johannes Kepler was uncertain of how to best write the third chapter of a book on applied mathematics. A sage beard and graying mustache hung down off his face. "In what follows," he began, "the natural method will be somewhat damaged, so that the human mind, which often takes a quite different route, might find greater enjoyment."

The Theological Turn

The ultimate barbarism, for a theologian, cannot be that of the body but of the spirit. When the Italian church and state, upon the joint decision of a council of bishops and cardinals overseen by Pope Clement VIII, sentenced Giordano Bruno to burn alive, it could not have felt to Bruno a worse injustice than when they first stripped him of his every priestly title. He was but an elderly excommunicate, having done little with his life but trek about Europe with his books and quill. His seven-year imprisonment by the Church was not, apparently, as painful to him as would have been a false repentance.

Kepler had almost been a priest. Bruno was almost a philosopher, but always a mystic. His heresies were myriad, rejecting transubstantiation and the divinity of Christ, among other deep entrenchments of the Italian church-state. All combined into one semicoherent heterodoxy.

His militant endorsement of Copernicus was more of a footnote, then, fodder for his mantic view of the universe.

> Is it not the mother of follies that this endless space, with no visible limit but countless different worlds, which we call stars, each so essential as to be sufficient in and of itself, does on the contrary compose in relation to its proximity but one single continuous orbit about this point, rotating in equally endless circles in such a short span of time?

Giordano's universe, merely physical yet unfathomably vast, gave even more power to his singly immaterial Being, the omniscient, eternal, pantheistic creator-God, who so endowed creation. Hence, such a theology became laced with an unremitting carnality; the sacred body became animal, mere material. Reading his early works, coming upon a fable of a donkey and a lion having anal sex in a river, one is struck by the ribaldry, the wit, the Keplerian carnival of it, how similarly both men reacted to their age. Their differences were matters of extremes, sobriety, and luck.

As Bruno grew madder still, locked in the cell of his intractable trial, Kepler had begun his teaching and researches in Graz. Rich patrons made requests for horoscopes and astrological calendars, which Kepler obliged, with careful warnings of their fallibility. He would send his old teachers long, somewhat needy letters. After three in a row to Maestlin without a response, he justified himself in yet another letter.

> It's just that, the brilliance of my colleagues exceeds that of any other sort of person. Marvelous memories, gilded oratories, divers knowledges, boundless opinions: but the mind is always then troubled by sickness, sadness, simpleness, and is ever exposed to ever more things. This is what has shaken me so that I write you privately. But it may turn out to be of public interest.

Life was quite lonely in these northern climes. Kepler often complained of the cold. So few students attended his lectures on mathematics (some things never change) that he added sections on history and Virgil's *Aeneid*.

What students remained endured a frantic and loquacious teacher. His thought would drift, start, and stop without compunction. He would yammer on about astrological obscurities, which he relayed in writing.

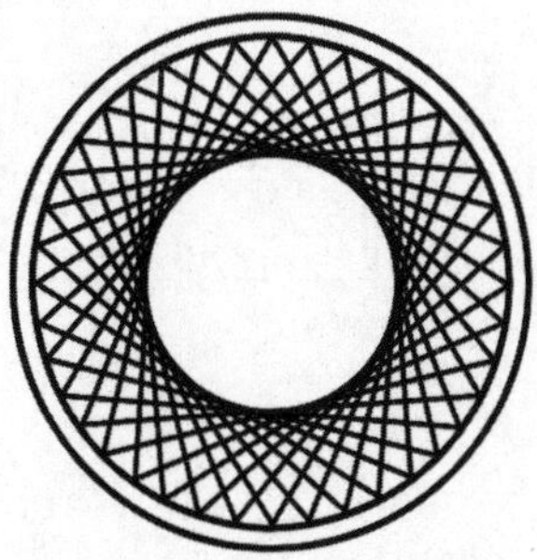

Kepler's drawing of the Great Conjunctions.

> About the 9th or 19th of July 1595, I was teaching what students were listening the way Great Conjunctions jump eight

> signs at a time, and how they progress step by step from one Trigon into another; I had drawn many triangles, or quasi-triangles, in the same circle, in such a way that the end of one was the beginning of the next. Thus the points where the sides of the triangles cut one another indicated a smaller circle.

Kepler stopped: new thought. These two circles were in proportion to one another. "This ratio," he startled, "appeared almost the same as the orbits of Saturn to Jupiter!"

Class was finished. Kepler rushed to check variations on a theme. He was enthralled. It felt as if he were discovering an entirely new world, like "the egg of Christopher Columbus." "It is amazing!" he wrote, "I no longer regretted lost time; I no longer tired of my work; I shied from no computation, however difficult."

First he called it his "little text," then his "elephant," and then, rather uncomfortably for Maestlin, their baby: "without your fire as midwife, I could never have given birth." His friend read the book.

Any unastronomical reader of *The Secret of the Universe* will be immediately disappointed. The title was irksomely literal, and Kepler's universe, unlike Bruno's, remained justifiably bounded. Apparently, its secret was three: "the number, shape, and motion of the circles," that is, orbits and the reasons for them. The reader may be further disappointed by its answer, if they disdain the fantastic.

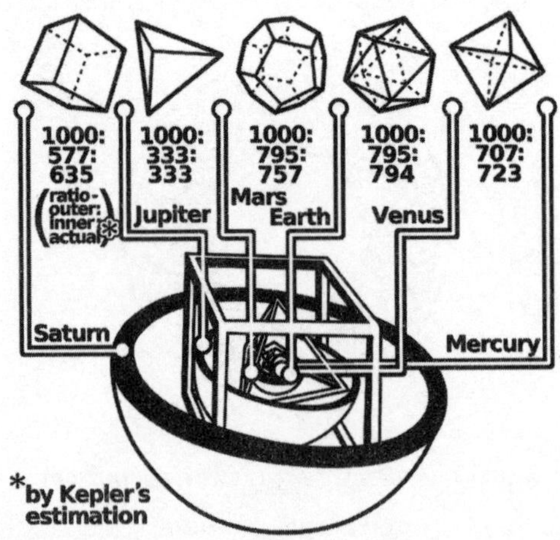

The orbits as dictated by regular solids, based upon an engraving from The Secret.

"The most great and good Creator," Kepler prefaced the book, "in the creation of this moving universe, and in the arrangement of the heavens, looked to those five regular, same-faced solids." These five regular solids are, he repeated, the "cube or hexadron, pyramid or tetrahedron, dodecahedron, icosahedron, octahedron. There cannot be more than this—see Euclid, Book XIII, scholium after Proposition 18."

Each regular solid, he explained, can be encased by a sphere. That sphere can then be packed inside a different regular solid, also encased by a sphere, and this can continue on up to the last regular solid. Given the right order, the circumference of these spheres defined the size of each planetary orbit. God had done this, according to Kepler, intentionally.

A "cutaway" of Kepler's hypothesis, from a drawing for Maestlin sent on September 14, 1595.

How could anyone lacking Kepler's disposition not be baffled by this ludicrous hypothesis? It is truly absurd, but people must try to understand each other.

Firstly, one cannot deny that this ridiculous idea predicts numerical ratios between the sizes of planetary orbits, which are marvelously close to the actual ratios, as measured in Kepler's era. Kepler's freakish leap of faith, which he would repeat many times, was to believe such a relationship was more than an entertaining coincidence.

Having the opportunity to make such leaps of faith was fundamental to Kepler's scientific method. It brought him profound joy to guess at the causes of things, and there was no one around who could deny it to him through a better explanation or more accurate experimental data. The poverty of human experience allowed him to believe.

As a child, it had brought Kepler physical pain that he had been denied the gift of prophecy. Personal, immediate, mystical experience of God was the form of it he most desired, but he was too immoral, he thought, had committed too many crimes to receive such a gift. He never acted like a mystic, but he hoped he might access mystical insights through mathematics.

So he knew, of course, that "if the measurements do not agree, the whole of the preceding work has been a delusion," but this evidence came after he had formulated his conclusion, a most irreverent, quasireligious method. *Fides quaerens intellectum*, wrote Saint Anselm five hundred years earlier: "Faith seeks understanding."

Kepler was not the first to ask his three questions about the number, shape, and motion of the planets, and his sources are clear enough. God himself, spake the Bible, "hath ordered all things in measure and number and weight." Though Kepler drew on his study of Greek, Euclid, up through Copernicus, no book weighed more on his mind than the Holy Writ. If the transfiguration of science were to happen at all, it was not to be by science alone. "I wished to be a theologian," Kepler beamed, "for a long time I was troubled, but see now how God is praised through my astronomy!"

At times, Kepler would even step out of astronomy into theology pure. "God the Creator had mathematical numbers with them as archetypes from eternity in their simplest divine state of abstraction," he wrote, espousing the quaint belief that God and number are coeternal. In his belief in coeternality, he had company with another great lay theologian, Dante Alighieri, who early in his *Inferno* wrote that "Justice moved my maker on high." It was justice, Dante said, that existed alongside God, the first mover, provoking God to take Dante up through Hell to Heaven; it was through number, Kepler believed, that he had risen above human ignorance towards an understanding of God's creation.

In this most counterintuitive way, theoretical scientists may conspire with religion, if they like, whenever they practice their most necessary and abused method of faith. When Kepler asked these age-old questions, with his brand-new answers, no small part of the debt was owed to his priestly training, Lutheran upbringing, and unique faith. When he added to this empirical verification, religion and science were smashed together into a hideous chimera, which only Kepler, it seemed, found to be beautiful.

Kepler knew this, and expressed his fear. "I shall have the physicists against me in these chapters," he wrote, "because I have deduced the physical properties of the planets from immaterial things." This was true; Kepler had determined the sizes of planetary orbits using immaterial mathematics. But then, he flipped this on its head and forced something theologians had believed immaterial to be a physical property of a planet.

> Just as the source of light is in the Sun, so in this case the life, motion, and soul of the universe are assigned to that same Sun; so that to the fixed stars belongs rest, to the planets the secondary impulses of motions, but to the Sun the primary impulse. In this same way the Sun far excels all others in the beauty of its appearance, and the effectiveness of its power, and the brilliance of its light.

There was a confused gift for physicists latent in Kepler's sun. Already at Tubingen, Kepler "had then reached the point of ascribing to this Earth the motion of the Sun, but where Copernicus did so through mathematical arguments, mine were physical, or rather metaphysical." He was unsure which was which.

Throughout its ill-considered hybrid of religion, philosophy, and science, *The Secret* bursts with the pleasure of mathematical play. The casual audacity of it all does bring to mind a particular Kepler, immature, needy, and lonely. Only in the first brief chapter did he even bother to justify his belief in the widely criticized Copernican theory. So ennobled by inexperience, Kepler quoted Copernicus's student Rheticus more often than teacher Copernicus! Maestlin, to whom he entrusted the book's first printing, even chose to add Rheticus's *First Account* as an appendix. The teacher was quizzical, to say the least, but supportive. His protégé's numerical tabulations were so disorganized that he had to set the type for the printer himself.

Maestlin even revised *The Secret* directly, at Kepler's request. "You see me get bogged down," he read, "I've a hopeless lack of Copernican knowledge. I'm eternally grateful for anything you could possibly write. I may start to beg you, if you do not get back to me promptly." Now, what on earth was he to do with that?

Despite his leaps of faith, Kepler always stopped short of unquestioning belief. While Rheticus had dedicated paragraphs to the divinity of the number six, he held back, as "one ought not draw explanations from those numbers which have acquired some special significance from things which follow after the creation of the universe." The ninth chapter of *The Secret*, devoted entirely to relation between number and personality, is a comical error, but a very short one.

As the slim text winds to a close, the picture is completed of a talented, if overexcitable, postgraduate, his nose in too many subjects, just setting the stage for his future academic career. It ends with a devotional hymn, an extended paraphrase of the Book of Psalms, into which Kepler had inserted a line on "the five-fold pattern of the starry spheres."

It was the turn of the century. Giordano Bruno, having refused all recantation, was shaved bald and made to ride an ass to his funeral pyre. A spiritual dirge emanated from his rear as the monks followed behind. In the holy city of Rome, he was stripped naked and tied to a stake. Those in front held up a portrait of Jesus on the cross, wishing to monopolize even this, the last moments of a denuded priest.

Bruno would not look—out of disdain for Christ? Unlikely—out of disdain for the Church.

Though he did send a personal copy of his work to Tycho Brahe, Bruno never knew Kepler. His was but another ponderous mind in that burgeoning horn of plenty from which scholars feasted, in arguments and disputations until proven or disproven, just as Kepler would become to many others. "If you had discovered any planets revolving around one of the fixed stars," Kepler would write a future colleague, "there would now be waiting for me chains and a prison amid Bruno's innumerabilities, or quite rather, exile to his infinite space." Elsewhere, "I learnt from Wackher that Bruno was burnt in Rome; he claims that Bruno bore the torture stoically. Bruno asserted the vanity of all religion, and in his vexation flipped God and Man about in circles."

Judgment

At this stage, Kepler could and would be seen as both a weak and mighty man. Such things differ with the angle of approach. Granted, he was in upper echelons of society, stationed far above the millions of illiterate peasants consigned to the land, still working under an essentially feudal system of labor. Kepler would occasionally complain about these "course and suspicious peasants" of the countryside, but when they migrated to cities, he found them wholly more agreeable. "Those of the lower classes," he described the citizens, "are always in good number, with straight forward and active minds . . . I might well call them my teachers."

Kepler's name, though still laced with the faint traces of nobility, was not elevated by this preference for intellectual capital over financial. He rose just above those rare city-dwellers and craftsmen whose mercantile spirit was slowly seeping across the country. Perhaps the bachelor Kepler was not immune to the squabbling of these petty bourgeoisie, which shrewdly included even marriage, the most tender of tender. Perhaps it was a priceless matter of the heart. "Literary men love solitude," he had declared in *The Secret*, and then, quoting Virgil from memory, "woman is ever a fickle and changeable thing."

He sent a letter off to Maestlin. "Look at the comedy! In '96 I was led to my wife."

So it came to pass that Herr Johannes Kepler married Frau Barbara Muller on April 27, 1597, to the interest of the greater community, after some concerning negotiations over the suitor's inattention and lack of funds. On the one hand, the girl was a catch: a voluptuous heiress to a small fortune. On the other, she was twice widowed at twenty-three with a small daughter and an extended family expecting a conspicuous display of wealth and gift-giving from any prospective suitor.

"I should not, after a few years, need a salary any longer, if that would suit me," noted Kepler in another lengthy letter to Maestlin. Yet, because of his wife's holdings, "I could not quit the land unless a public or private misfortune arose." Such concerns were prudent, as the Counter-Reformation began sweeping through his country. "Not always will King Rudolf be so free and easy to us," he worried, "some incensed Lutherans find cause for war, out of which they openly condemn the state." "For what has Germany rent itself in half?"

Almost instantly did the Keplers break the bed, but the chance of new life would not resolve these conflicts so much as create more. Their baby boy caught sick and died in two months. "The passage of time does not lessen my wife's grief," young Johannes held strong to Maestlin, before breaking down himself. "O vanity, vanity, and all is vanity! If the father would soon follow, this fate would not be unexpected . . . I have glimpsed a little cross on my foot, the color of which turned from blood red to yellow . . ." Kepler's world swelled with a physical and spiritual sickness.

Kepler would bear out his little cross. The young newlywed took the publication of *The Secret* as an opportunity to enroll himself in communications with every astronomical mind worth writing, and then some. While such community untied many of his tensions, on the work they could not agree.

Maestlin, having helped publish it, kindly endorsed his disciple's little book. "Recently," Kepler informed him, "I have sent two copies of my (dare I say your!) little work to Italy, which to the greatest pleasure and satisfaction of my soul was accepted by a Paduan mathematician, named Galileo Galilei, who attested to enjoying it. He himself has been in the Copernican heresy for many years now." Kepler was besotted by the prospect of friendship; the somewhat older Galilei had only sent back a civil response indicating that he had skimmed the preface, and the work looked commendable. Kepler immediately replied to "The Italian," but got no return.

Kepler also wrote an ingratiating letter to that old nemesis of Tycho, Reymers the Bear. "Pride of Germany!" he called him, "so much so I value your judgment. I love your system of the world." For one blissful, temperamental moment, Kepler was truly oblivious of conflict.

Reymers would not respond for two years, until he noticed Kepler's book was still selling. He did not critique the work. "Not in the least a serious writer," Kepler complained to Maestlin, but offered Reymers further compliment "because he is the Imperial Mathematician, capable of helping and harming me."

Response, where it was had, was widely varied.

To the noble Dane Tycho Brahe, he sent his note with which their story began. The great judge of astronomy gave his judgment.

A Letter Sent

Most erudite and excellent fellow: your letter out of Styria from the Ides of December last year was given to me about the start of March by way of a messenger from Helmstadt, in which beyond the teaching and cultivation uniquely ascribed to me, which may testify to my benevolence for you, ignorant of my visage and so far away, for such honorifics I yet have thanks for you. I have previously seen your book. Insofar as my occupation has allowed, I have scanned it thoroughly. It pleases me that it is surely not inconsiderable, and a certain sharpness in you and acuity unhidden by zeal shines from here: to say nothing of the sparkling and well-rounded style . . . For there is no doubt that all the universe has a certain harmony and proportion which is alternately divinely related and ordained; it is by way of number that the shapes may so succinctly encapsulate this manner: this was in some ways prophesied by Plato and Pythagoras. For they applied this quality with even further nerve; and if you uncover every consonance, with not a part wavering, you will be to me as great Apollo. However much I have the power to aid you in the matter of this arduous endeavor, you will by no means have trouble, especially now that I abide in Germany, and have left my fatherland with my whole family, so that my great astronomical treasury, prepared by the cost and labor of so many years, does not dissipate. At any time may you drop by, and personally confer with me on such sublime events in a pleasant and agreeable manner. Farewell . . .

T. B.

To that most excellent and clever man,
Mr. Johannes Kepler,
From the noble province of Styria,
Mathematician exceptional at the School in Graz,
And his dearly beloved friend.

Brahe's letter was actually quite long-winded. Between the first ellipsis lay multiple pages of critique, intending primarily, of course, to convert Kepler to the Tychonic system of the planets. Kepler was wisely warned against the dangers of unwarranted speculation. He scrawled his defense in the margins: "Everyone loves themself!" After the second ellipsis was an equally lengthy postscript, where Tycho added, with surprising calm, "I am indeed amazed to see you attribute my system of the world to Ursus, when in both your letter and book you attribute it (quite justly) to me. You seem to have written such things rashly." Kepler was caught! But how did Tycho know?

Why, Ursus the Bear had published Kepler's letter to him, without permission, as proof of his own brilliance! Hastily included, it was attributed to one "Mr. Johannes Repler," squeezed between a torrent of insults which Tycho had read to a growing ire. Kepler immediately recanted all praise of Reymers Ursus, offering many, many defenses of his first intrusion into the conflict. The most remarkable he wrote to his trusted Maestlin.

> That whole thing, you see, is poetic, taken from a poet, and uttered in a poetic spirit.

Whether the poet was ever forgiven is unclear, but the mathematician had been invited. Tycho Brahe had opened up his sanctuary.

The Need for Harmony

> Dearest Professor: the time that studies with you made Urania clear to me was also the time that would determine the death of my son: yet not once between us has there ever emerged the slightest offense, nor alteration in sex. Now a daughter is born to me . . .

In times of deep emotional strain, Johannes Kepler would write. Today he was drafting a novella for Michael Maestlin.

> . . . of Ursus and Tycho . . . Ursus betrays himself like a squealing rat, unmasked by his own words . . . of my mother and the news . . . in a business transaction I was shown a copper candelabra. Out of its tallow shines a maternal kindness, which I cherish . . . of discussions about gods . . . this work is most vapid, most annoying, and you have tossed me nothing of use, with which I might honorably adorn astrology . . .

Kepler's industrious mind was often faster than it was artful. If he ever slipped into incoherence, his circumstances justified it. A year had passed. It had been grating.

> . . . of me privately . . . I beg you, most gentle professor, fortify me with your counsel, whatever might be of use, as I will soon be forced out of this region, or pushed out by so many misfortunes: for which principle reason I am obliged to bear my soul. For if my life goes on, I do not see how I can stay here any longer. Nor am I any more respected in setting about my work . . .

Outside the window of his study, it was an easy and tepid summer. Kepler recalled the burial of his firstborn. In the cemetery, he had seen older people, who were dying in prison. Now even he had been fined for refusing to baptize his daughter as Catholic.

> . . . no matter what fortune may befall me elsewhere, I am certain that it will be no crueler than that which is lain out for us here, so long as the present government endures. The banishment of civilians now blows in the wind, which is unheard of: to show they are not yet committing seizure, after the citizens have begged the Prince, the response slows down. Out of court they drop mention of the gravest menace: if their voices be prophetic, never has something been committed of this magnitude: whether Lutherans may stay in this city, or have right to freely leave, carrying, exchanging, selling their goods. They would be full up on deception, the crime of treason would implicate all the citizens, and violence would have the embroidery of justice . . .

Inside the house, his zaftig wife would soon prepare dinner. She was best at salads and cooked tortoise.

> . . . I do not perceive that any private hope is possible in public calamity. There is never salvation at home with a riot outside . . .

Their daughter had died last month. There had been a large purulent abscess in her head. She was thirty-five days old.

Kepler was speechless with despair; he could only write.

> . . . of my studies and findings . . . now is the time, to return to you an account of my studies. If I were still in Tubingen I would be contemplating something new, a short book. Truly, my Maestlin, it is so right that you will be the one to say something similar, but without me you can compose nothing. For if I do not survive my debut with Tycho, I will not manage to craft these four chapters on cosmography. The title is The Harmony of the World. Prickle your ears—*eureka.*

Out tumbled the consolations of his imagination. Planets were plucked out of the quintessence, which knew no time, and placed back, recalibrated to the pulse of the human heart. The welkin grew thick with beats off the drum of reason, the parliament of fixed stars announced their fires in anticipation, the cosmos let out the ethereal drone of orchestral tuning. The universe began to play, conducted by Kepler's theory of harmony.

Each planet kept a mechanical watch upon the others. At the proper angles each sound their note, all arose a chorus, an effable chord which only mathematicians could hear. This was the music of worldly affairs, which were celestial affairs; war foretold peace, absence made presence, man begat woman, all was together in unity forever since the first day, when every planet had aligned in perfect conjunction: one clear, euphonious, sempiternal worldsong. It was serenity, it was bliss, it was light, and it was an extremely confusing theory, revealed too early.

> . . . I do not know if Tycho's little house of cards will cave under the pressure of this immense heavenly choir. I do not yet wish my words to be unpopular with great men . . .

Young Kepler had far too much life left unlived for his harmonies to be anything but aspirational. But he was afraid of death, and this made him precipitant.

Even the Catholics had been bewitched by this frenetic babble about world-harmony, but Kepler would never apostatize. One supportive government official gave him a courteous nudge.

> Tycho Brahe is circling about the court of the emperor in Bohemia, seeking a convenient place to fix his seat, and freely observe the heavens. He has received a salary of 3,000 florins (if the rumors be true) for services rendered. I would certainly desire such fantastic conditions for you, and who knows what fate has in store?

By the time Mr. Brahe sent along another letter of his own, Kepler was gone.

The Eyes of the Bear

Bumbling through varied German quarters, Kepler reached Prague a month later. He had fallen ill, but was received warmly at the house of a local baron, friend to both Kepler and Tycho. Tycho himself had received a castle in the more remote Benatky from Emperor Rudolf, still smiley with Lutherans, whereupon he immediately attempted to build a second Uraniborg, tearing out walls and adding new rooms.

The two were caught in the midst of an administrative hullabaloo. The plague had scattered the imperial court, forcing both King Rudolf and Tycho to move about constantly all last year. Tycho had been out questing to gain Kepler a place on the imperial payroll. Kepler himself had remained uncertain that this was his only course, until a final, dooming letter from Maestlin arrived in December. "I truly have no counsel for you," Maestlin stilled, "I can offer you nothing." A brief note on the solar eclipse, his family life, and a prayer to God was all.

In this fog of troubles, embattled by a quartan fever, Kepler marched upon Reymers the Bear. "I concealed my name, for fear that if he knew the matter was being raised by Kepler in person he would start an altercation." In such costume he told Reymers,

> Given that he had seen fit to drag me, who had written to him as a disciple, against my will into the judge's seat, he would therefore suffer me casting off a disciple's modesty, and assuming the role of judge in this quibble over words.

With that, he revealed himself and left in peace, to await Tycho's reply at the baron's house. He had chosen his camp.

So Kepler could truly say, "I said it to the Bear himself, to his face."

Reymers, on the other hand, had only just returned to Prague. Upon losing his livelihood to his enemy, he had fled the city.

Two Families

Kepler's calculations, he was learning, had not been accurate.

Longomontanus was Tycho's most trusted assistant. He had arrived in Prague two weeks before Kepler, right behind Tycho, and had already served eight years back at Hven. Almost ten years the elder, he smarted Kepler with the sort of knowledge only experience can possess.

Kepler had made an amateur blunder. In one calculation, he had inverted the relationship between Mars and Earth, scaling up rather than down. Left with a nontrivial bit of trigonometry, he had aggravated the problem by decomposing the shape into right triangles.

At this time, Kepler was working directly under Longomontanus. The preferred answer was rewritten concisely, running down the left-hand margin.

Soon enough, Kepler would master *Tycho Brahe's Triangle Practice*, the flimsy workbook given to newly arrived geometers, featuring a list of "dogmas" for just such use cases. But now, he was only just beginning; he was not yet thirty.

It had been this way since arrival. In early February of the new century, Kepler finally confirmed his invitation with Tycho, newly appointed Imperial Mathematician. Fetched by two of Tycho's cohort, Kepler had several hours in the carriage ride up to Benatky to size up his future coworkers.

Tycho's son was, at nineteen, by far the youngest in the compartment, a token Brahe, well-educated but, unlike his father, more politically than scientifically minded. Such aspirational families are a welcoming lot, but sport a brand of protectionism that can turn antagonistic.

Between the two sat Franz Tengnagel, twenty-three at the time. He was much the same way, if not even more so for lacking a blood relation.

The ambitious nobleman had been moving to change that, carrying on a tryst with one of Tycho's daughters, but he presently cut a rather pathetic figure. Both men routinely found themselves on such apparently banal errands as this.

These two, with Longomontanus, formed the core of what Kepler came to call "the Tychonics." Their namesake was waiting in yonder castle.

What apprehension fills the actor upon their first performance! "His wit and habit please me thoroughly," Tycho chirped upon first contact, "I think him worthy, loving and advancing all the best stuff." He had tantalized Kepler with work on the observations of Mars he had established years ago, and Kepler was only too willing.

Both would have begun in earnest but, as Tycho noted, "several difficulties insert themselves." The two were beginning to know one another.

Tycho Brahe was leading a posthumous life. His days were consumed by his legacy, dueling with the Bear over their old invention. His legendary Uraniborg was no more, but he guarded the leftover observations with avidity. The castle at Benatky was ersatz: instruments were strewn about outside the manor; the tiny mess hall could hardly contain all the voices; construction was ongoing, rooms were left half-formed, open to the elements.

Kepler was immediately disabused. He had arrived hoping to interest Herr Brahe in his theory of world-harmony; only now did he discern the validity of Tycho's own project.

As ever in times of conflict, Kepler sat down to write a long, and this time quite lucid, exposition. It was mostly about escaping poverty. "My skull is thick with deliberation," he wrote, "Perhaps I will become known to His Majesty, and arrive at a most honorable position. I had hoped that being near Tycho the Dane would provide for my discoveries, but there are more important (and easier) matters. Tycho has the best observations. If I stay in Bohemia, I will easily be able to learn—or rise to a higher office. My wife has no guardian for the journey; it has been difficult to bring my family together . . . "

On the next page, Kepler prepared his planned stipulations for working with Brahe. He had now recognized that what Tycho desired most from him was legal and scientific backing in the fight against Reymers, but "apart from astronomy, I must not be burdened with the necessity of other publications." "He will assign me no specific hours nor material," Kepler added, "In me, he must have faith."

Wherever Kepler kept this private list of howlers, it was not private enough. The Lord Brahe found some curious Latin and began reading. His aging eyes scampered over; his shaky hands grabbed a quill.

It was an insult of the cruelest variety, that unintended, extemporaneous kind, trifles of a conversation that linger for days, gnawing upon the whole foundations of one's self-image. In all his middling years, Tycho had been comforted by the idea that he was running a fraternity, not a business: for it is loyalty, not warrant, that forms the economy of a paterfamilias.

Kepler had written, "Tycho must concede to me my philosophical freedom."

"When have I ever prohibited that," Tycho scribbled in the margins, "which I do with vigor and zeal?"

Come April 5, the two sat down to forge a formal agreement. What was to be a colloquy between friends had become a professional transaction. Tycho brought along a scrivener, who carefully took the minutes.

Kepler could be guaranteed nothing, because Tycho had no guarantees to give. He would no longer offer up his own funds. Did Kepler want to live in Prague because of his family? "Never has this been done to me before. If it happens that the Caesar wishes to most leniently grant you that salary, to this end he is able to sustain this convenience of yours." Was Kepler expecting a steady income? "Concerning that yearly salary beyond provisions which I arrange for work in astronomical matters, as before, I wish to grant this, but have said that in this matter I am able to establish nothing certain."

This was legalese. Kepler had been stonewalled at every turn. At the end of the hour, he reluctantly agreed to absolute secrecy, whereupon he left Tycho reading over the minutes with his secretary. "Old age creeps in upon him," Kepler noted glumly, preparing to return to Graz. His only sanctuary was closing its doors.

Outside the horse-drawn carriage, at the precise moment of farewell, Kepler at last recognized the full weight of his error. He began to apologize profusely, lecherously—did he then, as he was wont to do in moments of sudden revelation, begin to cry? Did he recall John 9:4, "the night shall cometh, when no man can work?" Contrition filled his soul.

Three days later, Tycho settled upon his diagnosis of Kepler: "excessive arrogance."

Had Tycho turned the other cheek, his sanctimony might have been justified. But lately, he had become irritable and complacent, making misguided refusals and accusations. He thought Kepler was faking his fever; now that Kepler was an agent of Ursus, all was possible. Dear Kepler had become a "man on the wrong side of my every constant kindness," but Tycho himself proved far too rude for this self-ascribed constancy. "He has transformed into a rabid dog," he barked, "which he so often claims his stomach imitates (methinks it his head or heart!)."

The Dog, even more than the Woman, was Kepler's most beloved metaphor. He seems to have predicted this moment three years ago, in his autocaricature.

> His portrait is that of a delicate whelp. His appetite is alike: without class, he seizes whatever his eyes fall upon, though he drinks little and is contented very cheaply. He is similar in manner, ever insinuating himself with his betters (like a little housedog), leaning upon them for everything, and when he angers them he is desperate to return to favor. When the tiniest morsel is snatched away from him, he flares up and growls, as a pup. He is tenacious, persecuting those would-be crooks; that is to say he barks. And he bites, with a prickly sarcasm he has always at the ready. Therefore many hate him, and he is shunned by them, but his masters have love.
>
> And he hates baths.

These are not the words of an arrogant man. Back in Prague, Kepler let out a prayer to God and readied his atonement.

> O, these guilty hands, which strike swifter than the winds: which must start to soften, yet scarcely know how. For what shall I remember first? My intemperance? Of which there is a most bitter record. And you, noble Tycho, what of your blessings? They could not be enumerated, nor uttered with full honors. For two months, for no money, you nourished me freely upon your way of life; your whole family kept me safe, you celebrated my desires; what you hold dear, you have given me; never has anything I have said or done ever meant to hurt;

> in sum, not your freedoms, not your wife, not yourself, not anyone did you indulge, as you did me. All this, which I ought to have known, I now admit, and offer up this written testimony. I recall with great confusion how much the Lord and the Holy Ghost have permitted the intemperance and sickness of my mind . . .

Tycho knew the pains of an uncertain future. He forgave. The masters have love, but they remain masters all the while.

This rocky road would not smooth further. On September 30, all Protestants were legally barred from reentry into Graz. Nonetheless, when Barbara's father passed away soon after, Kepler made plans to covertly return to the site of his excommunication and attempt to produce some much-needed profit from his stepfather's land. There he climbed mountains and made first inroads into new studies on optics and light, but collected little by way of funds.

Such boded ill for his struggling wife. "What is good for her is unjustly scattered," the husband noted the year before. Kepler now bethought himself "a beggarly and contemptible man," and when that man wrote of her, which was often, it was no longer with desire but pity. She had become "simple and fat," "confused and perplexed," faithfully writing to her "heartlove" of the goings-on with "Diho Prei"—or "Brei," she could not decide which. She went out wandering in Tycho's gardens. Stray, worldly instruments loomed over.

Yet, of the people still present in Kepler's life, his family was all he could call dear. While back in Graz, Kepler asked after their wellness and finances, and questioned whether Tycho's daughter, Elisabeth, would "be a big bride." Young Tengnagel had placed the cart before the horse, impregnating the Brahe line before he was properly ensconced within it. When Barbara wrote back that yes, snow for the wedding in two weeks, but no, she had received no funds since he had left, the floodgates opened once more.

On June 13, Tycho Brahe opened a letter demanding, from this new employee already vacationing over a two-days' ride away, that his wife be paid remunerations. He could not be bothered to respond. "The betrothal of his daughter, Elisabeth, detains him," wrote back a secretary, "but so astonished am I at your words so rough and caustic." "To merit this fuss in the future, you want to use others."

The two exiles, so friendly at first, faded in and out this way.

"The marriage of Elisabeth is bound for a second day of festivities," the secretary reported, "but it is very cold." Such duality runs through life, with its "reigning loneliness of people," as Kepler called the Brahe manor. His own fever waxed and waned. "I desire for the better parts of my years to devour astronomy, here by these Tychonic riches," he wrote, "I would have long ago brought my theses concerning the harmonies of the world to their nexus, except that Tycho's astronomy has possessed me so, such that I nearly go insane . . . but Tycho will not produce his full supply for me! Only a little bit during meals, which leads him elsewhere, recalling now the apogee of this, the nodes of that." Kepler mentioned a recent study of magnets, adding to his experiments with light and optics, all bits and pieces of his grand harmony.

By this time, it was evident Kepler would never be a "Tychonic." Tengnagel was now a literal son, the obvious heir to Tycho's wealth in all its forms. Longomontanus, nearly forty, had retired from Tycho's service the prior August in order to marry and teach. He, at least, was a true mathematician, much milder in spirit, but forever cleaved to Tycho's technical bosom. A letter of his to Kepler professed that serious progress was being made on lunar theory, which was true, but attempted to persuade against any further theorizing. In the end, Longomontanus was the kind of ignorant sage who would spend the rest of his life hawking a method for squaring the circle, and Kepler had begun to steadily outpace him. They would keep a scholarly contact for many years to come.

Talk of Tycho's heir felt especially salient as the quinquagenarian trundled about the manor. "Tycho, who was great, was great before the year '97," Kepler wrote Maestlin, "big thoughts distract him; he becomes childish." "He always resembles a lost man."

More than one conflict was arriving at its sad resolution. "For so long have you been silent," Kepler began that letter to Maestlin, who had since stopped responding to Kepler's biannual treatise, each longer and more learned than the last. His quondam student, whether he willed it or not, had found a rough new teacher.

Tycho's beloved sister Sophie attempted to visit, but could not manage the long trip, even for her niece's wedding. Through emigration, Tycho had lost his truest family. Oddly, he saw Sophie's lover last, in 1598, as Erik Lange was skipping town on his creditors. Erik's obsession with alchemy had thrown him into insurmountable debt. He was constantly on the lam, leaving Sophie almost entirely longing and alone. From the bankrupt alchemist, Tycho forced further evidence of his enemy Reymers the Bear's betrayal.

This makes it all the less fortunate that, two years after Lange's testimony, a month before Maestlin's last letter, the Bear was extremely ill. When two ugly little men arrived at his bedside that September, he was not yet so foregone as to forget their employer. "When I was told that the Bear had fallen," Tycho wrote with glee, "I sent my attorneys."

> They arrived with a public notary to request that he retract his evil words, brimming with malice, as I prepared a list of his insults to be read back at him. He admitted several, denied more, but refused to recant. I noted this with pleasure, as I humbly obtained from the emperor four commissioners to try the case by law, but the culprit died.

Therein lies all that remains of the life of Nicolaus Reimarus Ursus, the Bear, proof that one truly can be lawyered to death. Tycho told his friends that Reymers had been struck down by syphilis. He continued to press for censorship, in the form of an old-fashioned book-burning of the dead man's words. It was successful; barely ten copies of Reymer's work survive.

"Struggles over priority are most bitter," a wise friend had warned Brahe at its start, "but this I surely know: if I wrestle with dung, win or lose, I am always defiled."

Kepler was told to prepare his tract against Reymers, but for publication, rather than trial. "I write of Ursus, and nothing else," he had complained to Maestlin. But whatever he was writing, all he could think of was Tycho's observational data. It was as the gluttons in Dante's *Purgatorio*, starving while tied to a fruit tree.

On October 24, 1601, the grapevine descended. Kepler would never be a Tychonic, no, but perhaps his jagged silhouette might fit—fit, foil, and reform—the subject of that rotting portrait Tycho had so vainly hung for his successor, long ago, at his little isle of Hven. By that portrait had been its label: Was Kepler to be the "Son of Tycho"?

*

Lunacy

Kepler recounts the episode in the finale of Tycho's observational journals. Tycho had gone to a dinner party, when "holding his urine past its due, he remained seated. Although he drank a little much and sensed tension in his bladder, he ranked illness below etiquette. When he returned home he was not able to get the urine out."

> He spent five days sleepless, when at last, with a most excruciating pain, he barely passed some urine, yet it was blocked nonetheless. Uninterrupted insomnia followed, intestinal fever, and little by little, delirium. His drinking, from which he could not be deterred, aggravated the harm. So it was that, on the day of 24 October, when his delirium had subsided for a few hours, in the natural way, amid the comforts, prayers, and tears of his family, he most peacefully expired.
>
> At this time, then, his series of heavenly observations was interrupted, and the observations of thirty-eight years came to an end.
>
> That night, which was to be his last, through the delirium, in which everything was most sweet, Brahe repeated these words over and over again, as if weaving a hymn:

LET ME NOT SEEM TO HAVE LIVED IN VAIN.

That family was surely his wife and children, although Tengnagel and Elisabeth were away on a year's honeymoon in the Netherlands, where Tycho had just been made a grandfather. Nor could there have been his sister, Sophie Brahe, who remained in Denmark and could barely comprehend the loss. "My mind winces at the word," she wrote, "My grief is too great to get the name across my lips." Decades later, she declared of Kirsten that "on account of the open and honorable life of both of them, she must be acknowledged as his wedded wife," but the good was for the children; Kirsten died after only a couple years of widowhood. The much maligned common-law spouse was able to live out her last years as in childhood, in the peace of a simple country home. The year after Tycho's lavish funeral,

Sophie finally married Erik Lange, who proceeded to leech her fortune in his search for the philosopher's stone, as she stayed home, lovesick, trying her hand at Danish poetry. She hired a physician to treat her for returning depression. She was never able to achieve her ambitions of fluency in Latin, the scholar's tongue.

The rest of the Brahe line were disgusted by Sophie's betrothal, but the woman could not be stopped. Tycho alone had acted with understanding. While he had respected Sophie's engagement, his fondness for Lange and alchemy had fallen sharply in his last years. In 1594, he had composed what may have been his last gift to his sister, an epic poem in her voice, a love letter from Urania to Titan, berating the god for his absence: "Your science is empty, leaving empty purse for money / And I, left with my marriage bed empty because of it!" In a subtler turn, Titan became the Sun, which was apt for Tycho, for whom the star was hardly the majestic center of anything. Urania, more intriguingly, was the Moon.

To make play at understanding this poetic metaphor demands a look at the scientific work. In the year of its writing, Tycho was also thrust into intense study of lunar theory. It was his last proper contribution to astronomy as an art.

With his improved accuracy, almost by accident, Tycho encountered more new anomalies than had been found in the past thousand years. He discovered evidence of nutation, the wibble-wobble of a planet's latitude as with a spinning top. In a very complex manner, he uncovered effects of the annual equation, a shift in the Moon's orbit due to the Earth moving about the Sun.

The most substantial discovery of Tycho's came from a close observation of eclipses, in an attempt to improve the predictions of ephemerides. The problems of present estimations made themselves woefully evident when Tycho himself missed half an eclipse over supper, believing it to come an hour later. He quietly passed over the mishap in his journals, noting tersely that "it is necessary to redo the examination." Of the eclipse two years later, he made better use; across two observations eight hours apart he found a ten arc minute lag in the prediction. He compared this lag time at conjunction to one of the Moon at quadrature, and another in between these two, and found a marked difference.

This change in the Moon's velocity, which Kepler dubbed "variation," is due to the Moon's orbit changing the Sun's pull upon it. In regards to it, Tycho noted his idol Copernicus "by no means saves the appearances . . .

which I now believe I have achieved," but whatever the achievement, it may have been appearance alone. The physical interpretation drew close enough to touch, its shadow colored every uncovered anomaly, but he never resolutely proposed how velocity changed from the data. Kepler, ever the student, friend, and opportunist, always gave the benefit of the doubt. "If Tycho's hypothesis for this variation is insisted upon firmly," he wrote, "the Moon is exactly as much faster in the syzygies as it is slower in the quadratures."

It is this slowly discerned moon, more multitudinous, cleverest in conversation, which Sophie was to Tycho, who first charted this new land of cosmic metaphor. The analogy of husband and wife to Sun and Moon was commonplace to the German poetry and culture, which Tycho had consumed in his youthful travels, but it had always been rote, unexamined, and Earth-centered. The Moon was held to be inferior to the Sun, weaker, without glow, but nonetheless corresponding, each having their own independent path among the stars. But by Tycho's hand, both the sister Moon and his sister Sophie were written in with a satellitic complexity, and the first slight preconscious of the learned passivity and forced subordination such a role required. Without her man, Sophie was "sunless, starless," Tycho wrote; "My art is better than yours," Urania whispered to the gold-obsessed Titan, "but whatever you like, I accept it." In such incidental flashes of insight, the poet does little but stand atop the scientist's experience and give form to his lazy thoughts.

Once revealed, the world of literature blossomed to the idea of Tycho Brahe. In 1605, a mere four years after Tycho's death, Cervantes published the first volume of *Don Quixote*, an absolute satire of dying medieval social customs, in which the titular character promotes himself to knighthood and goes on promoting his romantic ideals past the point of sense. But to call Tycho quixotic would be a mistake; whatever his ambition, through his obsessive observations, he became the patron saint of practicality.

Peel back that outer layer of comedy, and a different author is revealed: it is the Englishman, Shakespeare. At the end of his life, critics might suggest he was King Lear, but all before, in his constant consultation with death, alienation, and privilege, Tycho reminds of Shakespeare's Hamlet. This is almost literal; their deaths may well have been penned on the same day. Tycho's King Frederick built Elsinore Castle, the setting of the play, and Tycho's distant cousins, Rosenkrantz and Gyldenstierne, made a notable impression when they encountered the aspiring playwright. And how close to that famous quotation does Tycho strike with his own oft-writ motto, "Not to seem, but to be"?

Back at Hven, Uraniborg soon fell into desuetude. After Tycho's death, parts of its brickwork were stripped for a lesser house. His many instruments were sold off. Grave robbers broke into his tomb and stole his nose. The bare foundation of Uraniborg was later enshrined, a museum built by its side. Stellaeburg was excavated, and its exterior reconstructed. It is grand, but it is only a seeming.

Yet Tycho was, in life, never so depressing as to fully permit this Hamlet; he was only ever, and aggressively, his strange self. Out in Prague, drunkards still say "I don't want to die like Tycho Brahe" when they leave for the urinal. It is a difficult art, he wrote in a 1592 letter, to "drain a tankard, down to the very dregs"; "live well, and drink to my health!"

Reversals of Fortune

Kepler was disoriented by himself: "Am I rejoicing, or am I grieving?" At last, he was free to abandon his polemic against Reymers, which he refused to publish. That particular rivalry was now the dust of history. Strangely optimistic letters flooded his household. Only the more considerate allowed for his even stranger ambivalence. Thus, one message contrived, "I heard of Tycho's death not without great sorrow: to me, a brilliant hope for mathematics is at hand."

In yet another unanswered letter, he informed Maestlin of fortune's power.

> You have heard whispers of Tycho's death. The Kaiser has decreed that the administration of Tycho's instruments, documents, and unfinished studies shall fall to me. Using others' advice, which was to keep up my mask, I began to unite for the Caesar's judgment the very best, the true perfection of Tycho's works, a mighty tome given life by the name *Rudolphine Tables*. To complete it, I have requested half of Tycho's annual sum, that is, 1,500 florins . . . Now, what stands out most of all Tycho's work are his observations, the same exact books which

> he lorded over for all the years he labored. From his Prelude bursts pure ambrosia. I hope to publish for a wanting market. I do declare, I have been given hard work . . . Tycho did it, as did Hipparchus, suffering the greatest toil, which formed the foundations of the structure. But no one can do everything. Just as Hipparchus needed Ptolemy, who built the remaining five planets on top . . . Always goodbye, my cherished professor, and if you love me still, testify by writing a letter.

He acted quickly, and without hesitation.

> I do not deny that upon Tycho's death, because of either the absence of his heirs or the poverty of their practice, I boldly and perhaps arrogantly seized the observations he left behind, against their desires but nevertheless in keeping with the unobscure command of the emperor. Since he had entrusted the care of the instruments to me, I interpreted the mandate broadly and took especial care of the observations.

There were twenty-four books in total, and Kepler felt each one held hidden treasures. The pages on Mars were worth even more, beyond all measure, for it was them which Tycho had first promised, before the two yet met.

Kepler was now Imperial Astronomer, in possession of the greatest naked-eye observations that would ever be produced, and, relatively speaking, famous. He had a magnificent work, this vast star catalog called the *Rudolphine Tables*, to forever secure his new status. It is curious to imagine him in some podunk province, drafting an aspirational fan letter to his present self, but five years apart. What could possibly hold him back now?

In the middle of 1602, after an extended, genteel honeymoon, the stepson of Tycho Brahe returned home to find his life in tatters. Franz Tengnagel was now the de facto patriarch of a massive family of exiled nobility, and the twentysomething was utterly unprepared. His orphan wife offered up her house to all four of her spinster sisters; with his infant daughter, he found himself heading a gang of high-maintenance women. The new father had to lean upon the inheritance from his stepdad.

He converted to Catholicism, sold off Tycho's instruments as his own, and named his firstborn son, like Kepler's tables, after Emperor

Rudolf—anything to get ahead. Once he started marching about court, declaring his right to complete the *Rudolphine Tables*, nay, Tycho's entire *Theater of Astronomy*, Kepler began to fear for his livelihood.

When Emperor Rudolf recognized that his Imperial Mathematician had been stripped of his entire reason for being, Kepler was dragged before the court. "I was ordered to name what works will come from my study, which I would undertake to earn my supper."

"And you will witness the power of my industry," he wrote, almost menacingly, recounting the troubles.

Already, Kepler had completed an extremely rational and even-handed tract on certainty in astrology. ("Most of such pamphlets time and experience will expose as worthless," wrote the believer.) Next, as if pulling tricks from a hat, he pooled together his notebooks on optics, which were published as the genre-defining *Optical Part of Astronomy*, or *Optics*, in 1604. Now, he gathered together his latest triumph, the Mars notebooks he had begun in Tycho's service, "to be called *Commentaries on the Theory of Mars*, or *The Key to a Universal Astronomy*, or whatever else the name may be . . ."

This was the project which had been keeping Kepler awake at night for years. It was a fine replacement for the *Rudolphine Tables*, and no one, he assumed, could reasonably deprive him of the work he had been assigned by Tycho himself three years ago.

"More quarrels have arisen between me and the heirs," Kepler sighed.

> While I look out for the public good, they see Tycho's work as private property. There is shame in this battle, to which the cavils of spoliation cling: and so I yielded to an unfair contract, wherein my work could not be published without the consent of the heirs; finally this led to the crux of the matter, that the many observations would fall to me second, only if they gave them of their own free will. According to this contract, I do not claim the protection over Tycho's lucubrations for myself; but have cheerfully returned them to the shoulders of a nobleman, Franciscus Gansneb, called Tengnagel the Westphalian, who unwilling relinquishes the good parts of the work.

"He is as a dog in the manger," Kepler whined, "who eats of the hay, but offers it to no other."

Tengnagel now owned every bit of Tycho Brahe's legacy. The young protector had won the battle of politics, but in no time at all, realized he had lost the war on astronomy.

Alone to Kepler, he would mumble, "This is not my profession." The boy soon gave up his dignity and sent his court rival curt and perfunctory messages asking for help with observation. Kepler obliged. Together, they brought out the second part of Tycho's opus, the *Prelude*, from his leftover papers, but Tengnagel couldn't even get that right; the book was filled with errata and rudely failed to credit Kepler's contribution. The last third would be forever unwritten.

All the while, Kepler was exploiting Tengnagel's folly to privately work on Mars. His old notebooks were growing into a vast treatise, entire chapters were being rewritten, poems were being drafted for the introduction. But Tengnagel was scrupulously refusing to review the book, suggest changes, or give consent. The same emperor who had demanded Kepler produce was now withholding printing money. Life in publishing limbo was so demoralizing that Kepler started to consider distributing handwritten copies of his findings. This world of high-court politicking was repugnant to his nature.

By 1604, he was sending out letters to everybody who would listen, few though they were, bemoaning his inability to publish. "There can be no doubt," he averred, "so many things have been discovered that astronomy may justly be considered new."

He had found his title. This was, he knew, going to be one of the most reformative books in the history of science, if only he could get it published. He sent a worried letter to Maestlin.

"If you were Tycho, would you hate me? For what it is I dare to disturb?"

The War on Astronomy

Johannes Kepler's *New Astronomy* is the wartime epic of mathematics. All around Kepler, there were children dying of plague in the streets; there were peasant rebels hanging from the trees; there were Lutherans spitting on Catholic noblemen out on patrol. In his public life, Kepler was endlessly trying to resolve the hatred and confusion. As he worked in his private study, the tenor that his astronomical work took was, strangely enough, not so different. Kepler was trying to develop a model that would explain Tycho's observations of Mars. This was his war.

If this sounds excessive, it is an excess Kepler himself endorsed. His introduction maintained Copernicus not as a quiet Catholic canon, but a violent, abusive astronomer.

> A rumor has it that Rheticus (a Copernican disciple, first to dare yearn for a reconstruction of astronomy), when halted by wonder at the motions of Mars, and unable to untangle himself, sought the oracle that was his friend's wit. Copernicus, that cruel and irritable patron, seized the importunate inquirer by the hair and, over and over, bashed his head against the wall, leaving the body driven into the stony floor, adding the remark: this is the motion of Mars. A vicious rumor.

Kepler recalled all his fallen comrades. Tycho, in particular, had become a better friend in death. Several poems are composed in his honor.

> O noble one, give room, and do not disdain one who follows:
> All that I am is your gift, and all that I will be.
> All your monuments give me rest from my worries: without you
> Shadow I was; fathered by you, body I will be.

Then Kepler turned to the "difficult and strenuous war." He repeatedly quoted from Virgil, the mythic poet of war. The old gods warned him, "Cease, O Keplerus, to do battle in a war against Mars." It became *New Astronomy*'s leitmotif. At every corner, after every apparent success, there was a new battle to be fought. "Victory is futile," he sorrowed, just before the end, "and war is breaking out again in full force." Yet he would not relent.

New Astronomy is a massive, disorganized book,* with unforeseeable twists and turns. It is a mathematical odyssey. Along the way, Kepler encountered the mathematical equivalent of the horrible beasts, sirens,

* Kepler's *New Astronomy* is a complex mathematical work. Readers who want an unabridged account of Kepler's mathematical personality may prefer to turn to the appendix, "Seven Vignettes from the *New Astronomy*."

cyclops, and lotus-eaters from classical mythology. At one point, he produced, almost by accident, what might very well be the first ever graph of a function. He had no idea what this alien creature was, and played with it for just a few pages before sailing away. To solve an earlier problem about Mars's equant, Kepler had to invent a method so time-consuming that, as he explained it, he asked the reader for pity. "If this wearisome method has filled you with loathing," he wrote, "it should more properly fill you with compassion for me, as I have gone through it at least seventy times, at the expense of a great deal of time, and you will cease to wonder that the fifth year has now gone by since I took up Mars."

This was not false modesty. Kepler had spent so many years at war with Mars that he had come unmoored from all previous traditions of astronomy. Traditions be damned; ending war was all that mattered. In the fog of battle, he sacrificed the idea of circular orbits, and began to consider different types of oval for the orbit of Mars. So many of his attempts were dead ends that he called himself a "hasty dog, bearing blind pups." He tried an egg shape, even a pentagram, anything to end the war.

With these strange ovoid shapes, there arose a baffling problem of how to tell the speed of a planet in orbit. To answer this problem, Kepler, as he so often did, took an insane leap of faith. In fact, he took several.

The first leap of faith was surely his most insane. The subtitle of *New Astronomy* is "Based Upon Causes, Or Celestial Physics." The name of its middle section was "The Key to a Deeper Astronomy, Wherein There is Much On the Physical Causes of Motions." One important angle in his geometrical problems he named the "physical equation."

When confronted with the problem of planetary motion, Kepler invented astrophysics.

This invention was entirely his own, the single idea that metamorphosed Tychonic observations from Copernican astronomy into Keplerian world-harmony. Aristotle, holding that planets were metaphysical entities, had believed they were moved by angels, with distinct, individual souls. Kepler's vision of godliness was much more holistic than this; he believed planets were moved by a single physical force, emanating from the glorious Sun. He called this physical force the *species*, or *motive power.*

> There lies hidden in the body of the Sun something divine, which may be compared to our soul, from which flows that

> *species* driving the planets around, just as from the soul of a man throwing pebbles a *species* of motion comes to inhere in the pebbles thrown by him, even when he who threw them removes his hand from them.

This *species* is deserted territory. Not one physicist other than Kepler ever took it seriously. But what Kepler's *species* represented, the very idea that planets were moved by physical forces, was beyond serious. It meant an end to millennia of gobbledygook about Ptolemy and epicycles and equants. It meant astronomers were no longer subordinate to Aristotle's metaphysics, limited to the mathematical articulation of planetary motion, but could discuss the causes of these motions openly as a part of their science. It meant, in other words, that this metaphysics, the study of unchanging causes—what Aristotle had called "theology," "the highest science"—was, when it came the heavens, also a part of physics. The invention of astrophysics was not a war between religion and science, but their peaceful, loving, sensual conjugation. They were married together to end a different war, of human understanding against Mars and heavenly motion, and Kepler was their child.

This truly was a new astronomy. With it, there naturally followed the idea that the speed of planets in orbit corresponded to a physical law. Here, Kepler took his second insane leap of faith.

Because of physics, Kepler said, the speed of a planet in orbit was related to its distance from the Sun. Then he claimed, without any proof whatsoever, that these distances from the Sun were directly related to the area they covered over time. Proving that distances and areas were related to one another was not possible with Kepler's geometry or algebra; it would have required a brave new calculus. At one point, he even stopped his explanations, and begged geometers to invent a new mathematics. "Many first-rate mathsters labor endlessly on matters of questionable usefulness," he wrote. "I call upon one and all to help me here!"

How did Kepler know? How did he always seem to predict the future?

As a boy, he had yearned for the gift of prophecy. Kepler's final, decisive leap of faith was his least incredible yet, at the same time, his most notorious. There were hundreds of different ovals to choose for a path of Mars, most of them beastly and horrible. But all the most accurate ovals approached a certain peaceful, buxom figure. Here one version of an oval was "little different from an ellipse," he realized, and then for another, "it is just as though Mars's path was a perfect ellipse . . ." Oh! Oh. Oh, why not just throw out the oval, why not just . . .

"Let it be a perfect ellipse."

In one sentence, this strange yet elegant child cut the cord from its ancient parents: the ellipse born of the oval born of the circle was adopted as Kepler's theoretical foundation. There were no parties, no parades, just one boy, alone with his theory, struggling to end the war. "As if," he laughed through the heat of battle, "one were to squeeze a fat-bellied sausage at its middle, they would squeeze and squash the ground meat with which it is stuffed, outwards from the belly towards the two ends."

The ellipse can be defined simply. It has two foci: Take a loop of string and place it around two distinct points. Now, with a pencil, pulling the string back into a taut triangle, rotate it about the two points. This is the shape. In Keplerland, it was better understood as the closed planar intersection of a cone.

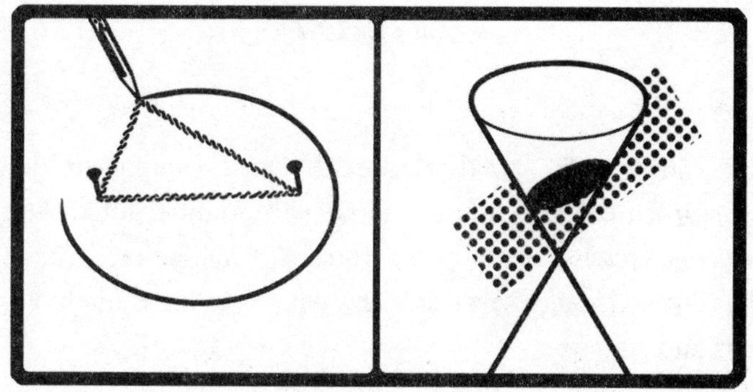

Two ways to generate an ellipse.

Finding an ellipse of the right width, Kepler tested it against his observations. "It was as if," he recalled the moment of revelation, "I were awakened from sleep to see new light." It fit the observations perfectly.

Kepler had the ellipse. He had astrophysics. He had the law governing the movement of planets. The war was over.

Dawn broke through the clouds, over this battlefield of pen and paper. Kepler had won. *New Astronomy* is the ballad of his victory, its heroes, its villains, its martyrs, its cowards, a song in sixty chapters.

Upon recognizing that peace had come at last, Kepler commissioned a special woodblock for the finale of the book. Urania, goddess of astronomy, rides her chariot down upon the last ever epicycle, with a laurel wreath in hand to crown Kepler, the pacifist, master of Mars, the god of war.

Kepler's First Law

The orbit of a planet is an ellipse with
the Sun as one of the foci.

Kepler's Second Law

Line segments from the planet to the Sun
sweep out equal areas in equal time.

Ascension

Dante's *Paradiso*, the last third of his *Divine Comedy*, is unlike its predecessors in that it is a poem about astronomy. Dante has ascended to the heavens. In what can now, for the first time, be recognized as historical, he travels through seven spheres, which he had thought controlled the planets and stars.

With my eyes I returned through every one
Of the seven spheres below, and saw this globe of ours
To be such that I smiled, so mean did it appear.

Scientific inaccuracy will alter the virtues of a poem. Dante did not much mind.

And I was in it, aware of my ascent
No more than one becomes aware
Of the beginnings of a thought before it comes.

Born into the provincial backwoods of a fractured Germany, Kepler had risen to the highest possible seat of astronomy in Europe. Rented out to the most infamous astronomer of last generation, he had outlived him, found the love in him, and untied the master's most gordian knot. In 1604, as if nature were celebrating its favored watcher, a second supernova broke through the skies. Even brighter than Tycho's nova, it would become "Kepler's Star." All that remained, it seemed, were the difficult and pleasurable decades of research, set before him like the bold fermata of a church chorale, yet the man felt incomplete.

Ascendancy is no inviolate joy. You leave people behind. Just before his ascent to heaven, Dante turned back to speak with the poet Virgil, who had been his guide. "But Virgil had departed, leaving us bereft: / Virgil, sweetest of fathers, / Virgil, to whom I gave myself for my salvation."

One day, as Kepler was in study, the essentials of *New Astronomy* already sketched out in his mind, a letter came across his desk. He opened it and read:

> Deliverance in our special savior Christ. For several years now, you brave man, famous mister, and genuine friend of honorable faith, I have been lazy in writing. Yet, although you have been carried to a highly esteemed station, whence you could look down upon me, if you wished, your steadfastness of mind, piety, and radiant love have not slackened but firmly clasp the foot of the bed. How firm you are I can only guess. Which hindrances thus interrupted my letters, I truly do not wish to recount, nor further excuse myself, with one exception: that I had no fitting response to your excellent mathematics.
>
> I am compelled to confess, your questions are at times too sublime to be satisfied by my own wit and erudition. And so, it is unavoidable that I grow dumb. But so much for that. I will henceforth, by the will of God, sir and true friend, offer up my exertions, to counterbalance that which I've neglected, with all I have in me, and that I am able. With what I know, I will respond: with what I do not know, I will confess to live in stupid ignorance.
>
> The new star is to be wondered at. I clearly remember a day on the 3rd or 4th I caught sight of some powerful light

breaking through the clouds. But then, by Jove, from where I resided it was concealed by another cloud.

How intensely I rejoice and revel, that the motion of Mars has been up to now captured by you, and that you have nearly finished the war! Truly, what you write, I confess I am not entirely able to follow, lacking the head which comprehends. From what I gather, you have written the same to me before, of their numbers, which you remember, having indicated to me the beginning and foundations.

Since that writing the facts have not returned to me, which therefore adds another reason I can hardly begin to speak. In the rest I recommend your own counsel. But you must say goodbye; God, may He calmly keep you safe, that which you do, and are happily able to conclude. Made in Tubingen 28 January 1605.

All yours, most loved
Prof. Michael Maestlin

To the brightest and finest man,
Mr. Johannes Kepler, of His Imp. Maj.,
Expert mathematician etc. Master
and His Friend in unyielding faith.

Dante was guided into heaven not by Virgil, but by Beatrice, the angel of his childhood love, conformed into a Christian girl, pure and chaste, perfect in every way. For Dante, it was not the physics but love, love of God and love of Beatrice, which gave the stars their motive power. It is not the beatific vision of God, not the picture of Eden sitting on the mount, not the animation of Satan chomping at the bit, but Beatrice, who has always formed the most impossible part of Dante's trilogy.

Barbara Kepler was a good woman, who had been given a hard life full of motherly toil and heartbreak. Her childhood had been spent under some wealth, and her husband could not reciprocate. "Did Johannes lower her daily wage?" rumormongers whispered of the couple, "Was she not supposed to eat?" He was always busy in study; she was slandered by city folk, who were repelled by her husband's religious independence of mind. They called her "Mrs. Stargazer." A dread shadow crept over Kepler's sketches of her: it was—dare he say it? (Yes! He must always confront the demon, face-to-face!)—melancholy.

Their marriage had become habituation. "Both of us well knew how our hearts felt toward one another," he wrote, but "not much love befell me"; "She was a woman entirely captured by love for her children." There were three of them now, plus three dead in infancy. When she had been pregnant, dozens of women would gather around her bedside. She had once been adored.

Barbara was no great theologizer like her husband, but she had grown beyond him in piety. Lutheran zealotry abounded; Emperor Rudolf, whose caprice was graduating to senility, militarized his own city to suppress the Protestant uprisings. The Counter-Reformation had arrived in Prague. Gliding over the blood-stained streets, searching out her despair, Barbara came to minister those stricken with plague, prayer book in hand.

She caught their typhoid.

Bedridden, reeking of sweat and sick, helpless doctors could do nothing but order she be given a new shirt. Her arms raised, dimly aware: "Is this the dress of salvation?" A small chapbook of her husband's poetry about her was sent to the printer. It held an old verse, which had charmed her.

Though for now you perceive but shadows received,
Endure life: the true incandescent essence
Your selfsame soul will forever perceive.
Eye blind to this light, at what do you tremble?
To lose all the fear, to reach something better.

"Kepler," another began, "married the stars."

A few years earlier, in 1605, about the time he was discovering the ellipse, Kepler had started to suspect he was dying. Recalling his childhood, he requested his old university in Tubingen accept a rough draft of his *New Astronomy* to publish, "should it befall me to die." "Verily," came the formal reply, "if you are unable to before death, we will manage this. We all hope that you may survive by your works and deeds." Throughout May and June he was stricken once more with intense fever, which provoked him to take a warm bath, to "purify the body"; it was, he suggests, the first time he had ever done so, and in the heat he found it difficult to breathe. "People like me are short-lived, as a rule."

Conceived in 1600, finished in 1605, embroiled in publication difficulties until 1609, *New Astronomy* arrived after unspeakable labor into Kepler's loving, living arms. His song was not yet over. It had just reached its final part.

"I think it to have happened by divine arrangement," he smiled.

There is no evidence that Kepler ever read Dante's *Divine Comedy*, although it is possible, of course—Latin translations were available. He had certainly read Virgil, Dante's guide, but this is secondhand. There is no direct link. It is just a metaphor, of his life running up, up, up towards its end, right in parallel with the Italian.

GALILEO GALILEI

The thorny second part, being very crowded and necessarily intertwined

DRAMATIS PERSONAE

Johannes Kepler	the emperor's imperial astronomer
Roberto Bellarmino	the reigning theologian of his day (& enemy of Galileo, villain)
Benedetto Castelli Gianfrancesco Sagredo Filippo Salviati Federico Cesi	members of a ragtag group of monks & noblemen known as "The Galileists"
Tommaso Caccini Niccolo Lorini Christopher Scheiner	members of a ragtag group of monks & nobleman known as "The Anti-Galiliests"
Rudolf II, Holy Roman Emperor	the aging King of Bohemia
Maffeo Barberini	a preeminent churchman
Virginia Galilei, Vincenzo Galilei, Livia Galilei	his children
Franz Tengnagel	the stepson of Tycho Brahe
Martin Horky	a wayfaring youth
Antonio Magini	Horky's able Bolognese teacher
Susanna Reuttinger	Kepler's new ladylove
Katharina Kepler	Kepler's mother
The Medicis	a royal family, born of opulence
The Popes	whose passings keep the time as the second hand of a clock

MAJOR WORKS

Sidereal Message • *The Assayer*
Dialogue on the Two Chief World Systems • *Two New Sciences*

YEARS ENCOMPASSED

1604–1642

Descent

In 1588, when Galileo first began to strive for a professorship, he was asked to compose a disputation on the exact size and shape of Hell in Dante's *Inferno*. When Dante was in Hell, he had been overcome by pity, but Galileo managed to keep his dignity intact. The sheer physical terror of the underworld found new expression in bathetic measurements concerning Lucifer's navel, the exact depth of the ice in which he was frozen. Because each level of Hell was a circle of linearly decreasing circumference, it was shaped like an inverted cone, 3,245 miles deep, just below Jerusalem. "It is easy to descend into Hell," Galileo warned, "and nigh impossible to escape."

By 1604, he had achieved a coveted full-time position at the prestigious University of Padua. Mr. Galilei became a teacher of men. He was a strong one. "From Signor Galileo," a student vouched, "I learned more in three months than I did with other men in as many years!" "Everything I am," cried another, "I am through having been your disciple." His recent lectures on the new star had been so popular that no hall at the university could contain them. With bravura, he performed them outside, where over a thousand boys jostled between one another for a view.

His teaching style was sharp, aggressive, even a bit violent, with a heavy visual bent. Early in his career, after he had solved a problem on centers of gravity with a clever but cryptic diagram, he sent the solution to several learned men. When one of them was confused, he simply drew it differently, and sent it again. "Then I saw where Signor was heading," and they were friends. After a brief contact with him, a poet recalls:

> What happened to me was what happens to men bitten by little animals, who do not yet feel the pain in the act of being stung, and only after the puncture become aware of the damage

received. So I, not perceiving that I was being taught, saw only after our discourses that I had a somewhat philosophical mind.

Aristotle had discussed philosophy with students while walking beneath the colonnades of the Lyceum, for which his school became known as *Peripatetic*, or "The Walkers." When one of Galileo's old students became a teacher in his own right, he would stroll up and down the banks of the Arno chatting with young mathematicians after class; it is not impossible to guess from whom he picked that habit up. The students Galileo left in his wake referred to themselves as "The Galileists." "His conversation was full of wit and conceits," one wrote, harking back, "rich, grave wisdom and penetrating insights. His subjects were not only the exact sciences but also music, prose, and poetry. He had a wonderfully attentive memory and knew much of Virgil, Ovid, Horace, Seneca, and among the Tuscans, Petrarch almost whole, and the poems of Ariosto, who was his favorite."

Despite his failing constitution, the forty-year-old never took sick leave. Instead, on weekends he ferried into Venice, to drink, laugh, and be merry. "I am blessed by God to be born someplace so beautiful," a newfound friend would say, guiding him around the city, "Our freedom infects every mind. It is unique in the world." This was Gianfrancesco Sagredo, noble gallant, lightly unshaven, the most happy-go-lucky man alive. It was his delight to expose friends to the Venetian "casino"; when Galileo discovered a courtesan was to have his child, he arrived back home with her by his side.

This woman, Marina Gamba, lived a five-minute walk away, as close as her admirer preferred. Galileo, who was now making quintuple his starting salary, moved into a not inconsiderable urban villa just before the birth of his daughter. Two years after the birth of his second daughter, he purchased an adjacent plot of land for a larger backyard. There, Livia and Virginia could play, aged three and four, dark and tawny hair flapping in the cool Paduan breeze. They were mostly reared indoors, though, near to the hirelings and watched over by his serendipitous girlfriend, soon to be gestating his third child, a son. Yes, Galileo loved his queer little family, no two ways about it; he even, in a rare case, turned to astrology to divine the personalities of the girls. These were "the best eighteen years of my life."

By middle age, Galileo had become a very busy man, but he could always find time to tend to his garden. In a soft leather smock, he seeded patches of cantaloupe, melon, quince, and a personal favorite, tart citron, out by trench beds of rhubarb, spinach, aloe, and rose. Herein lay the poetry of life; he was beginning to memorize the works of Ariosto, the Italian songster who had coined the word "humanism" eighty years back. Galileo thought his description of a garden to be very poignant.

> You will observe, anywhere else, how all in one day
> a flower will be born, live out its term and die,
> dropping its head on its bereaved stalk,
> for it is subject to the changing seasons.

Here, though, everything remained a verdant green.

"He would trim and bind the vines by his own hand," an adoring student recalled. "Agriculture provided him with philosophy."

Galileo dug deeper and deeper.

Upon Leaving the Top of the Arc

How is it that things fall? Certainly, it is slow at first and then ever faster, but it is not at all clear at what moment a fall begins, never mind its exact shape, or, Lord help us, its cause. In every case of interest, it is preceded by a rise.

Galileo gave his life as answer to such questions. He worked it out in his study, which was near enough to a draughtsman's quarters, with bits of scantling, all its woodworking tools, measured lengths of gut string weighted at one end, tiny brass balls and sheaves of parchment, and, most important of all, a quill and ink for record-keeping and analysis.

He was an extremely careful researcher, perhaps too much so. By midlife, he had published virtually nothing. As a joke between friends, he cowrote a trashy closet drama about the new star, but chose to release it under a pseudonym. "I realize how weak my arguments are," he apologized, "and how unworthy they are to come into the hands of others." There were some further writings on motion and a small tract on Copernican astronomy, but nothing that held to his exacting literary and scientific standards.

Slow research is not necessarily bad research, no more than a quiet life is worse than a noisy one. On this day in question, he recalled, "experiments have not been neglected."

> A piece of wooden molding was taken, about 12 cubits long, half a cubit wide, and three finger-breadths thick; on its edge was cut a channel a little more than one finger in breadth. Having made this groove very straight, smooth and sanded, we then lined it with vellum, smooth as possible, and rolled down it a hard, polished, very round bronze ball. This was done by placing the molding in a sloped position, one end some one or two cubits above the other. We noted, in a manner presently to be described, the time required to make descent, and repeated the experiment more than once for accuracy.

For any researcher in physics living before the refinement of mechanical clocks, timekeeping was the bane of experiment. Galileo, as was his way, made time into a physical quantity. "We employed a large vessel of water," he explained, "to its bottom was soldered a pipe of small diameter giving a thin jet of water, which we collected for weighing in a small glass for each time of descent."

In the hearts of the public, it was difficult for this sort of basic and honest study to gain traction. What the people wanted were novelties: fun, lucent, useful artifacts—what they had begun to call "natural magic."

> We then rolled the ball for a quarter the length of the channel, and measured the time of its descent. Next we tried other distances, then other inclinations, to no discrepancy in the ratios repeated a full hundred times, we always found that the spaces traversed were to each other as the squares of the times.

Now here was something new. Galileo, as a rigorous mathematician, was disquieted by his own experiment. "Of the accidents I have observed, I lack a totally indubitable principle to take for an axiom," he declared, and began to seek out a legitimate theoretical proof.

These internal disagreements had been growing like weeds across his life. He wanted more money; Sagredo tried use his nobility to snag Galileo another pay raise, was dispirited when he failed to help a friend, then thought of it no more. Virginia and Livia would have to live on in mere comfort. Meanwhile, Galileo's public teaching was guarded by university law, which obligated him to stick to the classics: Ptolemy one year, Euclid the next. Privately he was working further and further off the book, messing around with a primitive thermometer, playing with lodestones after reading about magnetism. But aside from his very best students, whom he loved dearly, the warmest reactions came from Venetian gentry and those of noble blood.

By the middle of the year, he was in contact with a duke and confessed a desire to become a "person of the court." The duke had asked Galileo about a certain doctor with medicine he needed. Galileo knew the doctor's son, Baldassar Capra, who was "about twenty-four years old, studies medicine like his father, and also astronomy and judicious astrology, about which he has much practice and exquisite judgment. This is as much as I can present to His Lordship. Should he request I penetrate further, I am at his beck and call." Such were the powers of Galileo, but they were not enough.

The court would not match the pay he received from teaching. He was hard put declining royal largesse: "If His Lordship is concerned with my spiritual integrity, I believe none can recognize but the purest sincerity, yet for my blindness I do not recognize the faults which those with clearer eyes see. May he pardon me, and excuse my weakness."

Court life answered to favor and fame; Galileo was a literatus, with no works to his name and a family to consider. Perhaps there was something worthwhile in this recent experiment, then, some clear application. A squared vertical motion, when combined with a uniform horizontal motion . . . it was just as a friend in court had suggested.

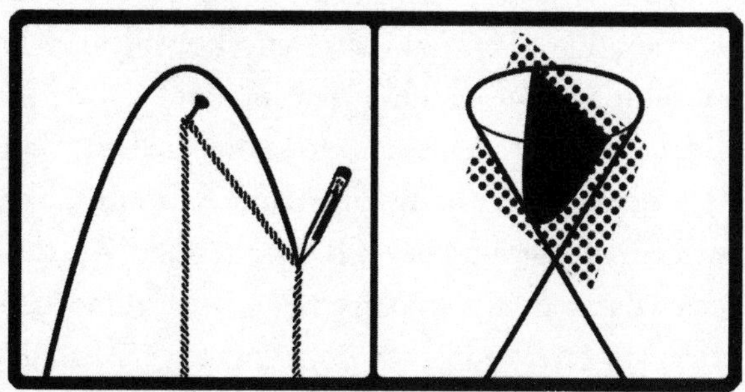

A parabola is a type of open planar intersection of a cone. It can be thought of as an ellipse with its foci an infinite distance apart.

> If one throws a ball with a catapult, artillery, by hand, or any other means, it will take the same path in falling as in rising; it is a line, which is in appearance, similar to a parabola.

Though left unpublished, this was Galileo's first great physical discovery: "The Law of Free Fall."

The Law of Free Fall

The distance traveled by a falling body is
proportional to the square of its time in fall.

Pupils

Not all students were grateful. On April 21, 1604, Galileo's scribe of eighteen months felt it necessary to comport himself, rather awkwardly, with the local inquisitor. "Galileo never makes confession or takes communion," the student professed. "On weekends, instead of attending mass, he visits the house of his Venetian whore . . . and he calls his own mother a hag." "Concerning matters of faith, though, I believe he is a believer." When the Venetian bureaucrats heard this complaint, they laughed it out of court; why, if Galileo had such difficulty seeing his family, perhaps they ought to pay him more.

Galileo, for whatever it was worth, was so well protected that he never even heard the accusations. He could carry on, shrewdly dancing the tightrope between teaching, tutoring, child-rearing, and research.

It was in private that the students came to know him best; Galileo, in turn, came to know the new diversity of his fatherland. Savoyards, Slavs, irresolute half-breeds, and an inordinate number of Bohemians added to the Italians that paid him for personal lessons in cosmography, optics, arithmetic, and whatever else he could offer. Most were military men, trying to get a handle on the basic geometry such fieldwork demanded, but some had a real taste for scholarship.

Benedetto Castelli had become a Benedictine monk when he was seventeen; a decade later he was helping Galileo with experiments after hours.

He was pudgy, with a large cresting forehead and not a trace of nobility to speak of, yet Galileo recalled, "he adorned every science with his light, and was full of virtue, religion, and holiness." Fellow monks had taken to sitting in on Galileo's lectures.

A friend like Castelli was the best Galileo could hope for. Most students were rather more distant; Galileo taught fortification to Tycho Brahe's nephew, who left lessons early. The stepson Franz Tengnagel had met Galileo himself, on voyage to Italy in search of a noteworthy astronomer to write the late great Tycho's biography. Galileo was not the man, Tengnagel realized; he was an "envious pettifogger, an unknown little boy, a baby mathematician who for his ignorance will publish nothing before he dies." Tycho had been a latter-day astronomer with a prelapsarian physic. Galileo did research in physics, and his astronomy . . . well.

In 1597, the young Johannes Kepler had sent him a book entitled *The Secret of the Universe.* Galileo had written back:

> Not days but hours ago did I receive your book; as the same messenger is returning to Germany, I thought it would surely be taken for rudeness if I did not offer my thanks for your gift. Thus I offer them, and in return offer greater thanks, for such proof that I have been deemed worthy of your friendship. I have looked at naught but the preface, yet understand your intent, and congratulate myself for having met a kindred truthseeker, and a friend to truth itself. For it is wretched, how rare are the students of truth, and how few do not follow some defective philosophy. But this is not the place to deplore the miseries of our world. I am rejoicing with you for the beauty which is to be found in the confirmation of truth, and I promise to read your book with a level head, as I am certain it contains many marvels for me. This I do all the more eagerly, as I came to the Copernican opinion many years ago, out of which I figured the causes of many of nature's workings, which I doubt the commonplace hypothesis will explain. I have worked out many calculations and overthrown many opposing arguments, which, however, I have not dared bring to light, for the fate of Copernicus, our teacher, terrifies me. Though he has gained immortal fame in the eyes of a few, there is an infinity (for infinite is the number

> of fools) who laugh him off the stage. I should publicize my thinking, were there more like you, but alas: there are none. I refuse the pain.

So much for astronomy. Galileo's growing notoriety as a private tutor was not as a stargazer but as the arbiter of an especially useful tool for Euclidean geometry, known as the sector. It was his cottage industry, for which he employed a live-in craftsman and copyist to reproduce the instruction manual. Around 1602, he had met the young hopeful Baldassar Capra to explain how it worked.

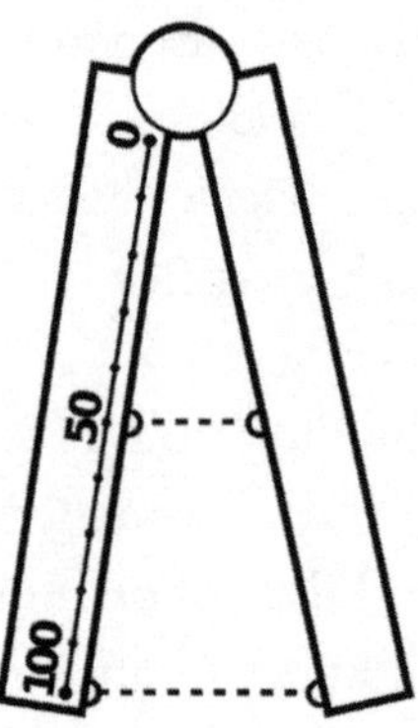

Bisecting a line using the sector.

A large swath of applied geometry can be characterized by two concepts: magnitude and proportion. An ordinary compass could easily take the magnitude, or length, of a line segment just by opening out to the ends of the line, but there was no clear way to take a ratio, because this would lose the original magnitude. The sector changed all this, almost like a second compass; it gave proportion an actual sense, a physical presence in space.

Two fat metallic boards, hinged at one end, rested flat on the page. A wide variety of numerical scales were engraved on its face. By opening the sector out to a line and associating that line with a number on its scale, a proportional change to that line could be found just by looking for the same change to its number on the sector. Measurement could replace calculation by taking the length of the new line.

The scales were not only arithmetical but also geometrical, which allowed for the easy extraction of square roots. Further scales were calibrated for the volumes of solids, the weights of metals, and even squaring a circle. The device helped set military flanks, transact exchange rates over coinage, and measure out gunpowder for cannon fire. Galileo had never fought a man in his life, yet he was not averse to warfare.

Baldassar left the lesson with a sector and its instruction manual, written in clear, sober Italian, with frank asides for "those with little schooling." "Be happy!" ordered the preface.

Capra was just so for two years, until he published his first work, a pamphlet on the new star criticizing Galileo's lectures. Galileo was "most excellent, with witty teachings," but "seems not to have taken much care on mathematical matters." The work was dedicated to his uncle; Capra was a noble youth, and his family had high hopes for him.

When Galileo received a copy, he channeled his outrage into insults. "His parents are beasts," he penned in the margins; Capra was branded as "my ox." Yet he declined to enter the forum—for infinite, he thought, were the number of fools.

In 1605, word of Galileo's prowess as teacher had spread so far that the obscenely wealthy Medici family called upon him to privately tutor their son. "I am late responding to Your Lordship," Galileo wrote, "out of a bashful deference, not to show a single note of rashness or arrogance"; "I long to be called your devoted servant!" The sector's instruction manual, five years out, was conveniently given a first printing, small and private, dedicated to his famous new pupil. He had published something in his name after all. This Galileo character just might make it in the world.

After twenty years of this slowly brewed success, in 1607, he discovered that a Latin variant of his manual on the sector was being sold outside of Venice, with added details on its construction, attributed to none other than Baldassar Capra. It was both infuriating and heartbreaking. "I enticed Capra to this enterprise with my own connivance," Galileo lamented, "for I dissimulated his earlier slanders and impostures against me." "It strips us bare; what remains of the lives of virtuous people is not just a mutilated infertility, not only an indigent beggarliness but a stinking corpse, despised, evacuated, abused." For the first time, he could not be silent.

The Medici family received a 30,000-word jeremiad about this young novice, "inflated by vain folly, promoted by arrogance, dumb with audacity, poisoning all the senses and more with his tongue." None could be allowed ignorance. Baldassar was taken before a special committee, examined, reexamined, cross-examined, and promptly expelled from university. His book was banned; his career was over.

Not all students were grateful.

This new century was not one of charity. In 1606, the Catholic Church placed the free city of Venice under interdict, for being rather too free with its armies and ships; more to the point, Pope Paul V excommunicated the entire Venetian Senate. Galileo was present that night to watch the accessory churchmen forced to leave state, each boarding the transport ship with their brief candle held aloft in the dark. His mellow Venetian friend, Gianfrancesco Sagredo, began to develop a serious enmity for the Catholic order, although he too was out of state. After two years as acting treasurer up in Northern Italy, Sagredo was promoted as consul to the least Venetian place on earth: Aleppo, Syria. Not even the most slaphappy man-child can refuse work forever. A most loyal student Benedetto Castelli had left too, ported several hours south to Cava, where he would send Galileo their first letters, of great length and learning.

Still, Galileo found an excuse to vacation to Venice without his best friends. There was camaraderie across his entire intellectual subculture. In June 1609, he was once again drifting across the soothing waterways, imbibing good drink and making good talk; this month, the best talk. While there, Galileo recorded, "a report reached my ears that a certain Dutchman had constructed a glass, by means of which visible objects, though very distant from the eye of the observer, were seen distinctly, as if nearby."

The idea kept hold of his imagination all that afternoon. Next morning he had figured it out. By the end of the week, he was at it in his study, a fitted wooden pipe by his side.

Galileo kept almost all his old letters. An unanswered reply from one Johannes Kepler still lay there, twelve years old, moldering in the stacks.

> How truly I wish that you, so blessed with a keen intellect, felt differently. Although you are wise and hide, your example of, as you say, withdrawing from general ignorance, never rashly inveighing or opposing the madness of common teachings, following the way of the first teachers Plato and Pythagoras,

> in this age there is Copernicus! Then many others, even the smartest mathematicians, setting about this immense work of moving the Earth. It is no longer so strange. Perhaps it will succeed, with the support of community, urging this speeding chariot towards its turning point, so that, because small crowds may be swung by reason, more and more we begin to overthrow the authorities. Have faith, Galileo, and come forward!

He took the pipe outside to test its magnification on the Moon. From his garden, the watchful spires of St. Anthony's Basilica stood just out of view.

A common spectacle lens was not strong enough; he would have to grind them himself. One of his favorite little tricks was teaching students how a rough surface could be made reflective, just by polishing it up to change the incidence of light. In the convex glass he could see his deformities pronounced: balding at the temples, red fleeing from his beard, very tall with giant ears. Magnitude. His right eye was off-kilter. A sense of proportion.

Horky's Odyssey

Martin Horky was an extremely German man, that is to say, born and bred in a house divided, and though he left it early, such stayed with him wherever he roamed. "I visited Tubingen in 1608," he recalled, barely eighteen.

> I was in Strasbourg, Heidelberg, Altdorf, Basel, Freiburg; then I left for Paris, in the company of a noble Silesian, to see that beautiful French capital. After a year there, I passed by Venice, lady of the sea, or should I say, the dens of Venus, where all was of enviable beauty, and I gave myself to not a few pleasures: I spent three full moons in her company. But she did not suit my spirit, praying to commerce

> rather than knowledge, so I went to Padua, which was the same. With a hardened mind I sought out the mother of study: Bologna. I shall stay here for a few years to study medicine and mathematics.

Quick as a whip, fully versed in Latin and knowledgeable of Greek, Martin had much to recommend him to Antonio Magini, an esteemed college professor, who took him on as a personal secretary. Had Magini not defeated the immature Galileo Galilei twenty years ago for this very seat in Bologna? Was he not sole inheritor of the astrological findings of Domenico Novara, the greatest teacher of the second greatest astronomer in living memory, Copernicus? Had he not been personally asked by the great Tycho Brahe to write his biography? True, true, all of it, yes, again, true: Antonio Magini was a name humanity would never forget. Together they sat down to review the latest work by the most famous living astronomer, Johannes Kepler's *New Astronomy.*

Horky was a great fan of the master mathematician; he had brought Magini his personal copy of Kepler's first book, *The Secret.* After careful study, he prepared to write the legendary author.

It was an extremely calculated letter, formal, while still drawing upon their shared German heritage; full of praise, but with enough critique peppered throughout to demonstrate independent thought. Only once did Horky lose himself in euphoria, writing, "You alone, Kepler, are the light and leader of my country." He hinted what a pleasure it would be to come to Prague, to meet Emperor Rudolf.

A month later Horky wrote Kepler again, of how "I am delighting in studies at Bologna, setting my eyes to the highest haven, that I may be able to reach it." His employer had also reached out to Kepler, by rather more detached language. Magini was terribly impressed with Kepler's *New Astronomy*, but was unconvinced, continuing to make his ephemerides in the Tychonic style. Kepler's mathematics were extremely imposing. The response came briskly, inviting Magini to Prague and with the postscript, "I will respond to Wenceslaus Horky at the nearest occasion. Send him my greetings." Martin Horky was embittered. His hero could not even remember his name.

He was a voracious reader. The University of Bologna had a vast library. Faculty had finer private collections, and all were up to date on the latest releases in their field. When Horky heard that the thwarted professor Galilei had finally managed to publish a book for public consumption, he was roused; when all 550 copies, not an insubstantial number, sold out by the end of the week, he was rankled; when he got his hands on the thing—oh my God. The old man had gone mad.

The tiny pamphlet was named *Sidereal Message*. Its title page proudly declared:

AN ANNOUNCEMENT CONCERNING THE NIGHT SKY

in which GREAT, FARAWAY, ASTONISHING spectacles are expounded,
perhaps for all, but especially Philosophers and Astronomers, by
GALILEO GALILEI
PATRICIAN FLORENTINE
public mathematician at the University of Padua,
with the aid of a device lately reinvented by him, the
PERSPICILLUM,
in which is observed the
SHAPE OF THE MOON, INNUMERABLE STARS, MILKY WAY, NEBULAE,
and above all else the
FOUR PLANETS
revolving around JUPITER at unlike intervals and periods,
with extraordinary speed, unknown to anyone until this very day,
so newly captured by their first author, who decrees they shall be
named the
MEDICEAN STARS.

There were more words inside. They did not matter.

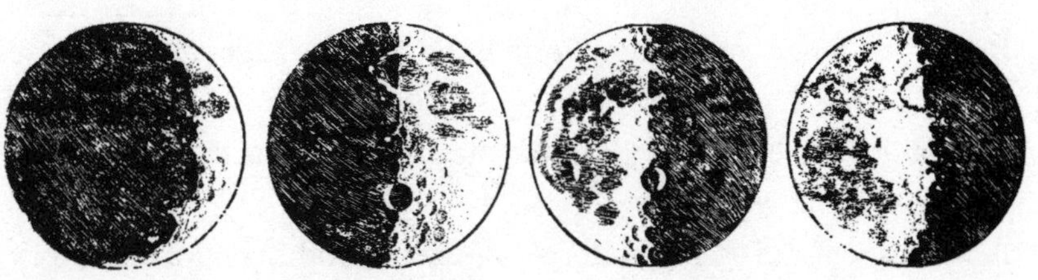

Galileo's observations of the Moon.

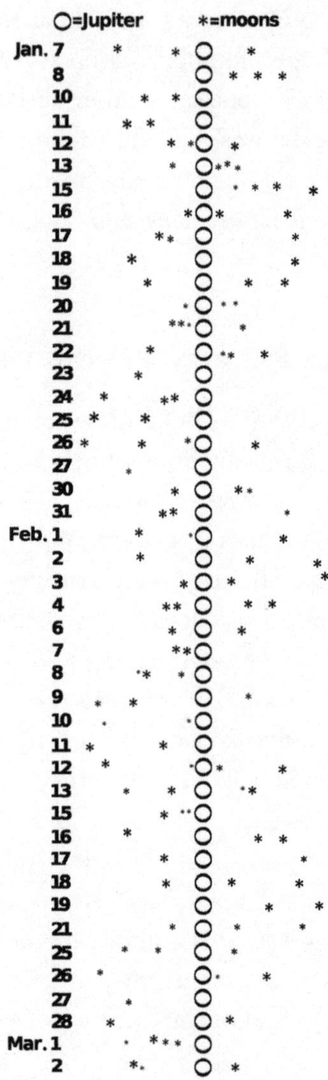

Galileo's observations of Jupiter.

Even an illiterate could palm through the first few pages to the pictures. They were large, half-a-page each, conspicuously of the Moon, yet unlike any humanity had ever seen. "It is like the face of the Earth itself," Galileo declared, and that was that.

A few pages later were comforts familiar to both Magini and Horky, the Pleiades and Orion Nebula, but they were hidden in a pullulation of dots, barely recognizable if not for their captions. "With the *perspicillum*,"

Galileo instructed, "you will detect such a crowd of stars which escape natural sight that it is hardly believable." What was old was new again.

"Now it remains for us to reveal what appears the most important in the present matter: four planets never seen from the beginning of the world right up to our day." Nearly half the book was spent on simple pictures of these moons, named after the Medici family to win their favor. Each had a blistering orbital period, less than a week, which Galileo demonstrated with multiple observations across a single night. No fraud, it seemed to imply, could so extensively fabricate such petty details.

Galileo's *Sidereal Message* was, quite literally, a comic book, excepting the most important trait. It was dead serious.

"We have in these moons an excellent argument to remove the scruples of those who, while tolerating the revolutions of the Copernican system, are so disturbed by the attendance of one moon to the Earth that they conclude this constitution of the universe is impossible." Galileo was openly defending the Copernican system. There was not a joke in sight.

Horky was in shock. Like nearly every reader, he had no *perspicillum* with which to verify any of this. He could only ask the thoughts of the best authority he knew.

> Always to you, lord Kepler, never from you. If you have followed up on *Sidereal Message*, and might send as much to me, I wish that this be done. If not, I must inform you on the matter. It is miraculous; it is stupendous. True or false, I do not know . . .

Professor Magini expected that Galileo's *Message* was a "trick," a self-deception made "by the reflection of the Moon," and dictated as much to Martin, his amanuensis. Every astronomer Martin met was going completely insane over Galileo's book; they seemed to him to be "howling wolves," petty and fierce. As a Protestant living in a Catholic stronghold, Horky was afraid, and the academy was teaching him paranoia. This new book had made everything mysterious and unstable. And so, for his own sanity, Martin jumped to drastic conclusion. He made a private note that Galileo had invented "four fictitious planets," and that instead of working as a scribe, "I would rather be writing the truth about them."

On the April 20, 1610, Horky was ecstatic to receive his first response from Master Kepler, bringing the letter to his lips. "I love it, I kiss it," he exclaimed, but at first had no time to respond. For Galileo, the one, the only the grand mountebank, was coming through Bologna on a book tour, and would be sleeping in the house of his hospitable rival Magini. Horky was preparing for the event. He would be there, in the middle of it all, in the very same room as Galileo the fraud.

"The honorable Magini provided for Galileo a feast, luxurious and indulgent," he recalled. Italian cuisine was the only one on the continent to favor lightness, all sweetmeats and fish; rest assured, they drank wine. Over twenty guests arrived that night, but Horky could tipsily make out Galileo somewhere across the table. He did not look at all like the savior of Italian scholarship. "He's bonkers," Horky wrote, "mad in the head."

> His hair is falling out; his skin is bursting with syphilitic sores; his eyes are broken from obsessing over Jupiter; his hearing, taste, and touch are withered away; his hands are inflamed from having nearly destroyed the wealth of mathematics and philosophy; his heart palpitates because he has sold everyone an astronomical fiction; his insides are fit to burst because illustrious men are no longer tickled by his games; his feet are screaming the gout, for he is an eternal wanderer. God bless whatever physician could nurse this feeble messenger back to health.

That night, Horky got his paws on Galileo's spyglass. "I did not sleep for all of April 24th and 25th, but tested the instrument a thousand different ways," he informed Kepler, "Down here it works miracles, but in the heavens it deceives." Next evening at a second dinner party, Galileo attempted to demonstrate his instrument, but "all acknowledged that the instrument deceived. And he became silent, until the next day, in a foul mood, he took his leave from Prof. Magini in the early morn. The wretched Galileo gave no thanks for the favors."

That settled it. Horky secretly readied his intellectual debut, an unmasking of the greatest scientific hoax this century. Around the beginning of June, he told Magini he wished to see the city of Modena with a traveling companion, where he brought his book to press. It was called *A Brief Peregrination Against the Sidereal Message*, a work just like his life—nomadic, a bit silly, and not at all long. Kepler was writing back to him sporadically, applauding his search for truth, but suggesting that any difficulties with the *perspicillum* were likely due to his own eyes. The warning arrived too late.

Perhaps there was some small truth in Horky's final grievance from his *Peregrination* that "these four spots around Jupiter are useless, and serve only to bring Galileo his golden fame." There were many valid complaints in what he had experienced these past months, but taken together, they were not enough to validate this, the first public condemnation of Galileo.

Galileo did not respond to the attack directly; he certainly wanted to, but wrote with a slow, methodical style, and simply lacked the time. A no-name student of his took up the honor. Dear Master Kepler sent Martin a letter with these brutal opening lines: "I read your *Peregrination*. I am unable to maintain the purity of my reputation alongside your friendship."

Horky would have been devastated, were he there to receive them. But upon his return to Bologna (rumor had it, that very morning), Professor Magini called him forward, provided some very strong words, and told him to pack up whatever he could and leave town. Whatever doubts Magini had about Galileo were doubts, not rejections. Only a fool would confuse a siege for a duel.

Appeasing whispers filled Galileo's mailbox that Horky had become a wretch, a deplorable bum lost on the streets of Bologna, Modena, Pavia, Milan. None knew quite what had become of him. It was nearly three months later that, as part of their rekindled friendship, Johannes Kepler wrote to Galileo Galilei, the court philosopher of the Medici family, a humble professor no more.

> Come to me yesterday was Martin Horky, returned from Italy, having gone hither and thither ever so much, in a journey comprised of delays. A marvelous and remarkable occurrence: he wore an exultant expression, and addressed me as though I agreed that he had triumphed over Galileo. Thus I responded as I had in the letter, that I had renounced his friendship. This much and more confused him; he knew not my renunciation, nor much else. After much debating he finally became open to the errors of both sides.
>
> His opinions have been greatly damaged.

Before the year was out, Martin informed Kepler that he was chasing after a "little beauty." He decided to settle down and marry. His wanderlust was over. He had learned. He had been taught. Everyone was, in the end.

Their Rekindled Friendship

In autumn of 1609, the Frankfurt book fair, oldest and grandest in all the land, had announced *New Astronomy* as available for purchase. "From that time on," Kepler recalled, "like a general who had won glory enough through a strenuous military campaign, I took some respite from my pursuits. I supposed that among others Galileo too would discuss with me the new celestial physics I had published, and would resume our interrupted correspondence, which had begun twelve years earlier."

When Kepler discovered just what author had taken next season's fair by storm, he was deeply struck by the coincidence of life. "Rather than perusing some foreign work, my Galileo has busied himself with a most unusual thesis!"

Galileo had sent Kepler this thesis, with an exhortation for the man's illustrious opinion, but he could not have expected the response to be nearly as long as his own *Sidereal Message*. Kepler had taken only five days. "Perhaps I seem rash in accepting your claims so readily," he vouched, "but why should I not believe? His is not a bid for vulgar publicity. He loves the truth, and bears the jeers of the crowd with equanimity."

By next month, so many had besought the Imperial Mathematician Kepler his opinion that he edited the letter for publication. His *Conversation with the Sidereal Messenger* immediately sold out, was bootlegged, and sold out again. It was his most read work.

While Galileo had written with an uncanny deadpan, Kepler wrote that he "regarded humor as more pleasant." "I shall design a lens shaped like a tit," he decreed, as a means to improve Galileo's *perspicillum*. "All rays converge to a common focus; such is the function of a hyperbolic nipple."

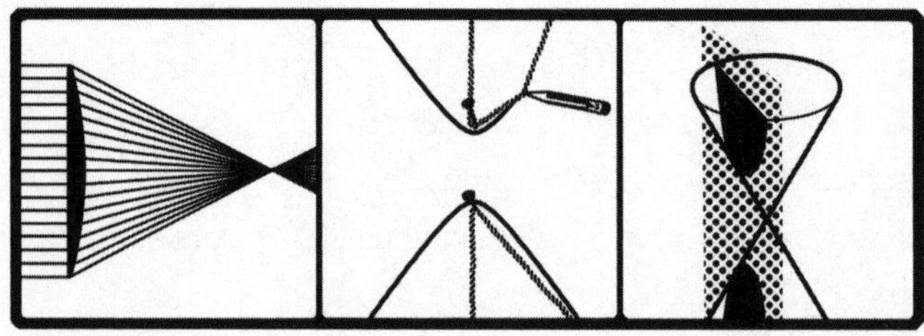

The hyperbola is the last type of conic section. It can be thought of as an ellipse where one of the foci has "wrapped around" infinity to the "other side." It refracts parallel rays about a single point. Kepler thought it looks like a nipple.

This joke was the start of a brief, intense exchange, political at one end, intellectual at the other, almost mythical in its effect: Galileo and Kepler, romancers of the stars. "It is not without reason that those whose explanations anticipate the senses attain greatness," Kepler wrote. "Let Galileo take his stand by Kepler's side. Let one observe the Moon, turned up towards the image of heaven, and the other study the Sun, turned down towards his writing desk (lest that '*spicillum* burn his eye)."

"You were the first," Galileo teared up to Kepler, "the first, and perhaps the only, to believe with but slight inspection. You did so nobly, with sublime talent, proffering complete faith in my assertions."

The Naming of Things

"Tnisoidohversesoagarowtivtonovmbullse."

Kepler stared at the trim scroll of paper before him. It was an extremely important piece of gibberish. Dead Lord Tycho Brahe had once talked about using these sorts of ciphers for his alchemy. It was an anagram—an anagram! What a curse upon Kepler, so incorrigibly obsessed with the meaning contained in every little enigma.

This quality he shared with his Emperor Rudolf, who had received the puzzle, written by Galileo, from the Medicis. The king had managed to read *Sidereal Message* all by himself and became so obsessed that he would call upon Kepler just to babble about its meaning. Rudolf had a chance to play with a spyglass from Galileo that same month, without realizing it already belonged to a cleric. When it was taken away, he threw a fit. "These priests want everything!" he moaned. Protestant cabals were plotting his assassination, Catholic authorities were preparing armies at the border, yet his mind was fixated upon a *perspicillum* to call his own. He demanded answers, objects, anything to which he could attribute the fantastic experiences he was hearing about. It was not enough to read about; he had to possess.

Spanning four rooms across the lower level of Prague Castle, Rudolf already possessed a host of astronomical machines in his splendiferous "cabinet of wonders." It was a physical encyclopedia, a protomuseum: Turkish sabers, scimitars, stabbers hung off the wall, won by war with Ottoman janissaries; colonial dicker had brought ultrahot Indian spices, desiccated island berries, ceramics, metals, jewels, and gems from bizarre bazaars; the Egyptian vase, Persian plates of lustrous nacre, and stately domestic wares; copper crockery, silver dishes, golden beads, topaz, agate, jasper, crystal, pearly nuggets in all shapes; exotic instruments like the lute, the lyre, the sitar, spoons; great engravings by master hands, alabaster carvings; taxidermy birds, alien creatures, horns in every genre, even the ass (unicorn); other balderdash: human face carved on mandrake root, bells to summon spirits, bezoar stones; marvelous clocks, scary timepieces, big ones, pocketwatches, all sorts, inaccurate always; distant writings, migrant texts, many words, glasses in a thousand frames, convex, concave of variegated fillings, all the better to read with, to see it all. Would that the king could own people, too!

His pet astronomer Kepler was a wonder and a wonder producer. When, from a passing diplomat, Kepler was finally able to borrow a rare *perspicillum* made by Galileo, he verified his friend's discoveries beyond doubt. He immediately wrote a second pamphlet (in which he invented the word "satellite"), triumphantly proclaiming, "Truth is the daughter of time, and I am unashamed to be a midwife. Thus, I have made an inspection of the uterus, and it has not been abortive." But he was lagging behind the Italian. Kepler's best solution of Galileo's anagram was, he admitted, a "semibarbarism."

"Saith too soon: view Mars, revolving doublets."

Perhaps Galileo had discovered new moons around Kepler's favorite planet Mars? No, Galileo knew, playfully ticking off each letter against his own. "I am not so austere as to be revolted by jokes," he would one day remark, and as Kepler had joked, he, too, now had his fun.

"Signor Kepler has exhausted himself," the messenger for the Medici wrote, "He is imagining a thousand different things, and says he cannot still his soul." Who was this elusive Galileo? What were his plans, what was he thinking?

But Galileo had not been thinking, not with what self-awareness that implies. He was lost in a dream of glitz, glamor, and fame.

The upstart astronomer was being feted, hated, contemplated all across the continent. Even before he had published *Sidereal Message*, he was asked to present his *perspicillum* to the Venetian senate. Aged governors escorted him up the stairs. He was meeting the doge. He was being hugged by the procurator. He was being given tenure for life.

When he slept, it was feverish; when he woke, it was not wakeful enough. His susceptibility to illness was increasing with age, which he attributed to having spent "the better parts of winter's night outside, exposed to the open air." By his own accounting, Galileo would work until four in the morning, even in Venice, where "the gondola was hard to find, and all was wet with shadow."

Well over a hundred *perspicilla* had been manufactured by his own hand, yet only ten were up to his high standards. These, he decided, must be sent exclusively to royalty and a few high-ranking members of the clergy. He had obvious plans to reprint *Sidereal Message*, "in the Tuscan language, as it is highly sought-after here. I have a great desire that this second edition be nearer the grandness of the Medici than the weakness of their servant." This reprinting was never completed; time flitted by, moments too surreal. He had been touring across the country, and everyone who was anyone expected him to come to dinner, sit still, smile, look pretty for their friends. He was bleary. His back hurt.

The sole relief came in the form of a very long letter, "from Kepler." Galileo was quick to brag, "the imperial mathematician, confirming everything I have written, down to the last word. It is now being printed in Venice, and soon you will see it." Two months later Galileo was written personally by the Grand Duke Medici.

> The eminence of your scholarship and fullness of your courage, alongside your singular mathematical and philosophical skill, and the most affectionate obedience, vassalage, and servitude which you have always demonstrated, have made us wish to have you near . . . a salary of a thousand scudi shall begin the moment you arrive in Florence to serve us.

At last, a court had matched his present salary. He quit teaching and straightaway prepared for Florence. "Concerning employment," he had written a friend, "I despise no work but that of whoring, of exposing my labor to the arbitrary price of any customer. Now to serve under some grand Prince or Lord, this is the sort I shall always lust after." Less than a year after he had left Padua, he learned that the Queen of France, upon receiving his *perspicillum* with a low mount, dropped to the floor like a common wench; it was actually he that was bringing royalty to its knees.

The very last personal letter Galileo would ever send from Padua was the only he found time to write directly to Kepler, a man who he now equaled in social status. He mentioned his salary twice and was careful to conclude with his new title (granted per his own request): "Mathematicus and Phylosophicus to the Glorious Etruscan Prince."

"How I wish," he mourned, "that we could laugh together at the extraordinary stupidity of the crowd. For you would verily cachinnate, could you hear, dear Kepler, what is said against me." "Ah, why can I not laugh with you forever?"

Upon reaching Florence, Galileo at last revealed the solution of his anagram to the Medici, which King Rudolf, of course, begged to see.

I have observed two moons, lo, vigils to Saturn.

Galileo *thought* he had made out two moons by the blurry, distant planet. He had actually seen its rings.

The trip to Florence astride a hemorrhoid-inducing horse had been very tiring. He took two whole months to set up house (the drawbacks to bachelordom are, at times, quite fierce). Upstairs there was a lofty terrace from which to view the stars. There was even time to draft a few letters to some neglected friends, breaking his long silence, before he collapsed.

The New Man

The cypress trees of the country house were well-ordered; the essence of olive wood sang up from the beckoning grove. If one had a fine spyglass, they could follow that grove sloping out the yard of the sylvan mansion, down the hillock into port, where ferries came in to rest, juments carried their cargo into town, and men and women languished on the opposite side of the river, staring out, a sort of natural television. Galileo breathed in heartily and felt his spirits replenish. "I immediately knew the fairness of the air, and thus the detriment of the city, such that I have come to think I may move to the mountains." "At present I find myself at Le Selve, the villa of Signor Filippo Salviati."

Mr. Salviati was the first friend Galileo had made in court. As a boy he had been raised to be the perfect gentleman, trained in archery, fencing, equestrianism, falconry, and singing. He knew how to ballroom dance, but only lead, guiding his young bride about the floor their first night together, when he was only twenty. Next year he was a father; seven years later, he was not. Returning to Le Selve after his young daughter's funeral, Salviati passed by the four guardsman, met his twenty-three servants, all for a household of two. As loneliness is not so much a longing for company as longing for kind, he exhorted all like-minded nobles to vacation on his rural estate.

These like-minded nobles were few, for when Salviati had turned twenty-four, something very strange had happened. He began to read. Three years later, he was fluent in Latin and the works of Aristotle. After studies of Euclid, philosophy no longer seemed to him a mere tool for conversation so much as one and the same. He became a different, inquisitive sort of courtier. When he heard that the famed Galileo had moved to Florence, he, of course, wanted to meet him. When he found Galileo bedridden, exhausted from his trip, he offered him full recovery at a room in Le Selve without hesitation.

Salviati's father had died when he was thirteen. Galileo was eighteen years the elder. The long hours of confab which followed were as that between family, which by the very nature of its intimacy seeks to elude history. Salviati had lately been reading commentaries on Aristotle, particularly those by a Greek named Simplicius, last of the pagan philosophers. After a chat about this, Galileo introduced him to the works of his favorite philosopher, the "divine" and "superhuman" Archimedes, who had long ago dedicated his life to bridging the gap between mathematical abstraction and lived experience. Salviati gave himself over to Ptolemy and, soon enough, Copernicus. The men began to carry on experiments together.

A year later Salviati would give words to his revised worldview, which had as many names as it did adherents—he called it *the liberal philosophy.*

The liberal philosophers were profligate with money, a thing only good for what virtues it could access; they were extremely sociable, especially between one another; and each was a Catholic with a sharp distaste for self-denying Christian altruism. Charity and knowledge were not altruistic for them, but selfish, almost physical pleasures. They invited women to speak at their academies (some of them even listened). They sprinkled the word *new*, or *novus*, or *nuovo*, on every thought, as if it were sugar for the mind. Most were in open rebellion against their parents.

A great many of Galileo's close friends were like this. Contact with Gianfrancesco Sagredo had been crippled by the latter's move to Syria, but in a rare letter, Galileo learned to expect him back in Italy at the start of 1612. For the rest of his life, Sagredo would reminisce on his foreign excursion. He told Galileo all about it after their reunion.

> After my arrival in Syria, by the grace of God, I made a very happy life. There was no one to command me. I had no business with my father but greetings and stories.
>
> If Signor could have seen me in my studio, he would have laughed, as I, drawn by curiosity, leafed through one book, while in my heart wishing another. Fearing I would soon have to leave the house, I was overwhelmed by the desire to take it, then another, then another, until I was loaded up like a right jackass.
>
> My rules for good health are: leave the table a teeny bit hungry, drink a decent amount, eat what is tender, crumbly, nourishing, and delicious. Vintage wines are to be excluded as routine, but it is good to drink them after a bit of fruit. If anything is even a little spoiled, I do not let it near my friends. For them I save everything good . . . I have become convinced that this world has been made to serve me, and not me it.

Sagredo called his way of life *the true philosophy*. Galileo was seven years the elder.

Benedetto Castelli, the ex-student of Galileo (fourteen years the elder), was of a separate strain. He was not a nobleman and was fervently religious. His letters to Galileo were quiet meditations on a life of reading and scholarship. The love both men felt for one another did not demand any affinity greater than the intellect, and that undeniable humanity, which can be seen in anyone looked at with the right pair of eyes.

When Galileo published his *Sidereal Message*, Benedetto was among the first to read it, "and reread it, over ten times, with absolute wonderment and a sweetened spirit. Having well understood this profound doctrine, the deep thoughts, the learned speculation, I will hold it with the dearest care. I thank you for having deemed me worthy of such a treasure. Now when I see this beautiful, ingenious work full of curiosities I will spend only my laughter, and just go read it off the bookshelf, without having to pay a penny." Benedetto then applied for a transfer to Florence, hoping to return to work with Galileo. His Benedictine order had not yet granted it.

On December 5, 1610, Benedetto caught Galileo, just before he left for Le Selve, with a particularly insightful message.

> Since (I believe) Copernicanism is true, Venus spins around the Sun. Thus it should, at times, look like a horn, and sometimes not. I long to know if Signor, with his marvelous *occhiali*, had noticed any such thing.

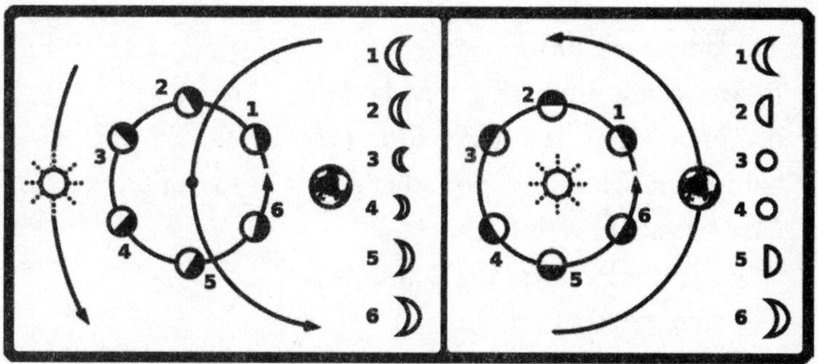

The Ptolemaic phases of Venus, left, and the Copernican phases of Venus, right. In the Ptolemaic system, the epicycle of Venus is always on the line between the Earth and Sun.

The hypothesis that the inferior planets had phases like the Moon was quite brilliant. Galileo had been observing Venus for some time. He had some very smart friends.

Johannes Kepler was one of them. Galileo had ordered a copy of Kepler's *Optics* before his sojourn to the mountains, and the work exposed just how intensely outlandish this German fellow was. The rector of Kepler's old university had stated the problem clearly enough: "Never in my entire life have I encountered anything in the field of mathematics, I near say all of philosophy, that is so difficult. I cannot tell if I have understood even the smallest part of what you describe."

Kepler, at that moment, in light of all these new visions, was writing up a sequel to his *Optics*. He was made sore when this troublesome work was interrupted by yet another anagram.

As I oft collect these darn immaturities vainly,
roughing men . . . oh no!

"I look forward to what Kepler has to say," Galileo taunted. He received a reply almost immediately.

> I implore you, do not continue to hide from us, whatever this is: for you see, it is as a matter between brothers. I am impatient, sifting through your many secrets. I have had no success.
>
> Five out the Sun is rotating a melancholy, monolithic red smearing.
>
> The others are incomplete:
> I have seen a simmering moody Sun curling etc.
> Momma Solaris is a fiery, cutting red, on the go, etc.
> Saturn and Mars have yet other goons in the culling etc.
> Mercury adheres to flame etc.
> The star-theater gyrating, a fountain etc.
> The Sun gyrates etc., etc.
> See what misery you thrust upon me with your reticence?

Damn this Italian! "In time," Galileo wrote three weeks later, "I make clear the truth of the matter, when there remains not a shred of doubt." And he did so then.

Hello. Carnal mother Venus imitates
Moon-God Cynthia in its figure.

Castelli's hypothesis was correct; Venus had phases like the Moon. The Ptolemaic system had been handed an epic defeat. Kepler hastened to translate the letter into Latin and add it with an extended preface to his new book on optics.

Galileo returned from his first trip to Le Selve, well-rested, at the end of January. He had managed to survive the most fabulous year of his entire life, thus far. This old, large, sickly man, having endured such an improbable fame, had been revived with a fresh resilience. On March 19, he sent a brief letter to the secretary of the Medici family concerning a request he had made late last year.

> I am anxiously awaiting a litter to convey me to Rome, which has not yet arrived, nor have I heard anything new of it. It upsets me, how the time flies by. It means I may not be able (should there be a bit more delay) to get there before the Holy Week of Easter, as I was much hoping, yet for other reasons I am still set on travel. It seems to me most necessary that I shut up the mouths of those who wish me ill.

Their Dying Friendship

Rome was the ancient still-beating heart of the world as Galileo received it. It had been the center of all known politics for two thousand years. Galileo arrived "in good health" on Holy Tuesday, which raised his spirits ever further. He slept each night a block away from the oldest racetrack in the known world, where the masses had once chosen to consume their bread and circuses. From there it was a skip to the Pantheon, a jump to the Forum, left to this, right to that. The monuments were endless, and they were making more. Galileo could barely turn a corner without scraping the facade of some brand new basilica. Even the Roman artifacts, by juxtaposition, became testaments to Catholic rule (except the Colosseum, it seemed, the gladiatorial ring that dwarfed every church). Pope Paul V slept each night like any other person. A thousand feet away, in his Sistine Chapel, where the paint of Michelangelo had dried its hundredth year, important men would come and go, talking.

Galileo was now one of these men. Rome was his mecca. Even little children in Russia knew his name, a letter informed him. For nonspecialists, there was sincerely no one living in the entire world whose opinion on astronomical matters held more weight. For specialists, there was Kepler, who had just released a new work inspired by his famous friend.

No less than four full letters from Galileo are quoted in Kepler's *Diopter*, right at the beginning, with complete Latin translations of each, so that everyone knew exactly what they said. The cover advertised this fact, as if it were a selling point.

A "diopter" is an old-fashioned word for sighting instruments which Kepler, as usual, reinvented for his own purposes. In *Sidereal Message*, Galileo had explained virtually nothing about the *perspicillum*; Kepler set out to explain virtually everything, "to open up this new playground, by way of the inherent force of mathematics." His model was Euclid, working entirely with definitions, axioms, propositions, and porisms. "This is not an easy book to understand," he admitted.

In his *Optics*, Kepler had established a model of the eye as an instrument, just like the *perspicillum*, which was acted on by light, an instantaneous outflow from a single point. He speedily built upon these definitions towards refraction, convexity, and concavity. By proposition 86, he had introduced systems of multiple lenses, in order to explain Galileo's spyglass.

Imagine, to begin, a single biconvex lens. For simplicity, take Kepler's desire, that each side be a hyperbola, focusing light perfectly about a single point.

The human eye sees an object as blurred when light diverges or converges too strongly. This can be seen easily by trying to focus on an object very close to the eye.

If the eye is just before the focus, the image will appear blurry, because the light rays are converging too strongly for the eye to handle. But if the eye is immediately behind the focus, the image is blurry, and inverted, because the light rays have crossed one another and are now diverging too strongly. If the eye is positioned a proper distance before the focus, this is a magnifying glass.

Kepler noted casually that "by a nearer, contrary convergence, the excessive divergence may be healed," and thus, "two convex lenses enlarge and define the image, but flip it around." With that slight remark, alongside a healthy exposition, the modern version of the *perspicillum* was defined.

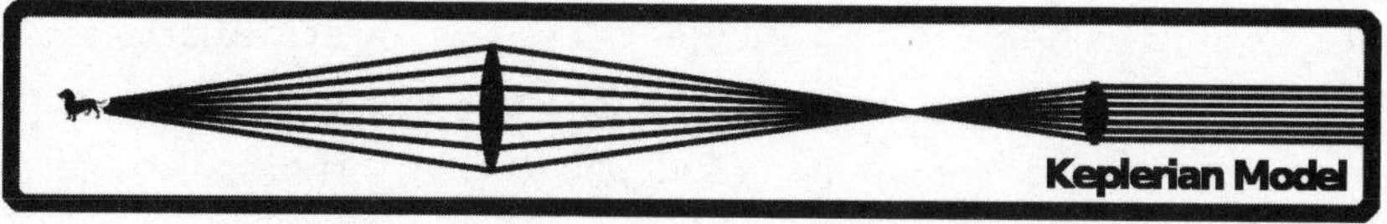

Keplerian model with two biconvex lenses. Plano-convex lenses can also be used to focus on a point at infinity.

Galileo had not provided the world with many details of his own design, but he did give all that was really necessary: "First I prepared a lead tube in whose ends I fitted two glasses, both plane on one side while the other side was convex and concave, respectively." The optical lens was before the focus, as the rays were still converging. This kept the image upright, but offered a much smaller field of view, which had caused Galileo serious difficulty when observing the Moon. Kepler developed a method with a third lens, which would keep his image upright, too.

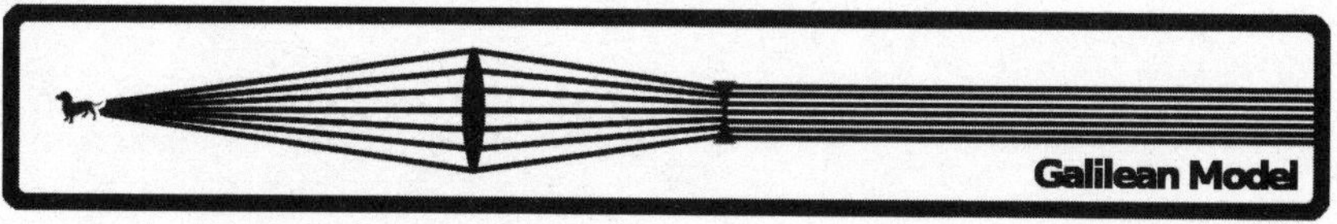

Galilean model with biconvex and biconcave lenses. Galileo used plano-convex and plano-concave lenses to similar effect for objects located at virtual infinity, such as the stars.

Kepler's task, in *Diopter*, was not to improve on Galileo, but to understand him. After his variation, he spent a much longer time considering Galileo's own design. He sent Galileo a message, more than once, letting him know about the work, in which "I judge that nothing has been left out by me, if it might be proved from its causes."

Galileo received several pleasant recommendations from his friends, at least concerning the preface, but he never gave Kepler a personal review. The two had offered one another mutual aid—career development, theoretical novelties, high sales figures—and that was enough. Kepler had clearly extended the possibility of a genuine, long-lasting intimacy, but let it pass unconsummated, in the way a circle dancer with an unwilling partner simply moves on to more playful bodies.

Nearly four years later, at the tail end of 1614, a young French Copernican was traveling through Florence. He was an avid diarist. "On Wednesday morning I saw monsieur Galilei, a very famous philosopher and astrologer, who was to be found in bed, due to some sort of indisposition. I had not wished to pass so close without seeing him, and discussing with him the new phenomena."

> After some conversation, I asked him about refraction, and how to shape the crystal of the *lunette* in such a way that objects grow and approach as desired. To which he replied, that such knowledge was not yet well-known, that no one had yet treated of it, outside of perspective, except for Jean Kepler, Imperial Mathematician, who specifically wrote a book about it. But it was so obscure that not even the author seemed to understand it.

The Renaming of Things

Ever since Kepler had published his *Conversation with the Sidereal Messenger*, people had come to call Galileo, and his little book, by that self-aggrandizing title: *Sidereal Messenger*. Life being so full of unexpected changes, it was an honor he would not refuse. About October 1613, Kepler was chatting with a fellow astronomer at the town carnival (he had a habit of taking colleagues out to lunch) when he suggested each Medicean moon take the name of a lover of Jupiter: Io, Europa, Ganymede, Callisto. These names also stuck. In short time, these new discoveries were embedding themselves in public consciousness.

The first thing Galileo did in Rome was throw himself at the feet of holy men. The day he arrived he went straight to a prince of the Church to explain himself in loving detail. "The cardinal embraced me warmly and listened attentively, full of hope." Galileo spent his first weeks scurrying between the houses of prelates and the Roman College, where all the brightest Catholic boys went to learn their letters, numbers, and the Bible.

> I have been favored by many who have wished to see what I have observed. All have been satisfied. At our meetings I have seen marvelous statues, paintings, ornamented rooms, palaces, and more.
>
> This morning I was presented before the pope. I was about to kiss his feet, when the Ambassador told me that, because I was so extraordinarily favored, His Holiness would not have me speak even a single word while kneeling.

All of Galileo's observations were formally verified by the Roman College's premier astronomers two days later. They passed around recent publications denouncing Galileo, just to have something to laugh at. Earlier that month, Galileo had sat enraptured by a religious sermon from a charismatic preacher, warning against the folly of pride.

In coming to Rome, Galileo wisely brought with him some new discoveries to entertain the men in power. He found these by staring directly into the Sun. Small, dark blemishes could be seen on its surface. Never mind that both China and Europe had known about these sunspots for millennia—no one had seen them through the *perspicillum*. There were only new things under the Sun; the Sun itself was a new thing. Galileo attempted to show them to several astronomers at the Roman College, who believed him, but refused to look, fearing the Sun would burn their eyes. Galileo had no such fear.

The elderly clerics of Rome were more than what they looked. They were loud and full of life, throwing some of the best parties in the world. Galileo was so busy socializing, he told Salviati that "I have no time to write to my friends, especially patrons. So you will have to stand in for all," and then, sweetly, "I do not care to spite, even in one city, those who have spited me before the entire world. I would much prefer to gain a friend." He could not bring himself to anger or malice in Rome, where every day was a celebration in his name.

The most extravagant celebration in Galileo's honor took place in a highland vineyard on April 14, after he had received two full weeks of tireless praise. It was vespers, and skies were kind. One side of the yard looked out onto a staggering vista, the whole of Rome. Before dinner, conversation turned toward the philosophy of nature and, by popular demand, Galileo brought out his *perspicillum* for an impromptu display. There were at most twenty men in attendance, few enough for everyone to get a turn: one bishop had brought his full entourage; there were professors from all over, in botany and mathematics; a renowned peripatetic looked on in silence. But none were so affected by Galileo as the bishop's nephew, Federico Cesi, who had funded the whole evening.

Cesi was a twitchy, baby-faced aesthete, with facial hair of noticeably fine grooming. He was very rich. Like many of Galileo's friends, he hated his father. Unlike Galileo's friends, his father hated him back. Even the pope knew Cesi's father had cruelly denied him the right of primogeniture and probably also knew his father had spread the rumors that his son was gay.

These rumors were not unprompted. As a teenager, Federico had been, to be frank, a total freak. He had a weird passion for flowers, which he liked to mutilate, label, draw, and give fancy, sesquipedalian names. His apathy towards women was so severe it crossed the border into contempt. "Love for woman is profane," he wrote, "an enemy to knowledge." Marriage was "a meek place, an emasculating repose." He had no friends his own age.

Instead, he spent his time with three men, all about ten years older and considerably poorer, whom he brought traipsing around the woods collecting florae and dead things. Cesi gave everyone fantasy names: they all had to call him "Sky-Wanderer"; the man purported to be his lover he dubbed "The Illuminated One"; the other two were "Eclipsoid" and "Mr. Tardy." The four men agreed to form an intellectual collective, funded by Cesi, who named them "The Academy for Lynxes," because, wrote Mr. Tardy, they were trying to "penetrate the insides of things, to know their causes and internal workings, just as it is said that the lynx has eyes which see not just what is in front of it, but what is hidden inside." Out in the forest, they all complimented one another on their "lynxiness," or, on their better days, "brotherly love" and "confraternity." One member called it a religion. Each wore an emerald signet ring of a lynx to symbolize their purity; Cesi was adamant that entrance into the academy include a vow of chastity.

His father, to no one's surprise, shut down the operation the second he heard about it. Federico received a very stern lecturing. "I hate courtiers," he sulked, "All of them are traitors. I trust no one. I care not a whit. I laugh at them madly."

The reach of this esoteric gang was farcically long. The Illuminated One, who was run out of Italy by Cesi's father, went to Prague and met Kepler and Tengnagel, both of whom very nearly became Lynxes themselves. In one of his many curious adumbrations, Kepler wrote, nine months before Galileo was in Rome, that "without a doubt, I know this: Galileo has the eyes of a lynx."

Cesi asked to meet with Galileo personally. The boy was all of twenty-five now, still uptight, but less absurd. Genuine academies for professional philosophers to discuss nature were not ridiculous; there were several cropping up across Italy, with appealing names like "The Drinkers" and "The Lazy Men." He had better resources than any of them. Why not the Lynxes?

As he pitched the idea to Galileo, he echoed something like the charter the group had drafted five years ago during their forced separation. Chastity was no longer a requirement.

> A Lynx must desire, in addition to knowledge of the things themselves, to give themself over to the pursuit of nature, particularly through mathematics. At the same time, they must not ignore poetry, nor the adornments of language, which decorate all of knowledge like a beautiful dress.
>
> A Lynx must pass by political controversy, the lowest form of debate, in silence. They cultivate peace, and seek out quiet study no matter what.
>
> A Lynx must always give themself, heart and mind, first and foremost, in perpetuity, to the faith of Christ, and most especially, live in devotion to the Church. Praise God.

He then asked Galileo for his membership. Galileo said yes, obviously. Only good things happen in Rome. This was all after the dinner party, though. Right now, there was a small crowd of powerful people lining up to try out his *perspicillum*. His spyglass. The new spectacles. The perspective cylinder, the *fistula dioptrica*. The tube of long vision, the toy trumpet, the *visorio*. The Italian *occhiale*, the French *lunette*, the English *trunkes*, the Dutch *om verre te sien*. When Galileo was a proper geriatric—bad back, arthritic hands, sharp mind—he would call it his "old discoverer."

A voice rose up from the mob—it was that boy, Cesi.

"Telescope!"

First Signs of Night

I know that I am dead and damned," King Rudolf trembled, "I am a man possessed by the devil." He had been prophesying his fate for more than five years; he would blurt it out loud in court as a means to make his circumstance clear. The Protestants had rebelled against the crown. He had killed them. He had ceded them rights. He had started a propaganda machine for Catholicism. Still they rebelled. Now the ladder-climbing king of Hungary, his own brother, flesh and blood, had heeded their call. There was a battle coming, and Rudolf knew he would lose. Trapped in a maze, with every path a dead end, he turned to frantic solutions, finally attempting murder spells and black magic. The Holy Roman Emperor had lost his mind.

"1611 was a year everywhere filled with sorrow," Kepler wrote, despondent. "Rudolf, suckled by the vain hope of rapprochement, did not wish me to quit the court." Unpaid, full up on the bitterness of being subject to a one-sided love, Kepler could not bring himself to abandon his dying, senile lord. "This devoured my time and money."

The new emperor had enough nonchalance to retain Kepler as Imperial Astronomer. Because executing family members was considered gauche in royal circles, Rudolf was held prisoner in Prague Castle, forcing almost no change in his lifestyle except a complete withdrawal from politics. In a funny way, this was what he had always wanted. He never did receive a telescope, but he still had his cabinet of wonders. His favorite painting still hung on the wall.

It is a very famous portrait. Rudolf's head stares out, larger than life, against an unremarkable black background. The unoriginality of the piece is denied by a single gimmick, which had brought it fame throughout Europe. Even Galileo had heard of it; he thought that "important painters would laugh," if it were tried to be passed off as a desirable style. "Whimsical painters decorate a human face with the products of agriculture, just the fruits and flowers from this or that season. Such oddities are nice and pleasing, while they are jokes."

Was it a joke? Here was the emperor, stripped naked, with a turnip for a trachea. An artichoke opened out his left breast, his nose a big dumb pear; his beard was chaff and his crown was any sort of wine-making fruit. His eyes were black cherries, stalk intact, blind and signifying nothing. A sash of pretty flowers wrapped over his right shoulder: Rudolf was Vertumnus, god of the seasons, god of the fall. If it was a joke, the punchline was true.

Three weeks after new year Rudolf was dead. Kepler, a single father of three, had no friend-king tying him to Prague anymore. He left his brood with one of the many elderly widows who made a living watching over others' children and departed the miserable capital forever, in search of a new home.

As one inheritor to the Brahe dynasty ran free, the other was trapped. Astronomy had vanished from Franz Tengnagel's cares ever since he had found paying work as a wartime ambassador, under Rudolf's younger cousin. He traveled to courts in Paris and Spain, where he was regarded with constant suspicion, dismissed variously as a rube, a charlatan, and a coward, yet he always managed to worm his way into their confidences. He had grown accustomed to abuse.

When the soldiers of the new emperor arrived in Prague, with all the rapine and violence soldiers bring, Tengnagel tried to flee but—alas!—was surrounded. Learning they had captured a high-ranking political agent of Rudolf's, the men carried their prisoner right back to the city, up to the torturer's quarters, under a cloud of dread.

The torture rack, like any other virus, was of unknown origin, and adapted differently to any given country. It had survivors. In lieu of every possible failure of imagination, let one speak.

> Now what a rack is, is three planks of timber, stood up against the wall, interlaced with small cords from plank to plank . . . it is longer than a man.
>
> I was stripped naked by the executioner, brought to the rack, and mounted by him on top of it. Soon after I was hung up by the

> bare shoulders. Thus being hoisted to the appointed height, the Tormentor descended below, drawing down my legs, then, ascending the rack, he drew the cords upwards. The sinews of my hamstrings burst asunder. The lids of my knees were crushed. The cords held fast, and I hung so damned, for a large hour.
>
> Mine eyes began to startle, my mouth to foam and froth, and my teeth to chatter like beating drumsticks. O, strange inhumanity of man, you monsters, you manglers! O, I am innocent, O Jesus! Lamb of God have mercy upon me. My trembling body was taken upside-down, to receive several similar tortures, on six parts: the calf of my leg, the middle of my thigh, the great of my arm. My left hand is still lame, and will be forever.
>
> Then the Tormentor was carrying a full pot of water, the bottom of which had an incised hole, which he stopped with his thumb till it came to my mouth. He did pour it into my belly, the first and second time of which I gladly received, such was the scorching drought of my tormenting pain, like I had not drunk for three days. But then, at the third charge, perceiving these measures of water to be tortures—O, strangling tortures!—I closed my lips.
>
> Whereupon the Administrator, in a rage, split my teeth with a pair of iron retractors; and so, my hungry belly waxing great, grew protuberant. It was a suffocating pain, as my head was hung downwards, the water vomiting back up through my throat with force; it strangled and swallowed my howling and groaning.
>
> Thus I lay upon the rack, between four o'clock in the afternoon and ten, my body beggared with blood, cut through in every part. The truth is, it surpasses the capacity of any human to sensibly conceive, or for me to properly express, the intolerable anxiety of mind and affliction of body which I sustained.

Tengnagel knew these possibilities of torture. He told them everything he knew, voluntarily. His tormentors had a brief think and then tied him to the rack anyway.

It is said they stretched him ten centimeters, for the length of time it takes to say one "Our Father."

> Our Father, who art in heaven, hallowed be thy name. Thy kingdom come, thy will be done, on earth as it is in heaven. Give us this day our daily bread, and forgive us our trespasses, as we forgive those who trespass against us. And lead us not into temptation, but deliver us from evil. For thine is the kingdom, the power and the glory, for ever and ever. Amen.

Frans Gansneb Tengnagel van der Camp did not die that day, but his honor and human dignity did. For the rest of his years he would walk with a gimp and a pain in his side, a constant reminder of his political life.

Galileo Galilei had escaped the politics of Rome. He was back in Florence, enjoying a work of art. Here was a style of painting he could celebrate without qualification: portraits of himself.

The piece was commissioned by Salviati, as apparent thanks for recommending him to the Academy for Lynxes. Such a trade pleased Galileo enormously. He so liked the idea of it, that a work of art might preserve a friend for all eternity.

Cesi accepted Salviati as a Lynx, "of good mind, sense, and quality." The head of the Academy often conferred with his elder Galileo concerning potential members. Cesi's draconian self-restraint was loosening; he had fallen under Galileo's sway. Though he would forever refuse members of a religious order to join his academy, he allowed a special "Friend of the Lynx" title to be devised for Benedetto Castelli, who had returned to Galileo's side in Florence. Even Sagredo, that unchanging rogue, was back in Italy. They were all between them talking science and art: Cesi wrote Galileo about Kepler's elliptical orbits; Sagredo was investigating glassmaking for the telescope; Salviati cajoled Galileo to return to Le Selve that they might discuss a piece of theater he had seen.

There was even an important cardinal Galileo had met in Rome, named Maffeo Barberini, who had been so charmed as to try to stop by Florence to see him. Galileo hooked him onto astronomy as well, sending a lengthy letter of his recent research on sunspots. "It was a pleasure to see," Maffeo wrote, "your opinion is quite akin to mine. I have come to admire this letter, with your other work, as signs of a rare intelligence."

Galileo had brought everyone together. It was a sight to behold.

Salviati was the first to realize he could not stay. He was hardly thirty, and decided it was time to discover the world. He traveled to Barcelona.

Cesi sent Galileo the letter on April 26, 1614.

> The pain I feel at the loss of Signor Salviati is beyond words. All the company we shared, which today I have gathered together in part, have not been able to stop groaning. We loved him, held him in high regard, and know well what the world now lacks. We have thanks to give, to you who introduced us.

It was a freak illness. No one knew just what. As was customary, the Lynxes arranged a funeral oration. It took about an hour to read aloud; Cesi did so himself, before a congregation of scholars.

> . . . we can still see, dear listeners, his volumes of Aristotle, filled up in the margins with notes and annotations, sometimes admiring a powerful idea, sometimes expressing dissatisfaction, sometimes interpreting a great deal, sometimes adding his own findings, where parts of the reasoning seemed unclear. We are readily persuaded, this was a mind far from simple, an audacious mind, laying bare the meanings of works, and sure to go far.
>
> Such minds may disagree with commonly held beliefs, from those who either do not study so precisely or have too narrow an imagination, and even more who cannot manage geometry, who show no passion for triangles and circles. He clearly knew that only reason could satisfy him, and how beautiful it was, and how truly he was attracted to the philosophy of nature, as it appears to conform to idealizations which only geometry could reveal.
>
> And then, finding his thought in consonance with our esteemed academician Galilei . . .

The Animals

A papal advocate walked into the shop of Galileo's Venetian publisher. They were beaten over the head with a history of the Counter-Reformation. Some jokes write themselves.

The enemies of Galileo were various and many. They were philosophers, they were mathematicians, they were preachers. They were German, French, Italian, smart and dumb, old and young, Catholic, Protestant, all strangely unified around this one shiny noggin. The whole world seemed to be growing, with all kinds of new creatures revealed.

On November 1, 1612, an aged Dominican friar named Niccolo Lorini became ensnared, as all sometimes do, in a conversation to which he had nothing to contribute. Given a moment of contemplation, the old man thoughtfully remarked, "the opinions of this Ipernicus fellow seem contrary to Divine Scripture."

This conversation managed to reach the ears of its famous subject Galileo. Niccolo stumbled over himself to make it right. "I trust that all our nobility is very Catholic," he wrote Galileo, "I long to please you, and serve you as my master."

Lorini was a Dominican of a particularly dogmatic flavor, those who, punning off the name, called themselves *Domini Canes*, "The Hounds of God." When you heard one, you knew.

The congregants of the grand Florence Cathedral, having encountered another Hound of God, funneled out in a daze. This new preacher was a voice from another world: fiery, loud, Bible half memorized—and so youthful! "I rejoice to hear that Tommaso has preached in a proper cathedral," his brother wrote, "all the more, if it has given him the satisfaction that he pursues." Tommaso Caccini had joined the same religious fraternity as Lorini when he was only fifteen. His older brothers were in agreement that their little sibling had the potential for holiness, and took it as their duty to offer advice and support. They brought secular wisdom; the eldest had obtained some political offices, while the middle child had moved to the cultured city of Rome, spending time in the court of Maffeo Barberini.

After *Sidereal Messenger*, Tommaso had begun to speak ill of Galileo, mostly in letters, but also in public. His brothers were worried: "Friars, at times, have an excessive temper." But Galileo was not worried at all. "They do not tire of their machinations," he wrote, "all the more being that their enemy is so close. But they are so few in number, and, as may be recognized in their writings, are part of that league, so I laugh at them."

The "league" Galileo referred to was very real. They were informal and met irregularly, like a book club for suburban housewives, discussing how the new discoveries would impact their philosophy. They were not especially different from the Galileists, in that respect; they were not even necessarily hostile, on occasion meeting with Galileo at Le Selve to engage in some serious academic inquiry. He recalled one of their dialogues as a subject of genuine interest.

> The meeting was a discussion in a circle of gentlemen concerning heat, cold, moisture, and dryness. A professor of philosophy said something that is commonplace in Peripatetic schools, that the action of cold was to condense, and he gave the example of ice as condensed water. I questioned this, remarking that ice should rather be called rarefied water, for if it is true that condensation brings denseness and rarefaction lightness, then since we see ice float on water, we must take it to be lighter and rarefied. Whereupon I heard it promptly rejoined that ice did not float on water because it was less heavy, but because it had a broad and flat shape.

This encounter quickly escalated into a confrontational "game"; one member of the league ran around town showing people his experiments with flakes of ebony, which floated on the surface of water while other ebony shapes sank. It was a baffling result, but Galileo thought it broke the rules. He complained, "If you put ebony *into* the water, as per our agreement, it will go to the bottom, even if it is thinner than paper."

This was the first time in Galileo's career that a major dispute would go unresolved. The two parties never found the same terms; neither had any clue what ice was, and neither understood the tension, that the way things act on the surface are not as they are below.

Members of the league announced themselves as "anti-Galileists, out of respect for Aristotle." They were in turn targets of the derision of several of Galileo's followers. Their main organizer had the last name "Colombe," which sounded a lot like "columbary," a house for birds, and so, for a chuckle, the Galileists called them "The Pigeons."

Laughter may be an immense and delicious pleasure, a wholly sensual pleasure—but real laughter is beyond joking, mockery, and ridicule. Laughter that takes no object.

In Rome there was a man writing against Galileo who could not be laughed at, because he refused be named. He was a more serious intellect than all who preceded him.

He spoke through an interlocutor, first as a "certain friend of mine," and then, once it was clear his letters would see publication, as "Apelles, hiding behind the painting." In ancient Greece, Apelles had been a skillful artist who would crouch behind his canvas to hear the unbiased commentary of passersby. Galileo was passing by, but he was hardly unbiased.

Apelles sought something old-fashioned, to "liberate the Sun's body entirely from the insult of spots," but he did this by modern experimental methods. The spots moved across the visible half of the Sun in, at most, fifteen days; thus, they should take about fifteen more to reappear. Careful observation showed this not to be the case. As the spots showed no additional parallax, Apelles suggested they were shadows of new planets, from within "the heaven of the Sun."

The creation of new moons wasn't old-fashioned at all. Galileo responded, "It seems to me that this Apelles, moved by the force of novelty, begins to lend his ear to the true philosophy, but has yet to free himself from the fancies once imposed upon him." That true philosophy was, of course, Galileo's own, that "the dark spots on the Sun are not far off the surface, but adjoin it . . . Some arise and others disappear." "I hope that with what I have pointed out, this matter will be considered closed."

Galileo was able to begin prolonged observations of sunspots once Benedetto showed him a technique written in Kepler's *Optics*, using a pinhole camera to project the Sun into a darkroom, rather than looking through a glass darkly with the telescope. This new method made him nearly certain of his own thesis, and he readied a further response for Apelles. Cesi agreed to fund their publication.

Tone was, for Galileo, the most confounding difficulty of courtly life. It brought his writing to a standstill. He had tried to learn grace and social cues from the etiquette books, but they were never enough. He had the grammar of a perfect gentleman.

Apelles had published further letters, entrenching himself in his earlier position. Galileo asked all his friends how he ought to respond. Cesi had suggested "an unassuming mood, as sweet reason scorches worse than sour words," but another gave a stirring call to arms. He wrote, "Signor Galileo. When one foresees evil and escapes it, this is rightly called 'prudence.' But when people have so flagrantly exposed themselves, the time is up. You must to face the storm."

Galileo would make time. His sunspot letters are trenchant criticisms, abrasive, but never malicious, and they always end with a disconcerting request for friendship, for a fellow "lover of truth."

A year later, Cesi ferreted out the true name of Apelles. He was Christopher Scheiner, a respected Catholic priest out of Germany. Once unmasked, Christopher sent Galileo a few stilted letters. They clearly disliked one another, but their early rivalry was not personal. From a distance, it looked like civility. Christopher had his own distinctions; he built a much better telescope than Galileo ever did, after studying and understanding Kepler's *Diopter.* He had an idea for a massive tome on sunspots, to prove that he was the greater scholar. It would be called *The Red Bear.*

This was at the close of winter 1614. There would be vicissitudes, other seasons, and then it would return, to a slightly older man, a little less suited to its sting. The birds were migrating inland, and you could see one floating down the Arno, with a carp in its craw, red chest puffed up like it had just won nature's battle.

Wine and Women

The Germans have a word for excessive drinking, *saufen*, and they considered it a sin. Though not even the best men could always restrain themselves. Most common, Kepler recalled, were the "casks of wine, imported from Italy, coming through the Alps," but this year Austria had an especially plentiful grape harvest, so he instead saw the barrels running up the coast of the Danube. He decided to try a bit of the stuff himself.

Kepler was usually abstemious. He was only purchasing alcohol (he claims) "as a good family man," to "meet the duties of marriage." A woman had brought it out in him.

After he had fled Prague, Kepler headed straight for the city of Linz. When his first wife was with him, he had inquired about a teaching post there, specifically for her well-being. She had died, but the offer stayed live. Kepler went back to school. For a moment, he was twenty-five again, a bachelor searching for God, the names Tycho Brahe and Galileo elsewhere in time. His stepdaughter would have been seven years old in 1597; now she was nearly twenty-five herself, successfully married and moved away. Then his other children arrived, and he remembered what he was missing. He had become the imperial mathematician, a highly desirable widower.

How much must have changed, that he was received by no less than eleven suiting candidates?

The first woman had briefly met Kepler six years ago, under very different circumstances. Now she too was widowed, but when they met anew there were no sparks (and, Kepler added, she had "stinky breath"). To his surprise, he was offered up either of her daughters in her stead, and "I thus transferred my interest from widows to maidens."

A preposterous farrago of women followed thereafter. There were nobles, peasant girls, ugly hags, loving mothers, illiterates, well-read ladies, a frugal lass, gold diggers, virgins begging for children, matrons strong as oxes—these, in any case, were their first impressions. He was himself rejected several times, on account of his advanced age and controversial theology. Kepler was entirely repulsed by his sexual appetite: "Was it divine will, or my own vice which had me darting around these past two years, torn between so many parties?"

After he had confirmed his choice, Kepler sent for a booze merchant. Though he was almost forty-five, it sounds as though he had never encountered one before.

> Four days after several hogsheads were constructed in my house, a vendor came with a measuring stick, which was used to examine every single container, with no regard for shape, without reason or computation.

This timeless business practice struck Kepler as a practicality waiting for his support. He wrote a whole book about it. "You see, it is not unreasonable: new marriage, new mathematics."

The New Measurement of Wine Barrels is the best-titled piece of mathematics in the entire seventeenth century. It touches on all the major themes.

When he was a young punk, Kepler's writing teetered at the edge of babble; after years with Tycho Brahe, it crossed over. Works like *Optics* and *New Astronomy* are so brimming with content that coherent structure is lost, and they take on the feeling of a magical codex, hiding more than exposing its secrets. But *Wine Barrels* begins soberly. With it, and *Diopter*, Kepler first achieved his mature style.

From here out, he introduced all his works with clear, pure mathematics, until he arrived at a problem which no book had yet solved, as he always did. The shift from science to art was almost palpable. He turned silly and began inventing.

It was Galileo's hero, Archimedes, who formed the foundation on which Kepler here built. Even Kepler's elliptical orbits had been inspired by Archimedes. In *Wine Barrels*, the contribution is more evident; the first section is titled *Archimedea*. These were the beginnings of a great revival; Archimedes was being reforged into a legend, a rival to Aristotle, a second route along which science could travel.

Archimedes, Kepler reviewed, had proven some fascinating facts. He demonstrated the ratio of a circle's circumference to its radius was about 22:7 (neither he nor Kepler called it "pi"). He considered the rotation of conic sections about their axis—that is, spheres and ovoids—proving more obscure facts, like that a sphere and the smallest cylinder containing it have areas in proportion 2:3.

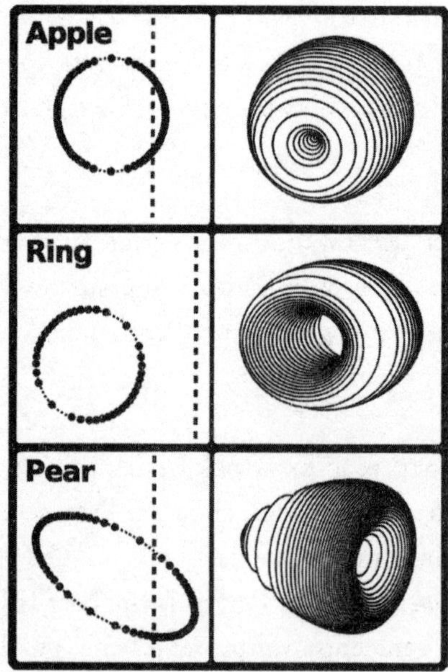

The shapes on the left rotate out about the dashed line, on a third dimension, to produce the shapes on the right.

"Until now, Archimedes and the geometry of the ancients have sufficed," Kepler wrote, but "we are searching for the true figure of a barrel," which was neither a cylinder nor a sphere. In search, Kepler let new shapes be formed, by rotating a conic section around not just its axis, but any line at all.

Kepler no longer had any idea what he was doing. No one did. He began by naming and organizing; the "conic solids" were sorted into no less than ninety-two groups. Without any easy means of definition, he tried to relate each shape to a familiar object; thus, certain solids were "apple-fruit," "fat plum," "pear," "olive." Rotated hyperbolas and parabolas, Kepler noted, had an infinite area, but sometimes appear "like Mount Aetna," or take "the form of a belly button," or, "part a big cave, part made thin and sharp, in the way the urinary bladder, as the Jews say of us Germans, is concealed by the glans penis with a foreskin."

The areas of certain conic solids remained subject to deductive geometry; the ring, for instance, could be "straightened out" into a simple cylinder, but such specifics ignored the true problem. Kepler did not uncover a general solution. The approval he gave for the wine seller's way of measuring was based on a separate, rather crude approximation of a wine barrel using two bottom halves of a cone. More so than any other work of his career, Kepler was running headlong into problems destined for future mathematics, and though he used the word "integral" several times, he meant it only as an adjective, to indicate the fullness and unity of a figure.

Nonetheless, it was a lovely start to a marriage, informal, yet honest and committed. A week before the ceremony, Kepler received a poem of similar character from a friend in Prague.

Out Regensburg, from Kepler, a letter is written
My dearest friend, my brother is smitten
With virgin SUSANNA, so new to the hour
Towards Austria rushes to fix the boudoir
Which has lately been so unproductive.

A wedding shall be consummated aright
A second, no doubt, in life's last quarter-light
For lightning arrives post-midday
And thunder is heard a great faraway
But we don't give a damn! The joy's unreducted.

Would that your guest could be present
To sing out the name of SUSANNA so pleasant
SUSANNA, who melts hearts majestic
A rose, and with roses, life sweetly domestic
Which lacking brings grumbles and grief.

But first the children, of whom you don't tire
By their lot a most faithful mother require
For you now are starry; Galileo doth settle
(Or this I reckon you up from the devil)
—what relief!

We send wine and books . . . there's Comedy in Susanna.

Susanna was young, too young, Kepler's stepdaughter scolded him. Having just made him a grandfather, he was marrying someone her own age. For whatever it was worth, Kepler chose his wife, an orphan, ahead of moneyed and noble ladies, because she was pretty, active in her loving, and appreciative of children—and she chose him, too, ahead of six other suitors. Circumstantial evidence suggests she enjoyed sex, and Kepler's old age would be ambushed by philoprogenitiveness, seeing his family triple in size. Her husband almost never wrote of her. It seems she was happy.

Whether she knew its full extent or not (and almost certainly not), Susanna had just been inducted into the difficult sorority that was the Kepler women. Publicly, their sisterhood was prescriptive and their burdens collective. When the tragic death of Kepler's stepdaughter left behind his three grandchildren, he made a scenic visit up the Danube with his now-eldest daughter, aged fifteen, where she stayed for several months learning to mother. Kepler himself had a dear little sister, deaf in one ear, over whom he was sweetly protective. She wrote occasionally and visited more than once, but after her marriage settled out west near their mother Katharina, caring for and being cared for by.

Katharina Kepler was about sixty-five (although not even she was certain) and had been a widow for over a third of her life. She had lived in the same small town, more or less, her entire life, though she did once take an exotic vacation to see her most successful child, the big hotshot astrologer Johannes Kepler, at the luxuriant Prague Castle. Kepler thought of her with the affection of a knowingly more educated son. To him she was a "wise-woman," an elder relict who accrued and distributed knowledge entirely on her own terms. All her adult life she had been a potion brewer, an art country wives often reached through their cooking. She tried to get a hold of her dead father's skull to turn into a drinking goblet, simply because she had it from a local pastor that this was a quaint burial rite among ancient peoples. In old age she loved to socialize and have people over for meals, and one can only imagine the officious trivia, true and false, which she forced upon her houseguests: the best use for rare orange zest, the proper boiling time of milk soup, how to keep a heifer alive through the winter. "Raw experience," Kepler wrote, "or, as physicians say, empiricism, is the mother who gives birth to science." He wrote this thinking explicitly of his own.

Katharina's last climactic role in Kepler's life began within the family, when another less thankful son spread eldritch rumors that she made food for Satan. He died soon after, without repenting, and Kepler was unaware of his brother's perfidy. The way gossip soaks a province, though, all a conflagration needed was someone to light the match.

It was the old town prostitute who did it. That is, according to Katharina's bitter testimony after the fact, Ursula Reinbold had been a prostitute. She was married now, but no course in life gave her the respect she deserved. Kepler's mother sounds, from the court testimony, to have had little toleration for such loose women. She was wont to give a moral lecture, but through the magic of brandy, both ladies could be happy in the same room. Offered a drink from Frau Kepler, Ursula awoke the morning after with her megrims worse than ever. Drinking once more with the town governor (everyone knew everyone in the country) out she sputtered, "That Kepler woman must take back her spell, before I die!" And so the prosecution began to assemble.

Frau Kepler had a habit of offering free drink to her many visitors. Watch the milk of human kindness spoil before her eyes! An art teacher's wife drank with Frau Kepler; she died; a hex. The town glasser had a swig; he went lame; another cursed spell. Soon her magic moved from libations to touch, curses, and then mere accusations, no method necessary. She slew a hog last week, kept three wives paralyzed, and what must have been, to a mother, the most unspeakable: she was a child killer.

This specific brand of cruelty—that against helpless old women—has a specific appeal for those who accept it. It pretends an easy solution to their own suffering. In an age of prolonged war and religious faction, if a county could burn a witch to stop the apocalypse, this placated many. Frau Kepler walked with the whole impotence of the countryside displaced upon her shoulders. When one grinning Lancelot pulled her aside, threatening her with his blade, her daughter could bear it in silence no longer. She composed a letter for Johannes, the golden boy of the family.

When Kepler received the news, he wrote, "my heart burst from my body."

Finding yet another lesson in the life of Tycho Brahe, Kepler turned to the law. He became the primary defense for his mother's case. To escape a lynching, she came to live with him in Linz for all of 1616; when she headed west again, Kepler rode with her. On the drive, he read a book on musical composition by Galileo's father, who had been a respected lutenist when Galileo was a boy. Even Galileo became thus involved in Kepler's grandest project, his world-harmony, to which he had finally returned after nearly twenty years. But for his mother, who had raised him up through a sickly childhood, he put it aside once more. For a long moment, he effectively quit scientific work altogether. Some truths are more important than others.

Kepler's months in court were best summarized by the prosecution, who wrote, "The accused appeared in court accompanied by her son, Johannes Kepler. Alas."

When weak evidence extends rather than terminates a trial, it has become a parody of itself. There is no doubt that, thanks to her son, Katherina's show trial became a real one, but there is also no doubt that, after six years of it, she ultimately saved herself.

Final examination in witch trials was frequently an interrogation under torture. In this case, the age of the prosecuted and strength of her defense led "merely" to a "frightening," an interrogation under the threat of torture. Frau Kepler was faced with the rack.

After a woman gave birth, it was believed her body waged a battle with evil spirits. Kepler's mother had done so seven times. She sold hay in her menopausal years, at interest, to support herself. When her own father was near death in 1595, she spent five years feeding her pop and cleaning his bedpan. Her own husband, when he was around, had been abusive.

She looked up at the Tormentor. "Do with me what you wish," she said, and thus secured her freedom. "Even if you were to pull out my veins, one after another, I would have nothing to confess."

She then fell to her knees in prayer, for the length of time it took to say one "Our Father." Without any forced confession, even the flimsiest legal proceeding had to let Katharina free. She gave in to senility and died hardly half-a-year later, overcome by this final strain from her hateful world.

Galileo Galilei also had a living mother, Giulia, but she really was guilty. She stole from him, spied on him, and fought with Marina, mother of his children, yet he would not refuse her place on his conscience. She had raised him up through childhood and taught him his first lesson in life. Neither marriage nor children had done anything to make her happy; there was only religion.

In 1612 Marina Gamba died. Galileo was not the type to talk about it. Without any emotional response from Galileo, her death appears to him entirely as a hassle. She had been caring for his little son, Vincenzo, who would have to be driven to his father in Florence. Both of his daughters, upon starting their adolescence, had already been brought near him. Galileo had been plotting their future.

Across the Arno River, about an hour south the city by foot, there was a small reservation where only women lived. Its population had many exquisite rituals. Each day, they all donned the same style of frumpy black robe and headdress. Their sick were dotingly attended, with oldfangled remedies of cinnamon water or liquor (wine was, therefore, primarily an anesthetic). They spoke to no men except the father confessor, who shared their vow of chastity, certain married employees, and visitors, who could only dine with them in the purgatorial "waiting room." Although girls from rich families lived best, the whole reservation was impoverished, so most goods were shared communally. Exile onto this virtual island was considered, by Catholic society, an even higher calling than marriage. For its inhabitants, this was not necessarily so. Lesbianism was uncommon but not unheard of; some poorer citizens, hysterical or just plain unhappy, would steal away a knife from the kitchen, curl into the corner of a stony cell, and end themselves.

In committing his daughters to a nunnery, Galileo lost one of them forever. Livia Galilei was only twelve when her move began, and she never learned to want a father. She did not feel at home with other women, and developed an eating disorder in her twenties. Galileo had done what he truly thought best, but had difficulty making the empathetic choice. He had no experience being Livia. If he were in his daughter's place, he would have preferred the cloister.

His other daughter managed the change of lifestyle in stride. Virginia knew how to deal with Papa. The clarity and reason with which she registered her new surroundings were coordinated with his own outlook. Suicides, laxatives, and plague were discussed with dispassion. News of deaths and marriages were always appended with an appropriately trite recounting of her emotional state. She told him exactly how she was to be loved; she was to be loved through physical tokens, the sending of food, alcohol, money, commercial goods, and customary statements of mutual affection. When she came of age, she returned the same, darning his socks, pressing his collars, and cooking him sweets. These were sent by courier that same day, almost every week, in a petite wicker basket with a fresh letter.

Most relationships Galileo and his friends had with women would not be so gainfully articulated, but this was certainly not for lack of trying.

In 1614, Federico Cesi became the very man his righteous teenaged self had despised, by falling in love with a beautiful woman, Artemisia Colonna. He put a baby in her, which killed her and the baby. With remarkable speed, he found a more suiting life companion in the cousin of the late Filippo Salviati, Isabella, which was about as close as he could legally get to marrying a fellow Lynx. Upon this collapse of his youthful continence, the visionary loudened, trumpeting the Lynxes as his "philosophical militia."

Cesi was not the only man transformed after friendship with Galileo. Benedetto Castelli literally walked in Galileo's footsteps, having succeeded his mentor's post at university. He adored teaching, tutoring even more, and wrote Galileo four or five times a month to express his thanks. Through Galileo, his reputation had reached the grand Medici, who invited him to breakfast at the royal palazzo on December 9, 1613.

Common practice had the wealthy wet a biscuit in a glass of morning chianti. Wine was like coffee to these people. This fat, abstinent Benedictine monk went on to have the most important conversation with a woman of Galileo's life, and Galileo was not even present. He had to receive it secondhand.

> The Grand Duke asked me about school, so I gave him a detailed account. Then he asked me if I owned a telescope; I said aye, and related to him some observations of the Medicean planets from last night. The Grand Duchess inquired about their position, and began to mumble to herself that they had better

> be real. So the royalty asked a friend, who answered that their existence could not be denied. I used this occasion to add what I knew about your wonderful invention. Another Medici gave me a smile so dignified that he must have been pleased. Finally, after much else, all treated seriously, the meal was finished and I left. But no sooner had I exited the palace, then the porter called me back.
>
> Now, before I say what follows, you must understand. All the while Her Highness's friend had been whispering in her ear, that although he admitted the celestial novelties you have discovered, the Earth's motion remained incredible, since it contradicted Holy Scripture.
>
> Anyway, back to the story: I entered the chambers of Her Highness, where I found everyone from that morning and more. After some questions about my own views, Her Ladyship began to argue against me using Holy Scripture. So, with the proper disclaimers, I played the part of theologian, with such deftness and alacrity that you would have been proud. A Medici assisted me, so that rather than being intimidated by the majesty of my company, I went off like a gladiator! The Grand Duke was on my side, and yet another came to my aid with an apposite quote from the Bible. Only Her Ladyship contradicted me . . .

Benedetto had become like an adopted son to Galileo. Their relationship was built upon the same kind of direct and open speech Galileo had achieved with his eldest daughter. It was understood that when Castelli spoke, it was in Galileo's defense. He was much more qualified than Galileo on theological matters, and his teacher was lucky to have such a canting advocate. Galileo returned thanks in a long, contemplative letter, with his own thoughts for future reference.

> Do scriptural interpreters think that, in a dispute about nature, someone who is right has the advantage over someone who is wrong? I know they will answer yes, because members of the true side have a million billion demonstrations to provide, and the false, nothing but sophisms and fallacies. But if the

> interpreters know they have such an advantage when only philosophical weaponry is used, why then do they whip out such a horrifying and insurmountable weapon as the Bible? The mere sight of it terrifies even the most stalwart champion. But, if I must speak truth to power, I think it is they who are terrified. They are unable to resist their opponent's blows, and struggle to find a way of stopping his approach. Yet consider what I have said. Whoever has truth on their side has a great, perhaps the greatest, advantage.

The full response is overwhelmed with this repetition and very genuine confusion, but Galileo's point was well-made. Why couldn't everyone just speak his language?

> "I have written rather more than befits my illness, and so I conclude. Merry Christmas."

Two Winters and a Spring

The Apostles gathered around Jesus and asked him, "Lord, wilt thou at this time restore the kingdom to Israel?"

And he said unto them, "It is not for you to know the time nor season, which Father hath set through his own power. But ye shall receive power, after the Holy Ghost comes to you, and ye shall be witnesses unto me in Jerusalem, in all of Judea, and to the ends of the earth."

After he said this, he was taken up before their very eyes, and a cloud hid him from their sight.

They were looking up intently at the sky when two strangers in white appeared beside them and asked, "Ye men of Galilee, why stand ye gazing up at heaven? This Jesus, who hath been taken up before you, shall come back down in like manner."

Then, on the day of the Pentecost, the Apostles were all gathered in one place, when suddenly came a mighty, rushing

> wind, and what seemed a cloven tongue of fire ripped through each of them. They were each possessed by the Holy Ghost, and began to speak in tongues.
>
> Now, they were dwelling in Jerusalem, where there were devote Jews of every nation. This noised abroad, a multitude gathered, and were befuddled, for each heard their own native language. Astonished, they asked, "Are not all of these who speak Galileans? How is it then that we hear our mother tongue? What does this mean?"
>
> Others, mocking, said they were drunk.

This passage from the Book of Acts was not the most alarming Tommaso Caccini quoted from the pulpit on December 21, 1614. That was a simple biblical poem in the Book of Joshua.

> So the Sun stood still
> And the Moon was stopped
> Till the nation had avenged themselves upon their enemies.

In a blustering oratory, Tommaso used poetry to condemn Galileo, Copernicus, and, it is said, all of mathematics. The poem was one the Grand Duchess had herself used against Benedetto in their dialectic, which Galileo, in his letter, had attempted to logically decipher in his favor. He used complex astronomical arguments, involving the "lunar month," "primum mobile," and "Ptolemaic world system," which a congregation of hundreds quite rightly did not care about. Their sense of poetry was immediate.

That Tommaso's and Galileo's sermons both carried this verse was, the former claimed, a happy accident. Be that as it may, men of letters traded epistles just like they wrote them. Castelli had given Galileo's letter to a friend, the friend to a friend, to a friend to a friend to a friend . . . this for almost precisely one year. When a writer wanted to remain private, they usually gave instructions that the recipient not be allowed to keep their words. Such was what happened to Tommaso, returning home two weeks later, still ruddy with aplomb. A letter from his cosmopolitan brother was awaiting him, out the court of Cardinal Maffeo Barberini.

> To a Most Revered and Venerated Gentleman:

> Your freakish antics disgust me beyond measure. Know that if I hear this shit again I'll make it so bad you'll wish you never learned to read. Know that nothing you could do could be thought of worse by those up top. Pray to God that you don't learn this in a trial!
>
> Don't even try to dress up in religious zealotry. We here in Rome know how you friars use it to cover up your filthy minds. Ugh! How thoughtless you are, to place yourself among these pigeons, these pigdicks, whatever the hell they're called!
>
> We were trying to help you, but you made it ugly.
>
> Brother Tommy, know that reputation runs the world. When you fuck up you lose it. I ask that you stop preaching. If you do not do this out of love for me, there is a second way. Maybe I'll find it. You've been warned. I wish you the best.

"Father Lorini was here," Benedetto told Galileo, "He regrets that the good Caccini let himself go so far." It is true Lorini preferred the proper bureaucratic channels. Nearly seventy, he and the Counter-Reformation had grown up together, sharing their lives through its ad hoc tribunal, the Roman Inquisition.

On February 7, he sent them a letter.

> Besides being the shared duty of every good Christian, us Dominicans are bound by an eternal obligation, since we were created to be the hunting dalmatians of the Holy Office. I have come across a letter, which is passing through everyone's hands here, originating from those known as "The Galileists." They follow the views of Copernicus. In the judgment of the many Fathers here, the letter contains many propositions which seem to be suspicious or reckless: for instance, that in disputes about nature, Sacred Scripture gets last place, that its interpreters are often wrong, and so on. You will see that, in the copy of this letter I am sending, I have underlined such propositions. I could have sent you further notes, but have refrained out of modesty, and because in Rome, the holy faith already has the eyes of a lynx.
>
> I believe that the Galileists are fine folk and good Christians,

> but just a little conceited. I ensure you that I am only taking this action out of religious zeal. I beg you most illustrious lords to keep my writing you a secret, as I am sure you will.

"The same Father who complained last year has hit me again!" Galileo moaned the next week. This time, Niccolo did not apologize. He returned to his friary, a saintly, subservient man, to live out the final days of his dotage in absolute peace.

"I wrote the letter with a quick pen," Galileo told a Vatican official. In actuality, he could hardly write fast enough to keep up with the revived opposition. "How harmful it would be, to take a doctrine from Holy Scripture for which there might be disproof!" Moving too quickly, he realized he needed a postscript.

> P.S. Though I have difficulty believing anyone would be so hasty as to decide to ban Copernicus, I know from past experience how bad my luck is. I have reason to be skeptical about the prudence and holiness of those with whom the final decision rests. They may be deceived by this fraud. However, I do not contradict my superiors. My upbringing is such that I should pluck out my eyes before I sin.

"I do not wish to quarrel with anyone," Galileo wrote. This was not how he wanted to spend his time. He had hitherto directed little energy toward what he called the "salvation of the soul," to him, the essential teaching of Catholicism. Now he was searching desperately through the weighty tomes of St. Augustine, St. Jerome, and St. Aquinas, mining for quotes, trying to learn. His righteous accomplice Dom Benedetto furnished him with a huge base of theological scholarship. Some hundred-odd pages of notes filled Galileo's study, none of which seemed publishable. To really do the job justice, he realized, he would have to return to Rome.

Unbeknownst to him, the preacher who had set this whole affair to flame had beaten him there. In the last days of winter, to repair his reputation Tommaso Caccini visited his brother, but when he arrived, he found a city entirely unlike what his brother had let on. The hardest truth of Rome was the part of his brother's letters which looked most vapid: "All cannot be written."

Tommaso soon learned that the Catholic Church would never dream of dismissing the claims of any Florentine vigilante. Rather, they absorbed the complaints and concerns of every sect, order, and individual who brought them, a lumbering, centuries-old giant with one eye and a hundred thousand limbs.

He met with a friendly Dominican cardinal to mollify his fears and was told he must report, immediately, to the examination hall of the Roman Inquisition.

It was a yawning cavern in three layers, with ornate decor, faux arches and baroque spandrels, frescoed ceilings, life-size statues, and a little man at the front desk to take all voluntary testimony. He was a persnickety type, submitting Tommaso to a rigorous line of questioning. He took exact notes.

A sentence was forming on the tip of Tommaso's tongue. He had just been asked if he had any preexisting hostility towards Galileo.

> I do not bear any hostility toward the said Galileo. Rather, I pray to God for him . . .

A wash of relief was over. Tommaso was no heretic. He was safe in the Church. This had been a dangerous game of chicken. After all this striving, he thought, it would be nice to settle down as the confessor to a local nunnery.

> . . . I do not even know what he looks like.

And then it was spring.

The Galiliests spent the first weeks of March pestering their Roman allies, trying to assess the damage. "Pray, do not form your fears," Galileo read, "nothing has been rejected."

> Cardinal Barberini, who, as you well know, has always loved you, told me last night that he would like to see greater caution, not pushing for arguments outside the domain of physics and math. For when new things are brought to theology, even by a genius, not everyone is sensible in their passions. One mouth amplifies; another edits; the words become so deformed that their first author cannot recognize them.

Budlets of oleander were outside Florence, flowering into a pink wood. On March 7, Cesi brought Galileo news of a convert from within the Church.

> I send Signor a book, which has just come to light, in which a Carmelite friar defends the opinions of Copernicus . . . this could not have come at a better time. He is now preaching in Rome.

The book was a boon, but revealed how much the ground had shifted. It was by a theologian, for theologians. Establishing heliocentrism as part of a rich community and long historical tradition was more important than the brilliant scientific arguments of any one man. The author even reminded himself through an ancient Latin verse.

> Never condemn the popular view
> For you will please no one, and fight quite a few.

One chapter contemplated how heliocentrism changed the location of Hell. It had the obscure title, "Hell is the center of the Earth, not the World."

Galileo, who had just begrudingly accepted his fifty-first birthday, spent most of 1615 too ill to travel. What he did manage that spring was to collect his theological notes into a short work for the Grand Duchess.

The letter was not, of course, actually for the Grand Duchess. She was an excuse for Galileo to courteously scream his opinion throughout Rome. In no time, it was seen by the highest rungs of Catholic power. They read:

> It absolutely must be prevented that an infidel hear a Christian raving about these matters, as though they were in accord with the Bible, while erring so deliriously as to be laughed at. The trouble is not the laughter, but that scripture should be thought in error, and thus rejected as ignorant, to the ruin of the salvation we stress about.

What a noble declaration! The scientific discourse was either irrelevant or already decided; this was about the fate of humanity.

There was a quirky unevenness that Galileo brought to this herculean task. He was sick, tired, and worn down. Misanthropy was leaking through.

> By nature, the number of people unable to grasp theology and other sciences far exceeds the number of intellectuals. . . .
>
> It is necessary to give the Sun motion and Earth rest so as not to confuse the tiny brains of common people. . . .
>
> Human nature is more inclined to (unjustly) oppress one's neighbor than help them up.

In spite of this, "I have the highest regard for how common people judge me." These social paradoxes scar even the body, over the years, of those who ceaselessly try; it is the bags under their eyes, the nibbles at their fingernails, the heavy wrinkles set like stone and tufts of hair which fall out early. It is not masochism, for they do not enjoy it; it is not martyrdom, for they do not accept it; they are sucked in, against their will but with their nature, kicking and screaming and fighting for breath like a failed Odysseus before Charybdis, losing grip on their given tree, staring down into her gaping, whirling, sable maw. Let it be heroism, in the most mundane, unyielding, daily sense of the word.

And then it was winter, again.

Galileo was, at the start of December, at last well enough to travel. The last time he had been in Rome, one cardinal commented that they ought to have erected a statue to commemorate his glory. Five years later, the difference was unfathomable. He had so many enemies. Even at friends' houses they sallied, where he would fend them off with his rhetorical prowess, fifteen, twenty, twenty-five at a time. The Dominicans spread false rumors that he had fallen out with the Medici family. Galileo recognized he would need to be here a very long time.

> I shall need as many days as my adversaries have had weeks and months to spread these opinions against my person. I hope my time will not be shortened . . . I have seen nothing in letters to assuage me of my fear. I beg you, shore me from this doubt, for I have no wish to leave here without the restoration of my reputation.

What Galileo was expecting was attrition. The Medici had given two rooms in their Roman palace for it, "to make life sober and secluded." There he could hunker down, to wait out the longest winter of his life. Something bad was happening to his eyes, he reported: staring into the fire, he saw it surrounded by a halo of false light. A languorous chill ran through the air.

It was a Thursday. Galileo had been in Rome two months. That night, he received word that a certain Cardinal Roberto Bellarmino, the most notorious scholar of controversial theology in his day, needed Galileo to come before him to discuss the reconciliation of astronomy and scripture.

The main portal out to the city was a dark arch of shaped lumber, split down the middle into opposing halves three times the size of a man. They turned about like Janus, the two-faced god, slowly exposing his unfamiliar person.

The Other Side of the Door

Deep into the wee hours, one, two, three in the morning, a ray of manufactured light would show beneath the entrance to young Roberto's room. He had made a pact with his eldest sister, his second-favorite member of the family, that if she snuck him a candle after bedtime he would spend it on a book and tell her its stories next day. Suppose that, towards morning, after a night of insomnia, sleep descended upon him while reading: he had only to lift his arm to arrest the Sun and turn it back on its course.

Such was the wretched boyhood of Roberto Bellarmino. It began bobbing on his mother's teenaged knee, absorbing, as if through her milk, that girlish submission, which would bring her twelve children in twice as many years, which, in male hands, would create one of the most ludicrous and involuntary rises to power in the history of the Catholic Church.

Poetry had possessed the boy before he could remember. The first he ever wrote was a prepubescent acrostic celebrating his V.I.R.G.I.N.I.T.Y., but he burned it after he was taught shame. None shall ever know what it stood for. He was forever in the little boy's habit of crying without reason. Though the pageantry of Church brought him to rapture, he abhorred all physical pleasure; aged six, little Roberto upturned an old soapbox in the family living room and, stark naked, climbed up to preach. He was loath to come back down.

From one sermon around 1570, when he was nearly thirty:

> Divine light is hidden by this insubstantial sack of meat. It is just like how we protect against the Sun; were we unprotected, we would be unable to look, as its rays would destroy our senses. Thus, we have words.

From another, around 1574:

> Other people look on drunkenness as a serious and despicable sin, but in a drunkard's eyes the real sin is to mix water with wine. Water is the foe.

From his most famous and astonishing sermons, on the "Four Last Things" (Death, Judgment, Heaven, and Hell), around 1576:

> All Christians have been persuaded that those who depart this life guilty of violating God's law are bound as convicts to eternal chains, and tortured without end, should they not sincerely repent before death. Yet we see daily how many people, unasked and uninvited, offend God of their own free will. What shall we say is the cause of this? My dear people, there seem to me three principal causes: lack of consideration, ignorance, and self-love. . . .
>
> Would that we could comprehend but little the eternity of the punishment. Think of the cross, suffered for the span of a single night, by a man suffering some grave affliction. How long does the night seem to him? What longing grips him for the dawn? How often does he inquire how much night is past, and whether the rays of light are starting to appear? If such a sorrow seems so huge to us, what will be the sorrows of those who lie, not about a soft bed, but in living and blazing flames, for the whole span of that night which has no dawn, and awaits not the light of day? O, deepest darkness! . . .

> Ignorant men do not realize that after death we shall have different eyes, and a different opinion, and a different reckoning of things than we have now.

Bellarmino's mother loved him special, more than all her other children, yet she was addicted to self-sacrifice. When he was still a toddler, a middle-aged canvasser came round to discuss a brand new religious order, specifically devoted to education, intellect, and Catholic apologetics. The first thing an authority ever suggested Bellarmino do, he did for the rest of his life. When he turned eighteen and left home to join up with this brand new religious order, his mother began wailing. "I am happier than ever to know I have dedicated him to God. Yet I cannot stop this aching in my heart."

Few would have such a transcendental failure as young Roberto, but the Jesuits would. Their world was the only place he wished to be.

It is not gratuitous to suggest that, within the scholarly community, the Jesuit order saved the Catholic Church from irrelevance. As anti-Catholic fervor spread across Europe, politicians and kings tried to beat it back. The Jesuits preferred to outwit it—or, worse yet, befriend it. Lutheran schooling was the first morsel of democratized education the underclasses had ever tasted. The Jesuit response ranked among the most radical positive reforms of the Counter-Reformation. They set up hundreds of schools for young boys which were free—*free!*—diverting corrupt Church funds back to the laity. Moreover, the education tended towards quality.

The problem with Jesuit reform, to the ultraconservative, was that a genuine education risked challenging received dogmas. Such a wildcard as Galileo found most of his earliest Catholic advocates in the Roman College, a Jesuit institution. His pen pal Maffeo Barberini had received his education there. Galileo's most respectable enemy Christopher Scheiner was a Jesuit through and through, constantly chastised for his brazen behavior. Even a figure as obscenely foreign as Johannes Kepler was contacted by astronomers of the Jesuit persuasion. A few over at the Roman College might dare pass for Copernicans.

Bellarmino, however, was not one of them. He had the freedom to choose, and he chose obedience. He read Aristotle, a few precis of Ptolemy, then turned back to the Bible and its commentators. No longer was he captivated by literary invention; literature was a constellation revolving around scripture, the unwavering book. After three years, he graduated with top marks and was sent to Florence as a Jesuit teacher.

Magister Bellarmino had liberated himself of private desire so totally that he seemed destined for middle management. In Florence, he was assigned to lecture on Greek. He did not know Greek. Rather than suggest a replacement, he puttered out before the students, explained that he would begin with the rudiments and learned the whole language that year. In his future sermons, whenever he had to give the true definition of a word, he assumed that Greek etymology held the answer.

A series of promotions brought him to a professorship in controversial theology at the Roman College. Since everyone kept calling him a theologian, he learned Hebrew, to read the Old Testament from its first source. The word was, for him, the essential thing. As a simple teacher, he was so distressed by the act of translation that he sent a letter to a cardinal superior, asking what it meant for the Church to accept a Latin version of Greek and Hebrew texts. He worried, "To say that for centuries the Church misinterpreted, that she has honored the misunderstandings of a certain translator as the very word of God, would that not advance a strange absurdity?"

He had recognized his world exactly. It was a strange absurdity.

To his immense discomfiture, Bellarmino discovered that such controversies were extremely popular. So many were attending his lectures that his masters bid him publish a textbook. "There's no escaping a book," he sighed, and proceeded to write out two million words by hand.

On the Controversies of Christian Faith Against the Age of Heretics turned Bellarmino into an international celebrity. In any land of dissent, England and Germany especially, a good Catholic lost in debate needed only run home, grab their copy of *Controversies*, and thwack their opponent upside the head with it. It's in there, stupid!

So estimable had Bellarmino become that, as the volatile seventeenth century opened out before the Church, he was risen to the cardinalate. "He has no equal for learning in all of Christendom," Pope Clement VIII declared. But Bellarmino had only ever wanted to be a Jesuit. He received the red in abject despair, with plaintive tears rolling down his cheeks.

From then until his death, Cardinal Bellarmino lurked in the shadows of nearly every Church affair grand enough to cast one. Already, an unseasoned Lutheran math professor by the name of Johannes Kepler had grown infuriated with *Controversies*. "Bellarmino is most learned," he conceded, "but he is shackled by greed and fear of disgrace." With sudden malice, Kepler spit. "Ech! How many obstacles surround the truth!"

In contrast, when an equally juvenile Catholic scientist named Federico Cesi was passing through Bellarmino's new diocese of Capua, he was housed by the cardinal and found the elder so supportive the two began a private exchange of letters.

By 1599, Cardinal Bellarmino found himself in an undesirable position as councilor to Inquisitors General. He was enlisted to read through the oeuvre of a Dominican transient named Giordano Bruno, who had been brought to court on charges of blasphemy. Giordano's books were a prodigal grab bag of poetry, mysticism, science, anything that might be contorted against Catholic doctrine. "We are all born ignorant," Giordano wrote, "and willing to acknowledge it."

> As we grow, we are brought up in the disciplines of our house, and we hear disapproval of those different from us. At the same time, they hear the same things about us. Thus, as we acquire zeal for our own ways, so do others for theirs. So it easily becomes habit that we oppress, murder, maim, and eradicate our enemies as a sacrifice to the gods. They do so too, and we all render thanks unto God for the light which leads to eternal life. . . .
>
> To Copernicus we owe our liberation from several false presumptions in vulgar philosophy. I do not wish to call it blindness—yet even he could not entirely remove it.

This was obviously heretical. Bellarmino edited the list of profanities down to eight propositions, which Bruno would be asked to abjure. He refused.

In 1605, at the death of Clement VIII, Bellarmino was recalled to Rome to join in a papal conclave. In the following weeks, he heard a rumor so scary that he locked himself in his room for days. He might be elected pope! Bawling inconsolably, pressing his rosary to his lips, he whispered over and again, "Dear God, save me from the Papacy." He was, but barely. Pope Paul V demanded he remain in Rome as his personal advisor.

When Paul confided to his beloved Roberto in 1606 that he was planning to excommunicate the Venetian Senate for defying papal authority, he received full approval. The Jesuits were first to leave the city, as Galileo

Galilei, watching them depart, felt his first incorporeal tug from this distant stranger.

Bellarmino was still in Rome at the start of 1611, when he was approached by this Galilei boy, who had been drumming up support from all the church prelates for his *Sidereal Messenger.* They had a conversation about a most curious subject: the interrelation between astronomy and theology. Bellarmino sent a brief note to his Jesuit colleagues at the Roman College asking for "definite knowledge" concerning Galileo's discoveries and received word that they were all true.

By his own admission, Bellarmino did not hear of the matter again for five years. Then, seemingly out of nowhere, a patrician coterie of Galileists was bothering him without end. Cesi was the first, reporting back to Galileo:

> Concerning the opinion of Copernicus, Bellarmino himself, who is one of the heads of the committee that deals with these matters, told me that he considers it heretical, and that the motion of the Earth is undoubtedly against scripture.

A few months later, a Carmelite friar sent Bellarmino his book attempting to reinterpret scripture through Copernicanism. Bellarmino wrote a long, considerate, and hackneyed response.

> It seems to me that you and Galileo would be prudent to speak hypothetically and not absolutely, as I have always believed Copernicus spoke.

After a year of follow-ups and interviews, including a full investigation of Galileo's letters on sunspots, the Roman Inquisition had cause for a private consultation concerning two garbled theses.

1. The Sun is the center of the world and completely devoid of local motion.
2. The Earth is not the center of the world, nor motionless, but moves with the whole of itself and also with daily motion.

Both propositions were discussed by a council of eleven church authorities, including an archbishop, several Dominicans, and precisely one

Jesuit. Cardinal Bellarmino was not physically present. They advised as follows.

1. All agree that this proposition is foolish, philosophically absurd, and formally heretical. It explicitly contradicts Holy Scripture, according to the literal meanings of the words, and the common interpretation of the Holy Fathers and doctors of theology.
2. All agree that this proposition receives the same judgment.

This advice was forwarded to Pope Paul. Paul called his advisor Bellarmino to his side. Bellarmino was to warn Galileo that night.

Bellarmino, as he had always done and always would do, wore the entire legal, political, and ideological superstructure of the Catholic Church as though it were his bald skin. It was his unremovable suit of armor, his honor and drowning burden. He called Galileo before him, to discuss this reconciliation between astronomy and scripture. He writ his name large on the *dramatis personae* of popular modernity: Roberto Bellarmino, enemy of Galileo, villain.

The tired, distressed figure who appeared before Bellarmino was one much like himself, in all the empty, pleasing, vital ways. Both men liked plain speech. They had both been teachers, and preferred wine to beer. Both were staunch defenders of free will, in keeping with the traditional Catholic doctrine of divine grace. Bellarmino, who considered himself married to his diocese, with its denizens his "flock," was a patriarch just as Galileo. Each had the cute beard of an old man. They had learned the lute and received training to be doctors when they were children. Bellarmino had even confessed, like Galileo, to suffer from "fleshly allurements." Both had trouble keeping time of their own accord. They were friends.

The ways in which Roberto Bellarmino was not Galileo Galilei were fewer, but they were significant. Bellarmino was twenty years older. He was far more worldly; Galileo had never left Italy. He cried so easily, while Galileo did rarely, if ever. He read more. More important than any of these, Bellarmino was very short. When he stood next to Galileo, he came up to his breast, and this had always been an embarrassment except during sermons, when he climbed up a footstool, taller than any, arms spread wide, voice booming out as he talked about God.

A Bad Memory

The man brought before Cardinal Bellarmino was an annoyance liable to produce a histamine headache. The cardinal's head was prone to ache from fasting and sickness. It ached so badly that he had fainthearted thoughts about death. But his suffering would soon be over, and his headache would go away.

Bellarmino raised his martyred eyes to the prisoner, who was down on his knees, where every guilty party was brought, to hear his sentence.

> Invoking the name of Our Lord Jesus Christ and his most glorious mother the virgin Mary, in the case and cases aforementioned pending at present before this Holy Office, you the aforesaid Fra Giordano Bruno, a felon examined, tried, found guilty, impenitent, obstinate, and pertinacious, by this our definitive sentence, on the counsel and opinion of the Reverend Fathers who are masters of sacred theology and doctors of canon and civil law, our consultants, we proclaim in these documents, state, pronounce, sentence, and declare you, the aforementioned Fra Giordano Bruno, to be an impenitent, pertinacious, and obstinate heretic, and for that reason to incur all the ecclesiastical censures and penalties of the sacred canons, laws, and constitutions, in general and in particular, as those that are imposed on such confessed, impenitent, pertinacious, and obstinate heretics. As such we degrade you in words, and declare that you should be so degraded.

This arcane branding was, technically, the only legal proceeding for the Catholic Church. Such a case was then passed to the secular branch of government, where it was completed with an appropriate execution.

Giordano met the eyes of his accusers with a low growl. These were his last known words. "Greater, perhaps, is the fear endured by those who deliver the sentence than those who receive it."

*

A Dove

I did not write last week," Galileo wrote, the week after being called before Bellarmino, "because there was nothing new to say." No one knew the specific dialogue of his special conversation between Roberto Bellarmino—few even knew it took place at all—because Galileo did not think it worth recording. When acting as scientist, Galileo recorded virtually everything, but here he was intentionally dismissive.

> They were about to make a decision on that business which I mentioned was of public interest, which my enemies have tried to drag me into no matter what cost. I mean the deliberations of the Holy Church on Copernicus's book and opinion on the motion of the Earth, which had been subject to complaint by that same friar who earlier declared them heretical. With his followers, he fought to make his declaration prevail. He has failed.

The Church had released their first public decree concerning Copernicanism. This was the only official opinion Galileo could read. Against the advice of their internal council, they did not use the term "heretical." Galileo noted as much with deliberate nonchalance.

> The Holy Church has only declared Copernicanism does not agree with scripture, and only prohibited books which explicitly state that it does. There is only one such book, by a Carmelite friar. As for the book of Copernicus, as I understand it, they might remove a word here and there.
>
> As one would expect from the nature of this business, I am not mentioned.

The Church had resolved into an official position, but Galileo was not ready to leave. His reputation remained in disrepair. Sagredo told him the Venetians thought "you have been violently pulled into Rome, and declared a heretic." Less ridiculous was the gossip Castelli said was echoing about Florence, that "you have abjured in secret at the hands of Cardinal Bellarmino."

Rumors were omnipresent, but they were nothing more. Galileo remained respected by those who mattered: Bellarmino, Barberini, and Pope Paul. The next week, on March 11, he went on a constitutional with Paul for nearly an hour. "He cheered me up," Galileo wrote, "and said I could put my mind at ease. I am safe as long as he is alive."

This ease was not shared by his employers. The Medici's secretary of state sent him the briefest of orders.

> You have tasted the persecution of these little monks. We fear that a longer stay in Rome might cause something awful. You have escaped with your honor. Poke not again this slumbering giant, and return!

In his final gambit, Galileo met again with Cardinal Bellarmino, hoping for protection against at least one rumor. He left Rome with a caring note in his pocket, by which to forever remember these months in contradiction.

> I, Cardinal Roberto Bellarmino, wishing the truth be known, declare that Signor Galileo has not abjured any opinion held by him, either before me or anyone else in Rome, or any other place that I know of. No salutary penances have been imposed upon him. It has merely been made known to him that the declaration of our lord pope adjudges the doctrine attributed to Copernicus, concerning motion of the Earth round the Sun, to be contrary to scripture, and therefore not to be held nor defended. In good faith I have written and subscribed this with my own hand on the twenty-sixth day of May, 1616.

Bellarmino's writing after middle age had gone soft. Even their titles alone, *The Mind's Ascent to God* or *The Art of Dying Well*, invoked a private sentimentality. As Galileo returned to Florence, Bellarmino was at work on a book called *The Lament of the Dove, or, the Value of Tears*. The presence of a dove, as a literary trope, implied not only a desire for peace but the immanence of God, because the dove was the most common form God took on Earth as the Holy Spirit. Bellarmino no longer contemplated the other form of the Holy Spirit, a tongue of fire.

Part of the book was dedicated to "the fount of tears," expatiating the many reasons a Christian fundamentalist had to cry for the worlds to come: lack of piety, souls of the damned, temptings of the Devil. "I am old now," the preface stated, "and close to death . . . thus, that psalm ever comes to mind, *O, who will give me the wings of a dove? For then I could fly away, and be at rest.*"

A Tongue of Fire

I am filled with joy by your safe return," Federico wrote his fellow Lynx, "but I fear that change in air may be damaging. Pray, take good care of yourself." "I think the season favorable, and believe the heat will be milder than here in Rome." His fears knew more than his hopes. Galileo spent most of the next two years in bed due to piles and a hernia that caused swollen testicles.

News of the edict from the Church slowly bled out across Europe. The Galileists and the Jesuits were disgruntled, but willfully obeyed, as any self-respecting Catholic would. "Those in Italy," Kepler wrote in 1619, "where Galileo has always been, are constrained by a horde of cardinals." He followed this up with the latest astronomical tattle. Three comets had been seen in the night, streaking after one another, "displaying near perfect form; first obscured throughout August; passing from Leo to Cancer beneath the feet of the she-bear."

Galileo was unable to see them. He could hardly even stand up. He was impotent; he was inflamed. For all his feigned disinterest, the truth was that he cared most personally about his Catholic Church. Since the day of the decree, his virulence had been irrepressible. "His unyielding will is to castrate the friars," the Tuscan ambassador had boggled, wide-eyed, as he helped Galileo escape from Rome.

The three comets triggered a slew of pamphlets on the phenomena, not least of which from the Catholic authorities. Galileo read the little thing, with a cover announcing it had been "presented publicly in the Roman

College of the Jesuits, by one of the fathers in that same society." This put a scowl on his face. He had just spent months with the Roman College. Why had the Jesuits not conferred with him? Instead, intellectually averse to Ptolemy but religiously prohibited from Copernicus, they had turned to the work on comets by dead Tycho Brahe, a man who had never even owned a telescope. This anonymous Jesuit kept referring to comets as "fires." "Buffoonery," Galileo scribbled in the margins. "Indolence," "ridiculous," "the ungrateful brute." He would show them what fire really was.

Galileo collaborated with one of his more ambitious assistants to publish a response. Even innocent Christopher Scheiner came under attack; with this nameless accomplice, "they aspire to be artists, though they are inferior to the most mediocre painters." When the Jesuit riposte arrived, hidden under a pseudonym, Galileo cracked. "It is customary for first-time offenders to receive less severe punishments than recidivists," he wrote. As he had labeled his first thankless student, decades ago, his new enemy became his "ox," "my tremendous ox," "a snake," "you absolute idiot!"

Galileo spent the next two years crafting a second rebuttal. It had to be just right; that he should continue the debate after this would be unspeakably immature. His judgments, he believed, were as fine as an assayer, a hypersensitive scale, and so he named the work *The Assayer*. By now, he had uncovered the true name of his enemy, but it was irrelevant. The Jesuit was an insect. "No lack of stings for me," Galileo wrote, "but I know the antidote. I shall crush the scorpion, and rub him in the wounds."

"You go about like a blind chicken," he jeered, "poking its head in the ground, hoping to find some speck of wheat." "You lack even that common sense through which everybody, even the merest idiot, can reach a conclusion." "You make use of very inappropriate ways of speaking." If Galileo had not been despised by most Jesuits before, he was now. A friend serving the pope in Rome eavesdropped on the monks. The word "annihilate" was heard, more than once.

The Catholic Church heartily endorsed this dick-waving competition. The official censor gave *The Assayer* a handcrafted imprimatur, praising "the deep and sound reflections of this author," with "nothing offensive to morality." Galileo had been cautious to avoid condoning the Copernican system.

The precise disputes carried off in *The Assayer* are minuscule and of no general importance to history. The material composition of comets, the shape of their orbit, and the proper use of the telescope were but a couple of the subjects in which no progress was made. "Possibly my opponent thinks philosophy is a book of fiction," Galileo wrote, moments before theorizing that comets are hallucinations caused by atmospheric vapor, "but that is not how things stand."

What distinguished Galileo's fictional philosophizing from that of his enemies was not its truth value but its character. Never before had a scientist this invested in the early secular state and this disaffected with their religion had the cause, time, and energy to express their view of the world so purely and plainly. No reader worth their beans has ever read any book, be it by Ptolemy, by Aristotle, or even the Bible, without a questioning eye, but to Galileo, authors had become obstacles to truth, rather than fellow travelers. None could be trusted, including himself.

> I do not have a perfect discriminatory faculty. I resemble a monkey who firmly believes he has a friend in the mirror. So live and real does the reflection seem to him that he does not discover his error until he runs round the mirror two or three times.

It was not enough to be a critical reader. Every work that civilization had ever created must be assumed false.

> Regarding certain knowledge as a mode of demonstration and discourse pursued by humanity, I hold that the more it approaches perfection, the smaller the number of conclusions it will promise to teach, and the fewer it will demonstrate; hence, the less attractive it will be, and the smaller its number of followers.

This was a method of doubt more pessimistic and universal than any healthy book reader could entertain. It was not humility except in absolute terms, closer to a rabid incredulity. What had happened to Galileo, to create in him such an unsparing, speciated vision of scientific practice? "I believe good philosophers fly," he wrote. "They fly like an eagle, not like a dove. Eagles are little seen and less heard, while birds that fly in flocks blot out the sky with their squeaks and squalls, befouling the earth beneath them."

Few would describe their fellow researchers in this way. Galileo had become an old man, in a cruel world, keenly aware of his growing isolation.

Death and the Garden

Not thinking about Copernican astronomy demanded tremendous effort. Galileo would pick up distractions wherever he found them. One recent poet-friend sent a pretty diversion.

> May Signor encourage me as I do him, by virtue of our shared friendship, in coming to listen to some of the compositions of my friends, decorated with the new style and Grecian vagaries I have previously mentioned. For just as he has studied mathematics and philosophy, and with such joy tried and arrived at new things, if he is not yet full he might compare thoughts with us poets.

The poet aimed to convince Galileo to "suspend his inclination towards the ancients, and not attribute so much veneration to antiquity." Among his literary friends, Galileo's adoration of Ariosto and Dante was legendary. They thought him a bit stodgy for it. "Signor has uncovered many shortcomings in Aristotle and Ptolemy," his friend continued, "so too might he recognize the imperfections of these Tuscans."

It was possible. Galileo was not much familiar with this "new style." Its foremost practitioner, Torquato Tasso, had recently died; his life had been ruined by a difficult mental illness, which the kids in those days were finding terribly romantic. *Jerusalem Delivered* was the name of his epic. By the end of its first stanza, Galileo was displeased. Tasso had called his protagonist "errant," but this was plain to anyone who had been reading and did not need to be stated.

Galileo placed his copy of Ariosto's *Orlando Enraged* by its side. "Over the course of many months and even years, I compared all the shared concepts of these authors, adding up reasons to prefer one to the other. I found far more for Ariosto." "Tasso's style is languid yet exaggerated," he criticized, "a bad sort of dense." He did not like the constant use of allegory, where nothing represented itself. He did not like the characters, who could not focus on their chosen task. He did not like the lovey-dovey sex scenes.

One description brought Galileo an unusual amount of consternation. The heroes had entered an island manse, where "a bonny garden is revealed from their vantage."

Its quiet ponds, flickering crystal,
Flowers varied, sundry herbs,
Suntanned hills and shady dales,
Timbers, caverns, from just this glance.

The beauty which works its wonders best
Is this art, which tries hard but appears not to.

"This is repugnant to truth," Galileo wrote, "such a palace would have to span hundreds of miles, yet this island is very small." The canto was a patchwork of psychical impressions. Its lies made his head spin. It was impossible, he thought, for there to be a place bigger on the inside.

At the outset of this work, on March 14, 1620, he received a terrible letter. He read, "My brother, Gianfrancesco Sagredo, who is now in heaven, was suffocated by a violent case of catarrh. He degenerated in unspeakable disorder for five days."

Sagredo had been his oldest friend. They had traded portraits just months before; his oleaginous face was hanging over Galileo in the living room. His letters, which Galileo kept, were endearingly selfish. In these latter days, he had opened up his own whorehouse on the Venetian strip. "Under my care is one blackberry," he described a Moorish girl, "presently with a pasty white teen." He was shameless and unflagging in the pleasure he derived from this "diverse service," driving his favored prostitutes out to the country to sketch them, next to plates of truffle and peach. "I avoid a singular taste in pretty pictures," he told Galileo, shipping facsimiles of rare classics to his house from Rome. To view them up close, he obtained a pair of "shortening glasses," a sort of inverted telescope, but he found they looked better from afar, "just like real bodies; women, when kept at a distance, appear much more beautiful." To match his private brothel, he had been collecting for a private zoo, filled with birds, martens, marmots, and wolves. Oh, to be so uncaring, so rich in character and everything else! "My idol," Galileo called him, and now he was dead.

Others discovered humor only as its sorry object. Cardinal Bellarmino lay in hospice, crying, his body festooned with leeches. The scientists had been overzealous with their cures. A mob had broken through the front door to rob this spiritual bishop of his worldly possessions, cutting up his cassock and miter for holy relics. They fought over the blood from his wounds, catching it in kerchiefs and cups. Had the Church not upped the guard for the funeral, thieves would have stolen his fingers.

That had been an ostentation. Without any fanfare at all, death crossed over from its Roman cathedral to a private Florentine domicile and put an end to a Galilei.

Virginia wrote her father, "I am most saddened by the loss of your cherished sister, my dear aunt, but my sorrow at losing her cannot compare to my concern for your sake." His daughter had grown up right, to his immense fortune, and had begun to care for him.

Galileo's mind carried more memories of the fallen than seemed able to fit. He recalled his sister's birth, when he was nine. He recalled running through the streets of Florence, chasing down silk cloth for her wedding gift, so pleased and energized by the kindnesses a family teaches a son. He recalled, after the death of his father, giving what meager earnings he had to pay off her dowry. He had been a giving man, and for this kindness, he was slowly losing everyone he cared for, everyone who understood him. His own son was off to college. Galileo's friendship with Benedetto had become so familial that he entrusted the monk with stewardship over the boy.

Federico Cesi was in poor condition, grieving for Bellarmino, for whom he "carried a particular affection." Running the Lynxes for so many years had proven unduly stressful, and he had entered into a libertine manner. "Neither Cesi nor his wife exercise at all," their physician recorded, "they sleep, eat, and drink too much, so they are fat . . . there is no easy remedy unless they want it." Maffeo Barberini was chipper. Despite his career in the church, he found time to write poetry. "Your virtues," he informed Galileo, "have provided material for composition." Who would go next?

Galileo looked at his own portrait. His skin was sagging, his beard ghost white, his vision distorted by an ocular cyst.

Then death came for the pope.

"Truly, father," Virginia went on, "it seems you have few others in this world left to lose."

The Changing Tides

Maffeo Barberini had enjoyed writing poems since boyhood, and they were perfectly adequate. After he was elected the new pope, none in Rome would shut up about their brilliance. New editions of his poesies were released each of the next five years. They included his celebratory poem about Galileo, or rather, a condemnation of diehard Aristotelians.

When the Moon shines in heaven
With calm unraveling its golden parade
Surrounded by glittering fire
What strange pleasures draw us in.
None more than the servants to Jupiter
Discovered by you, learned Galileo.

A singular focus invites its specific death:
Such a life happily grabs the sword
And plunges through the flames
Into battle, thinking only of triumph.
The good of peace is tangent to its arts
Beneath the duty of its own advancement.

Not always is there splendor within
That radiant outburst: consider the black
In the Sun uncovered (who would believe?)
By your art, your labor—*Galileo.*

The poem was titled, with a wink, "A Dangerous Adulation."

No one had anticipated that the cardinalate would spring for such a decidedly "neutral" candidate as Barberini, but it had been extremely hot that week, encouraging the archconservative and mere conservative factions into compromise. Such inanities gird every high position.

This changed everything. Galileo was sick in bed, again, "if only to temper my excessive joy"; "my hope, all but buried, has been revived." The glamor of the papacy made private communication taboo, but the last letter Barberini ever sent Galileo stated they were brothers. A sun was to be painted in the bedroom of the Vatican and considered for the interior of the papal carriage. Cesi immediately drove down to congratulate Barberini on becoming Pope Urban VIII, only to be interrupted mid-sentence: "Is Galileo coming? When is he coming?"

He came in April of next year. The pope gifted him a gold medal. They had long philosophical discussions twice a week. Galileo was supposed to have brought a copy of *The Assayer*, which had delayed printing just enough be dedicated to Urban, who enjoyed reading it over dinner.

Inspired by Sagredo's shortening glasses, Galileo also brought a device to view insects up close, which the Lynxes had named "*il microscopio*." Every time Galileo went back to Rome, he was expected to entertain rich people with these trinkets. "I am getting old," he sighed, "To be a courtier one must be young."

He returned to Florence in less than a month, and began, tentatively, to write about Copernicanism. He sent out a few feelers and found them warmly received. A friendly member of the papal court wrote Galileo that the pope read a large section from one of these apologies and "much liked the toy examples, the popular experiences through which you proceed." Galileo's apology ended with a curious advertisement that "you will be able to read about this topic at greater length in my *Discourse on the Tides*."

Back in 1595, when Galileo would talk shop with Venetian friends into the early morn, they would see the night tide rise up past a man's chest. It was not a phenomenon Italians could avoid. After more than a quarter-century of contemplation, he realized it might provide the first physical, earthbound evidence of Copernicanism and had plans to produce an entire book upon it. In anticipation, he wrote an extended outline for a sympathetic cardinal.

"Water can be agitated by the motion of its surroundings," Galileo proposed, "I am inclined to believe that the cause of tides is the motion of the seawater basins." "Each part of the Earth's surface moves with two uniform motions. Within a period of twenty-four hours it moves sometimes fast, sometimes slow, and twice at intermediate speeds." Galileo thought the daily and annual motions of the Earth composed so that the side of the Earth spinning away from the direction of its yearly orbit was moving slower than the side spinning towards it. When a body of water is long enough that its opposite ends are on parts of the Earth with considerably different speeds, it was like stirring a tub of water from only one end. The tides are this sloshing about of the oceans.

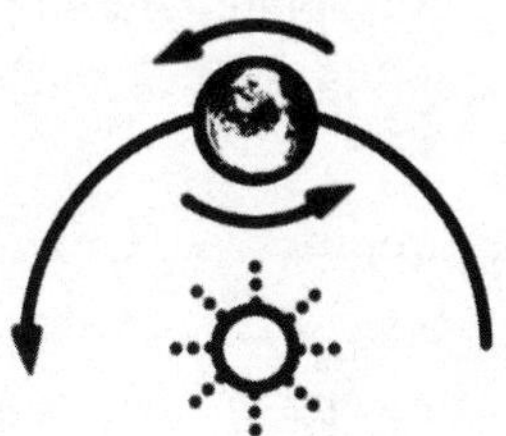

Galileo's tidal theory depends on the belief that the diurnal and annual motion of the Earth compose to affect its oceans.

Such a basic daily mechanism did not complete Galileo's explanation. Most tides occur at six-hour intervals, while this provided for every twelve. The remainder, he suggested, was due to self-excitation, whereby the tides fell back on themselves "by their own weight." Finer subtleties might be explained by the size of the ocean basin, in the way that pushing a bowl and plate of water, of equal width but different depth, results in a different sort of sloshing for each. The chaos of these individual contingencies was infernal, "infinitely puzzling to our minds," Galileo moaned. "How many days, how many nights have I spent on these reflections, despairing over ever understanding it!"

His hypothesis has several benefits. It allows for variation in the tides across different parts of the same ocean. It explains why small lakes have no tides. It extrapolates from immediate sensory experience. Nevertheless, it is wrong, and Galileo would spend the rest of his life in tragic irony, publishing the science that tells why. He concluded, "If this hypothesis should be declared false by better knowledge, then what I have written is worse than questionable. It is empty."

Though it was unexpected for Galileo to make this kind of mistake, it contained a part of him. This was his ideal vision of life on Earth, one single, self-contained, physical system, entirely independent from other planets. Meanwhile his experience of life, as a painful, unpredictable, interdependent series of highs and lows, was something entirely different.

Galileo wished to extend his work on the tides into a full discourse, but this fed directly into talk of Copernicanism, and then arguments against Aristotle, the phases of Venus, parallactic measurements . . . it would take him many years to achieve this. He still reviewed, from time to time, the gracious testimonial Cardinal Bellarmino had granted him in 1616, explaining that Galileo had never abjured the Copernican opinion, but that "the doctrine attributed to Copernicus" was "not to be held nor defended."

He could manage that.

There was a story floating around, from Urban's coronation, concerning an action strictly out of character. Barberini had descended the staircase of Capitoline Hill, about to ascend the Vatican, surrounded by Jesuits, Dominicans, and those too savvy to let their influence be known. Blood-red wine ran down the handrail of the balustrade out the newel, a carved lioness with a roar. He paused by her for an unnatural length of time, breathing in the adulation of the crowd.

There was also another, opposing story, after Maffeo Barberini had been Pope Urban VIII for six years. A friar was discussing with him the attempted conversion of two German Protestants, who were harrowed by the 1616 decree against Copernicanism. Urban winced and brought his hands over his eyes. "This was never our intention. Were it up to me, that decree would never have been made."

But it had. The Germans were lost.

Works of His Golden Years

Secret Revisited

"The seasons turn like a wheel," Johannes Kepler wrote, at the end of 1621. He was drawing up brisk annotations for the second edition of *The Secret*, an almost pitiable artifact of his twenties. "I was new in town, then," he recalled, "this work was brought out by an astronomical tyro." To meditate on one's own progress is a task likely to both please and humiliate the elderly. "This is just some astrological fun," he dismissed the ninth chapter, "if anything at all. It should not be considered part of the work." Two chapters later, "All this could be omitted. It does not matter." In a week, he held nearly as many pages of footnotes as original text. No longer did he believe the book's central thesis, that planetary orbits were defined by spheres circumscribing the regular solids (how could he, having discovered that orbits were elliptical?), but neither would he absolutely abandon it. With a perverse sort of pride, he realized, "Virtually everything I have ever published stems from this little book."

The Secret was a mad theorist's crazy dream, and its contents could hardly be less insightful. Upon its reprinting twenty-five years later, in addition to the invaluable contributions of his *Optics* and *New Astronomy*, Kepler had just published two landmark works: his textbook *Epitome of Copernican Astronomy* and his private obsession *The Harmony of the World*. How had he done it? What is it that tempers the spirit of a theorist?

A sage beard and graying mustache hung down off Kepler's face. He had reached this stage in life only by following his mother in his childhood, by learning from Professor Maestlin in his college years, by leaning upon wives and children for emotional support, and, most especially, by using Tycho Brahe's immaculate, meticulous observations. In the rare event that Kepler was forced to produce observations of his own, he called it a "ridiculous spectacle," but he still tried his best to produce them. It had been his attention to experience, of his own and, thankfully, of others, which had brought his imagination under control.

A year before the republication of *The Secret*, a professor of history had sent an inquiry to Kepler for his biography. This inquiry has not survived, but Kepler noted that, much to his surprise, "my friends would like to hear about it."

Mr. Kepler Writes a Textbook

Professor Maestlin planned to die in front of a chalkboard. He was a prolific writer without grand ambition; he had known all the greats, but never located the means to become one. The royalties from reprintings of his popular textbook *Epitome of Astronomy* kept him buoyant—forty happy years at the same university, day in, day out. Countless students had vanished before his eyes. Others were memorable, and a few distinctly refused to be forgotten.

A messenger arrived over the Swabian Alps with a stack of pages from his old student Johannes Kepler. "Cripes," Maestlin whinged, "This is really beyond expectation."

Immediately after *New Astronomy*, Kepler had recognized he would have to become his own popularizer. His attempt was the *Epitome of Copernican Astronomy*, a comprehensive tutorial "in imitation of that primer by Maestlin." "Its pages remained, tucked away in my writing desk, for seven years. I often reread them, making changes and editions." In 1618, the first ever textbook for heliocentrism was released, "in plain speech, with the tedium of proofs eliminated," Kepler hoped.

"This edition is not obliged to the uselessness of idly parroting back the doctrine of the spheres," explained the preface. Using a simple question and answer format, Kepler distinguished himself on the first page.

> Q: What is the relation between astronomy and other sciences?
>
> A: 1. It is part of physics, because it inquires into the causes of things, natural events: because among these subjects is the motion of heavenly bodies: because its final purpose is to investigate the structure of this part of the universe.
> 2. Geography and hydrography, the essence of nautical things, involve astronomy . . .

This was no longer an unexampled thought experiment. This was the most authoritative astronomer in the western world, telling parents how to raise their children.

The Catholic Church banned it straight away, on account of its open advocacy for Copernicanism. Kepler was caught totally off guard: "But all of me is Copernican!" He was clueless about the situation in Italy.

Kepler minded Catholic censorship a whole lot more than the average Protestant, for this he was not. Prescribed Lutheran doctrine disagreed with him almost as much as the Catholic one. His church regularly excluded him from their weekly rituals. "It breaks my heart," he wrote, "that religious faction has ripped apart the truth so miserably, that I must assemble it piece by piece, wherever I find it." "I seem to be a spiritual mutt."

The Jesuits, overhearing the discord, attempted, repeatedly across the next decade, to convert Kepler to Catholicism. His responses were flummoxing.

> How poor must my piety be, if I were only beginning Catholicism now! For I was baptized by my parents the moment I crossed over into this life. It is the same spirit, who here gives this one their doctrine, that now gives me my baptism. I cling to the Church, to the community of it, so full of God's love. Even while she thrashes and bites, so subject to human frailties, I cling.

During the years he was formulating this unique style of religious devotion, Kepler would often make the day-long trip out to Tubingen to borrow the ear of old Michael Maestlin. His favorite teacher was over seventy now, "my inveterate guide on the road of Copernican astronomy," as Kepler described him in the winter of 1620, recalling their discourse on his "new hypotheses on lunar motion." This was an important discussion for Kepler, who had decided such material was deserving of a second and third volume to his *Epitome*, for very advanced students, focusing specifically on celestial physics. It is possible he discussed with Maestlin his recent and dubious theory that matter was naturally compelled towards rest, a property he had named "inertia" (meaning "artlessness" or "without power"), but there is no record. Several months ago, Kepler sent Maestlin their last extant letter. There is nothing about it to suggest their conversation did not continue, and it is still easy to imagine these two old friends, in the middle evening, probably over a home-cooked meal, talking for several hours about the Moon.

World-Harmony

Kepler was not a creator, or at least, did not want to be. He wanted to be a scribe. He was trying to listen. He was listening for others, who spoke through their words and voices. He was listening for God, who Kepler believed spoke through his creation—the physical world—using mathematics.

The Harmony of the World is Kepler's most daring and awesome attempt to listen, for a property of the physical world he called "harmony." "Harmony," Kepler defined, "is a qualitative relation." That quality is "unity," the collaboration of parts to form a whole distinct from themselves. War and disunity,

Kepler wrote, belonged to "brute beasts rather than human nature." Kepler's understanding of human nature, despite the horrors he had witnessed over his years, remained one of potential good, because he held that human nature was defined by the soul, and the soul was defined by its ability to unify. He believed that absolutely everything in the world had the potential to be unified, in harmony, and he wrote *The Harmony of the World* to celebrate and share as much of this unity with his divided world as he could.

This fantastic optimism led Kepler to create an extremely funny and intensely strange book. "Two objects cannot be sensibly unified except within the soul," he wrote. When he proclaimed any two things to be unified, he was only sometimes stating necessary, objective truths. Other times, he was willing two clashing concepts into harmony, following the subjective dictates of his sentimental soul. He provided a fine example of both at the end of a discussion on the regular solids he had used in *The Secret.*

> I happened upon another figure with regular symmetry: the prickly dodecahedron formed by twelve five-pointed stars, which I have named the Echinus, that is, Sea Urchin or Hedgehog. It is very similar to the five regular solids.

Kepler's "hedgehog."

Obviously Kepler knew that his star polyhedron was not a hedgehog, but he relished the comparisons he saw between them. He had formed a subjective harmony, which he invited his readers to join. Meanwhile, he used this private eclecticism to describe a mathematical relationship, which was objective and universally true; this "hedgehog" shape really is like a regular solid. The two classes of shape really are in "harmony," and no geometer before Kepler had properly recognized it.

Kepler idealized subjective, unarticulated harmony as "passive," a quality he attached to water, darkness, emotion, and women. Objective, articulated harmony was "active," a property of fire, light, reason, and men. These were standard dichotomies of his era, but the true perception of any harmony, he declared, was "partly active, partly passive," in varying degrees, and the two could never be disjoint in lived experience.

Harmonies concerning single shapes, as with the star or regular solids, were the most perfect to Kepler, but he knew the world to be richer than that. Merely combining different shapes to fill a flat surface, like a jigsaw puzzle, already introduced a thousand complications.

This problem, practiced by seamstresses and textile workers for centuries, yet hitherto untouched by mathematics, is known as "tiling the plane." The simplest and most harmonious way to tile a plane is with the same figure, as with a square grid. Kepler, however, found tilings that were anything but simple.

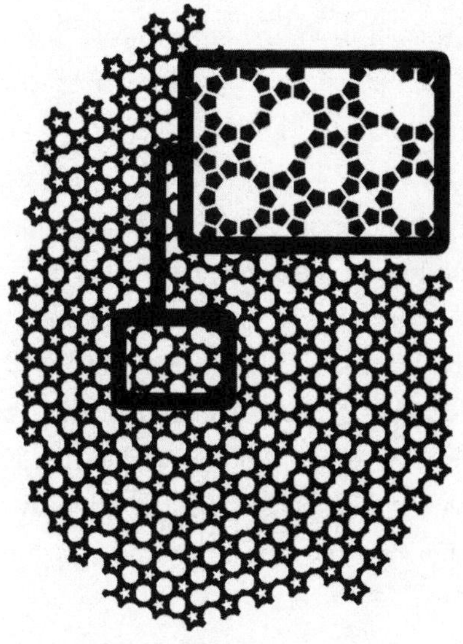

Kepler's nonperiodic tiling of the plane.

He was searching for ways to tile the plane with a pentagon, the most complex shape involved in the regular solids. He was having trouble. Building off of the star and decagon, he found a working tiling, except that "certain irregularities must be introduced. Two decagons must be combined, two sides removed from each." Although this had an evident harmony, it was difficult to classify what sort of harmony it was.

> As it progresses out, this five-cornered pattern continually introduces something new. The structure is very intricate.

Up to the third chapter, Kepler had considered geometrical problems on geometrical terms. Now, to make these studies truly "of the world," he opened up his playful soul to comparisons between every genre.

Mathematical equality could be in harmony with ideal jurisprudence, he wrote, "in which not only is each allotted his own, down to the last farthing, but penalties are imposed on criminals equally, without respect to person." Polyhedra could be placed in harmony with gender, classed, according to their faces and edges, into female, male, and even hermaphrodite, the former married off, "just as women are assigned to, and, in a way, subjected to men." Of the plentiful harmonies Kepler developed, some ridiculous, some apt, it was music that was queen, the richest and most receiving mother of harmony.

Everything could form a harmony with music. Since the age of Pythagoras, it was known that music had a basis in mathematical ratios, yet untrained ears, even animals, delighted in it.

> They rejoice in what they perceive of the harmony and proportion in a tune, and grow depressed when it is lacking; that which the soul desires (harmony) is called consonance, otherwise (no harmony) dissonance. If they also account for that other harmony, letting their voices fall long and short, then their bodies begin to groove, their shouts ring out; workers strike their hammers to the beat; soldiers change their pace: all to the same law. Everything is alive, while the harmony endures; benumbed, while it does not.

In the two decades since *The Harmony*'s inception, Kepler had read every book of music theory he could get his hands on. He had trained his ear and seems to have been able to notate a bar of music after one listen. Almost all the music he knew was religious in nature—the organum, the hymnody, the Gregorian chant. It was the new church style, the polyphony, with its many chords and voicings, which Kepler called "the splendor of harmonic form," where he found a comparison to astronomy.

Running the gamut: from low G (Gamma ut) *to high E* (E la).
The clefs (bass, alto, treble) indicate the range of a five-line stave.

"Voices and sounds do not exist in heaven," Kepler wrote. On this point he was extremely clear. The harmony between astronomy and music was not aural, but was made sensible through the human capacity for reason. Kepler was certain the creator-God had given humanity the ability to reason so that they could understand his creation, the world, as a unity. The synthesis of music and astronomy was, Kepler felt, "miserable worm that I am," his debased peepshow into this totalizing peace.

He began by reviewing musical notation. Musical notes cycle in octaves, from A to G, and the pitch of each note has a fixed ratio to the pitch of all others. The lowest note in the bass was referred to as *gamma ut*, contracted to *gamut*, which was then broadened to refer to all the notes. Thus, to play through an entire scale is to *run the gamut*.

The mathematics Kepler developed to map the planets onto the gamut were involved. The planets present potential harmonies everywhere: in their orbital period, their speeds, their distances from the Sun. It was very particular metric in which he discovered the greatest harmony: the amount of the planet's orbit that was traversed each day, as seen from the Sun.

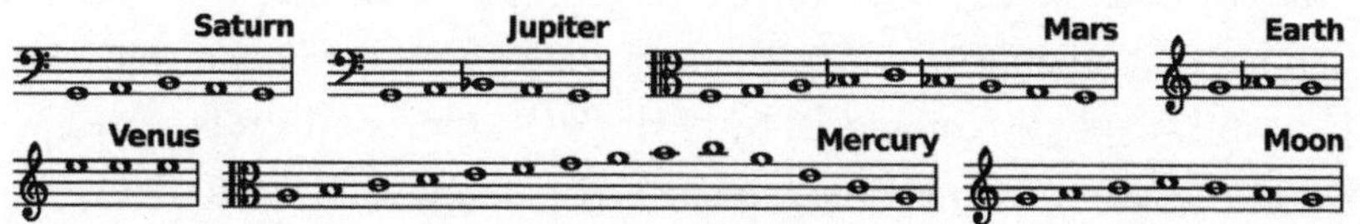

Kepler's attributed notes for each planet. The flat (♭) symbol lowers the note slightly, to about 15/16 its original pitch. Kepler abused notation here, indicating only the written musical ranges of each planet. The range between the Moon and Saturn, for instance, was well over seven octaves wide!

The musical interval associated with a planet was determined by the ratio of its metric's lowest and highest values, when the planet was furthest from and closest to the Sun. When far away from the Sun, the planet evoked its lowest note; then as it moved toward the Sun, it ran up the gamut to its highest note, then back around. Most of the time, this meant all the planets together were just evoking irrelevant, dissonant noise. What Kepler cared about were the rare occasions when some combination of planets evoked a lovely, consonant chord, of anywhere from two to seven notes. But it was infinitely improbable that all seven planets formed a consonance at the same time; Kepler wrote, "I do not know if it is possible to occur twice. It may be evidence of the beginning of time itself, from out which every epoch flows." Consonances of fewer planets, however, were fairly common, if people would only listen for them, through mathematics, and receive the divine pleasures such listening brought.

Kepler had sought a harmony between astronomy and music, and uncovered one of complexity and nuance, with extensive proof. His belief that this harmony was causal, that God had organized the planets to deliberately produce it, is more than wrong; it is the most hysterical thing a person could possibly believe. This happened because he lacked modern cosmology. But if Kepler seems to us unreasonable for his hysteria because of the earnest fervor with which he exploded every microscopic sensation of his inner life, then our pessimism is not a scientific but a philosophic attitude: that since the essential quality of harmony is aesthetic, it does not deserve exaltation; that geometry has nothing to do with jellyfish, or that men have nothing to share with women, or that astronomy has nothing to learn from music (or physics), even if both involve mathematics, and there is no joy to be gained by placing them in direct relation.

In 1619, the soul of Kepler's working life was completed. *The Harmony of the World* is no longer a book easy to comprehend, if it ever was. The premise that there is music in the heavens, *musica mundana*, was over a thousand years old, celebrated by both Pythagoras and Ptolemy. Kepler was the last astronomer to seriously write about it. In this sort of science, which is the synthesis of traditions, old threads may be rarefied, diverted, cut, burned up, or just as ably tied in a knot, so encompassing that it leaves nowhere else to go.

World-Harmony: Coda

Wait! Kepler was literally pages away from completing his *Harmony*, a day or two's work at most. On May 15, 1618, he was playfully interrelating the various properties of the planets, their orbital periods, their densities, their speeds, when he recognized an equation so upsettingly accurate that it had to be true. When Kepler had discovered planetary laws in *New Astronomy*, he had stated them obliquely, as if unaware of their presence, but this time he knew exactly what he had; it was "the sunrise of a most wonderful contemplation." But where ought he put it? He reached for the last chapter, and turned to the beginning of the end.

The Third Law

> KEPLER'S THIRD LAW: The square of the orbital period of a planet is proportional to the cube of its orbit's longest diameter.

Battle for the Soul of Tycho Brahe

Obligation bound Kepler to Earth as the ceiling does a party balloon, everywhere unyielding. It would have to be removed. Decades ago he inherited a contract with a dead emperor, through the designs of a dead nobleman. It was not something he enjoyed.

In 1607, "Tengnagel has given up hope of the *Rudolphine Tables*, but not me. Not while I live." In 1610, "I think before publishing the *Tables*, I shall write ephemerides for the next eighty years." In 1616, "The *Tables* need time. You will learn yourself, and others: mathematicians will report it." 1618, "I set aside the *Tables*, since they required quiet, and turned my mind towards perfecting the *Harmony*." Finally, in December 1623, "I can see the harbor. If the Kaiser keeps the peace, the *Tables* will soon see light"; "The *Rudolphine Tables*, which was conceived in me by Tycho, their father, has now been borne in my womb for a score and two years, day by day forming the baby. The birthing pains torment me."

There was no peace, no quiet. A mass of peasants rose up in the north, ending their military oppressors and proclaiming their freedom. Like a hurricane they spiraled southwards. The Brahes, in their last historic act of self-concern, felt certain Kepler was dishonoring Tycho's legacy. They threatened legal actions, which might prevent the *Rudolphine Tables* from ever being published. The situation in Italy was chaos. The Jesuits had effectively declared for Tycho Brahe, but this opposed them with the innumerable conservative Aristotelians in the universities. The Galileists heaped scorn on both parties, rejecting everything. Kepler was alert to all of this; by now he was reading the Italians and had learned how they talked. After a hefty Italian polemic named *The Anti-Tycho* was given to him alongside Galileo's *The Assayer*, the German decided to enter the fray.

The Shield Bearer for Tycho Brahe contains a malice unique in Kepler's repertoire. It is the "Anti-*Anti-Tycho*," dedicated and directed, most unsubtly, to "the famous and generous Brahe clan, of ancient nobility, having honor and resources." "I would be infuriated if even three people tried to extract meaning from *The Anti-Tycho*, all the more if they hurt themselves," Kepler roasted, and later, "I shall not teach the author how to read, the ass." Even Galileo, once a friend, came under tepid fire.

> Nothing is better for the advancement of astronomy than those observations of yours, Galileo. However, with your indulgence, I might state my mind: it seems to me that you would be better off to collect up your thoughts, when they go wandering off the course of reason.

Galileo had written, "From Copernicus and Ptolemy, we have systems of the world that are unified, with great sophistication. I do not see that Tycho has done any such thing . . . the system of Tycho is a nullity." Kepler preferred to write of the Tychonic system as a respectable halfway home to Copernicus, for those "too soft in faith to accept the motion of the Earth," but Galileo could not conceive of a weak transition. At that moment, he was writing his *Discourse on the Tides*, which had snowballed into a much larger *Dialogue on the Two Chief World Systems*. The Tychonic universe was not even up for discussion.

Fellow scholars had come to anticipate insults from Galileo, but from Kepler they were shocked. "My hopes deceived me," the author of *The Anti-Tycho* wrote, "you have really unbridled some repulsive slurs. I am unaccustomed to hearing such maledictions put upon my person." Kepler was the only one unsurprised; he had always been keenly aware of his capacity for sin. He stood accused, and plead guilty. "This is not a representative example," he wrote, "I desire to be civil, and attract mutual goodwill . . . I will send him an inquiry, proclaiming that this old man has gone astray, and, excusing himself, will bring back morality."

The Tables

Days before the end of winter, 1627, Kepler was forced to walk alone. He had been on the trail for four hours. An anal fissure prevented him from taking a horse; the cold air whipped him at every stride, and he had no fat left to protect his bones. A flake of snow skittered into his dress, and his fears were realized. The journey had been a fool's errand. He did not believe he could survive it and turned back.

Johannes Kepler had made a promise. In 1624, the peasant revolt had burned down his chosen printshop, consigning the galley proofs of *Rudolphine Tables* to the flames. The Counter-Reformation then swallowed Linz and placed his personal library under lock and key ("like a bitch being deprived of her pups!") He had escaped to the city of Ulm, chosen almost at random, where the printing industry proved inhospitable. Piqued, persecuted, he had tried in desperation to reach Tubingen on foot. The tables must be completed. He had made a promise.

The Frankfurt Book Fair received Kepler that year with their standard, hokey generosity of businessmen selling product. He had outlaid his private finances to push the work through publication. He had either soothed or evaded the Brahe family's every complaint. He had personally driven the books to market. He stood up a little straighter, the principal weight of life let off his shoulders.

The past few decades, Copernicans had been producing tables and ephemerides that were sleek and competitive. Kepler's *Rudolphine Tables* were not competitive. They were simply the best. It was mostly German astrologers who could discover this from a first edition, but these were Protestant and Catholic, teachers and students and hobbyists, and so too were there some French and English ambassadors, and refugees from all over. In their academies, middle-class abodes, monkly convents, holes in the ground, they opened to the first page.

PREFACE TO THE *RUDOLPHINE TABLES*

We scholars have eyes in common with the uneducated; in fact, as humans we have eyes in common with beasts. Although all of us, both educated and uneducated, behold the wonder of the stars together, we cannot perceive by our naked eyes their inner workings: the order, constancy, and eternity of the celestial revolutions. Here we need our minds, and the ability to remember.

Even now that the discipline of astronomy has grown up, it retains loving memories of its childhood, from the old computing tables of the Greeks, not to mention the ancient Babylonians, which have been out of use for so long that their name scarcely endures. Later, Hipparchus accepted these Greek observations, and comparing them with his own experience composed a rudimentary sort of table. This can be considered the puberty of astronomy. But Ptolemy was the first who, aided by the ancients, especially Hipparchus, edited a whole set of tables, thereby establishing a sort of young adulthood. Unfortunately, after the age of Ptolemy and his immediate successors, philosophy fell on hard times. From other regions arose new nations and empires, first the Huns and Goths, and then the Arabs. The first two were dumb barbarians, but the last were brilliant, if superstitious. Eventually, nine or ten centuries after Christ, those Goths and Huns lost their barbarity; Europeans gradually returned to the contemplation of God's creation, while the Arabs and Jews, taking pity on their failures, set about fixing them. Many books were translated from Arabic to Latin, even that stupendous work by Ptolemy, which Arabs called *al-majisti*, or "The Greatest." In this auspicious age, many new universities were encouraged by the few old ones. The Germans were especially sedulous, in schools in Vienna and Prague, with Peurbach and Regiomontanus discovering what was good and bad in existing tables, but also men in Italy, such as Domenico Maria of Bologna, worked hard to observe the stars. Although Regiomontanus would have been equal to the task, he died an early death, and was succeeded

> by Copernicus, a canon in Warmia, pupil to Domenico, man of the highest order, and (what is most important) a free spirit. He was succeeded within a few years by Erasmus Reinhold, who endeavored to finish the tables, a huge and disagreeable task; he called them the *Prutenic Tables*. Then Tycho Brahe arose, a Dane of outstanding nobility, who defied his peers and chose to restore astronomy on his own. He was an incredible man, the originator of the *Rudolphine Tables*, but I would prefer that readers learn of Tycho from the work of Christian Longomontanus than my own account. However, in 1601 I sent a letter on this subject to Antonio Magini, which he published in his ephemerides, reprinted in 1614. Eight years after the death of Tycho, I published my *New Astronomy*; thus, everyone and their mother has been stealing loot from the shipwreck of Tychonic astronomy. Finally, whoever makes use of these tables, be they students, philosophers, or even theologians, let them remember to attribute it to the munificence of my patron: Rudolf II, who gave me to Tycho as an assistant. They all merit thanks; may the reader pray for the whole blessed family.

Sunrise

Kepler's eyes fluttered open. It was dark and quiet, which was not unusual: of those beings made up of chords, which vibrate at the slightest zephyr, he had always been first to sound. It was unexpected, though, that he was still sounding. All his promises had been fulfilled, his aspirations achieved, yet he kept waking up. He kept cracking his back, kept clipping his toenails, kept using the toilet. He could outline his wife, in a dirty chemise, looking very plain, the twilit blues sliding off her up his left arm, into his breast. She could never make it alone, but their eldest son was good stock, training for a doctor's pay. Their eldest daughter had the will to help raise sisters; she had married a big strong man who looked after old ladies. The house of love will rule itself. Little paws were up and dancing now, thumping the floorboards of the nursery. Kepler could not decide whether to spend the morning with them or get straight to work. He turned over towards the window, a permanent smile etched across his face. Light was breaking through in force, but he would not look away.

A Family Man

Galileo's son was acting like the bastard he was. Presiding over his education in Pisa, Benedetto Castelli reported that "his obstinacy is stronger than ever. It is like bashing my head with a rock. I am stunned." Vincenzo was a brooding seventeen-year-old, locking himself in his room. His only solace was shopping for clothes downtown, modeling handsome cufflinks, waistcoats, monk straps, and bonnets. Thankfully, Castelli added, "I have not discovered any real debauchery." The troubles escalated when the boy moved to Rome, for there he met his cousin.

"The kid's got spunk," an overseer told Galileo, "but you wanted the unfiltered truth."

> Your nephew does not much like being corrected and has a bad memory. He studies so little that in a few years he won't know how to make money. He prefers promenades and prattle, and cannot resist taking this business home with him, I mean women, speaking dirty to them, withholding nothing in their presence.

Benedetto wrote, "Make no mistake. Your nephew is ungovernable. Everything goes in one ear and out the other. Upon the most severe reprimand, he bursts into laughter." For both these children, he had been bending over backwards, "which has led to a million hardships and horrors." Together, they broke Friar Castelli's foremost tenet: "What really spooks me, what makes me shiver, is your nephew's indifference on matters of religion"; "I am stunned by the words of Vincenzo. This is no mere aversion, but a poisonous hatred of the clergy."

Galileo was irate and had the upper hand. Financially, he supported his entire family. His nephew was promptly removed to Florence; his son, why, "he should consider himself lucky to have as many dollars at his age as I had pennies." Galileo could not imagine where they had learned such bad behavior: certainly not his sterling example. He had selflessly given his children everything he thought they needed. Not from his younger brother, either, who was a milquetoast with unquestioning deference. ("I dare not contradict you," he wrote Galileo, "I am very weak-minded.") Not even from his mother, since Galileo knew from experience she was best kept away from children. (Just before her death, his brother gave her final mention: "She is still so terrible.") No, no agent had planted these rude seeds; they grew up in the boys as wild sprouts in unkempt patches of lawn.

Vincenzo had recruited older sister Virginia to his cause. The nun had always been too understanding for her own good. "He stands in desperate need of more collars," she wrote father, or, "I want to offer a good word on behalf of my brother. Forgive him this once; blame his youth for such transgressions." In time, Vincenzo overcame his doldrums, though there forever lingered a vague coldness between father and son. Virginia, however, only grew warmer.

The poor girl was running out of teeth. She was not yet thirty, yet her cavities were rampant. "Send me some fatty mutton," she asked Galileo, "Surely I can manage that." "Send a warm quilt to protect me, else I shall grow numb with cold." "Send me more of your leftover meat, if you can. I enjoyed it last time." She was terrified her father would abandon her. But Galileo never, ever, ever would. He decapitated chickens for her. He fermented wine, repaired clocks, and lacquered windowsills. She wrote, "I suppose there is nothing to be said. You are our father—our sire—and we pin our every hope on you."

Nothing brought Virginia a more thorough and unobscured elation than the brief moments she was allowed with children. It was baby cousins first and, soon enough, nieces and nephews. (Vincenzo had made Galileo a grandpa.) But then they left, and she returned to God. She ate moldy bread and watched women die. Hers was a hard religion.

She kept up-to-date on her father's career, reading *The Assayer* and writing on the papal coronation of Maffeo Barberini with vicarious delight. Galileo told her when he was sick. She knew all his daily habits.

> I have learned, Papa, that you have returned to your usual labors in the garden, leaving me considerably disturbed. The air is still quite raw, and you are still weak from your recent illness. I fear this activity will do you harm, as came to pass last winter. Please Papa, do not forget so quickly the dire straits you were in, and have a little more love for yourself than your garden; although I suppose it is not for love of the garden *per se*, but rather the joy you derive from it, that you put yourself

> at such risk. If you will not do so out of love for yourself, then, do so out of love for us, your children, who wish to see you live to a ripe old age.

Galileo would not listen. He sent her fruits from his garden and flour. She baked him yummy pies. She sighed, "When he goes, I am alone in this world."

On the eve of 1630, Galileo completed his enormous *Dialogue on the Two Chief World Systems*, pending the corrections of his friends and, of course, the Catholic Church. His vision was worsening, so he was pleased to have finished, "before I lose my sight, and tumble into darkness."

The Dialogue

If the rumors that Simplicius was a dullard were true, he was at least a dullard with integrity. He kept a thumbed copy of Aristotle's *On the Heavens* in his trouser pocket at all times, so there was never danger of paraphrase. He was frank with himself, that "I confess to have an ordinary mind." Across four days, he visited a capacious Venetian palace overlooking the lagoon, where he, Gianfrancesco Sagredo, and Filippo Salviati developed a scintillating dialogue on popular cosmology.

"Bitter death," wrote Galileo in his *Dialogue*'s introduction, "has robbed Italy of these two luminaries in the prime of their years. As far as my feeble skills allow, I have resolved to prolong their lives on paper." He added parenthetically, "(Nor shall a good Peripatetic lack a place.)." Thus, he characterized the three actors of his scientific play.

"Quit your sneering!" Simplicius ordered Sagredo, who had a habit of turning Salviati's corrections into insults. Salviati, however, was an understanding teacher. "Good Simplicio is a man entirely without malice," he said; he complimented, "You are an Archimedes," whenever Simplicius made a basic inference. For Simplicius, calculation was "a subject in which I understand but little or nothing." He watched aghast as Salviati improvised fifteen pages of mathematics. "I make such calculations very speedily," said Salviati, the savant.

There were almost two millennia of commentary and apologia on Aristotle's natural philosophy to wade through. Though recent scholars had often forsaken it; in whole or part, no one before Galileo had attempted to publicly dismantle the edifice, brick by brick. Galileo's *Dialogue* was the literary equivalent of a single man trying to take apart a castle with his bare hands.

In his preface, he wrote that, after the Catholic Church had prohibited Copernicanism, "there were those who impudently asserted this decree had its origin not in judicious inquiry, but uninformed passion."

> Upon hearing such carping insolence, my zeal could not be contained. I propose to show in this work that much is understood on this matter in Italy, most especially Rome. It is not from failing to account for what others think that we have yielded to the assertion that the Earth is motionless. Rather, our reasons are supplied by piety, religion, knowledge of divine omnipotence, and awareness of the limitations of the human mind.

Halfway through the dialogue, Salviati added, "We Italians are making ourselves look like ignoramuses. We are a laughing-stock to foreigners, especially those who have broken with our faith." This was about Galileo's honor as a Catholic scientist. Religion can inspire men to crazy feats; honor too.

The shape of any entrenched institution is not a point or a straight line, but a nefarious maze. Medieval scholastics had apportioned every corridor of the Aristotelian castle into an infinity of specialized compartments, then built a garret for the Bible, all particolored walls, stairways to nowhere, and doors that led in circles. As with any thorough criticism, Galileo's *Dialogue* was forced to mimic the structure of its subject. He did his best to demolish the largest rooms concerning planetary motion, Biblical literalism, and stellar parallax, but this work had no aesthetic unity. It was a loose bundle of arguments, wrapped up in a closet drama, with only a vague set of guiding principles.

The most essential principle concerned "the authority of written words," that is, an innate trust of scholarship. Sagredo called this an "absurdity" and favored "sensible experience." Galileo's characters almost always preferred "reducing it to a little sketch," but a further principle expanded this distrust to even sensible experience, in favor of pure logic. "The very instrument of seeing has hindrances of its own," Salviati explained; sensory knowledge was of a human nature, while deductive knowledge "equals the divine in objective certainty." To convert those who disagreed, Galileo endorsed the Socratic method; Sagredo described it as "the pleasure of undermining one's companion, making things drop from their lips that they never knew they knew." He called it a "form of violence." Simplicius was the victim.

By the second day of dialogue, the Aristotelian was already bruised and beaten. Through Galileo's almighty pen, Simplicius had been forced to admit the possibility, even probability, of a mutable heaven, a movable Earth, and the depths of human ignorance. In retreat, he called up the most outstanding argument from *On the Heavens*. In Aristotle's own words:

> The Earth must be at the center and immovable, not only for the reasons already given, but also because heavy bodies forcibly thrown straight up return to the point from which they started, even if they are thrown an unlimited distance.

Simplicius then stood immobile, as Salviati and Sagredo chiseled the foundation of Aristotelian physics out from under him. To fit his usual style, Galileo had Salviati warp Aristotle's reasoning into a physical experiment.

> Drop a lead ball from the top of the mast of a boat at rest, noting the place where it hits, which is close to the foot of the mast; but if the same ball is dropped from the same place when the boat is moving, it will strike at that distance from the foot of the mast which the boat will have run during the time of fall, for no other reason than that the natural movement of the ball when set free is in a straight line toward the center of the Earth.

This experiment, Salviati declared, is nothing but an easy fiction.

> Have you ever made this experiment? Anyone who does will find exactly the opposite of what is written; that is, the stone falling from the ship's mast strikes in the same place on the ship, whether the ship moves or stands still.

> Shut yourself up with a friend in the main cabin below decks on some large ship, and have with you there some butterflies. Have a large bowl of water with some fish in it; hang up a bottle that empties drop by drop into a wide vessel beneath it. With the ship standing still, observe how the little animals fly with equal speed to all sides of the cabin. The fish swim indifferently in all directions; the drops fall into the vessel beneath; in throwing something to your friend, you need throw it no more strongly any one direction than another, the distances being equal; jumping with your feet together, you can pass equal spaces in every direction.
>
> When you have observed all these things carefully, have the ship proceed with any speed you like, so long as the motion is uniform. You will not discover the least change in all the effects named.

What an idyllic cruise! Whether it be at land or sea, when the frame of reference is shared between the viewer and the viewed, motions appear the same.

This invariance principle does not hold when the frame of reference is rotating; an Aristotelian could still argue that these butterflies, fish, and friends ought to be ripped off the spinning Earth and hurled out into space. Galileo became very confused on this point and attempted to adduce "a geometrical proof of the impossibility of extrusion by terrestrial whirling."

His confusion was far more respectable than he knew. The quest for a physical theory to underlie Copernicanism would fell many minds more learned than his. Such confusion, however, is especially painful to see on the last day of dialogue. Simplicius listened to Salviati reiterate Galileo's theory of the tides, that their cause was the combination of the Earth's daily and annual motions. Galileo's own invariance principle provided the first step in proving this theory wrong. The annual motion of the Earth each day is almost on a straight line, at almost uniform speed. The oceans travel with the Earth, as do humans, so this motion can barely have any visible effect at all, much less combine with another to cause the tides.

Oh well. Galileo had lost many moons to the tides and had thought many rich thoughts. It was time to release what he could to the world. "It would be excessive boldness for anyone to restrict divine power and wisdom to some fancy of their own," Simplicius concluded. He remained unconvinced, but intellectual disagreement did no harm to their friendship. "Come," said Sagredo, "let us go enjoy an hour of refreshment, according to our custom, in the gondola that awaits us."

Galileo laid his quill to rest. Writing had always been a difficult pleasure; publishing was just difficult. In years past, his system had been to hand the work over to Federico Cesi, who collaborated with other Lynxes in Rome to obtain an imprimatur from the Church and pay for printing. Galileo had no desire for this to change. The Lynxes were a powerhouse of scientific publishing.

Cesi's entire life was devoted to the Lynxes, and perhaps it should not have been. When he assumed the family finances, he discovered that his sanctimonious pater, having denied him his rightful inheritance, ought never have been trusted with money himself. Their debt was impenetrable; Cesi wrote, "Stoicism no longer helps me. I am lost at sea." He struggled to maintain his facade of wealth, until the stress ate him out from the inside. He was trapped in bed for days, refusing to make a will, until he was trapped in bed forever.

"With trembly hands and weepy eyes, I give you this unhappy news," Galileo read from a fellow Lynx. He was fast running out of friends. "I see our academy going to ruin," the letter continued. Cesi was split in half, longways, and found to have "gangrene in the bladder, tiny clods of meat stopping up the piss." He would not be printing Galileo's *Dialogue*.

There was nothing else for it. Galileo would have to print it himself.

The imprimatur system of the Catholic Church was deliberately prohibitive. When the Church certified Galileo's book to be printed, the license applied to one city only and required renewal for all future editions. To print in Florence, the censoring would have to be restarted; worse yet, the Master Censor in Rome was already invested in the work and demanded that sensitive parts be forwarded to him. The Catholic leviathan swallowed up Galileo's book and refused to spit it out.

"The work stays in the corner," Galileo mourned. "My life is wasted." His *Dialogue* had been up for administrative review for well over a year. He pressured the censors to let the book pass, sending pleading letters to his Roman contacts. When the censors finally caved, they requested significant emendations of the forward and epilogue, and specified that the title of the work could not reference the tides or any other physical aspect of Copernicanism. Before this judgment arrived, Galileo had already printed the midsection. He was desperate to see the work off, "while I am still alive, that I may know the outcome of my hard work."

Galileo was very much still alive on February 1632, when his *Dialogue* hit the shops. He would know the outcome the following year, after copies had made their way to Rome. He had settled down at a villa in the mountains, next door to the local convent, to spend his final years with his beloved daughter.

The Teacher

Immediately following the publication of his *Dialogue*, Galileo was purblinded so badly he could not read or write for two months. Afterwards, he apologized to several friends that, coupled with his failing memory, communication was no longer a manageable burden. Age wears on people. It can make them irascible.

At the end of August, Galileo received a warm message from a regardful disciple of over ten years, announcing that he too had written a book.

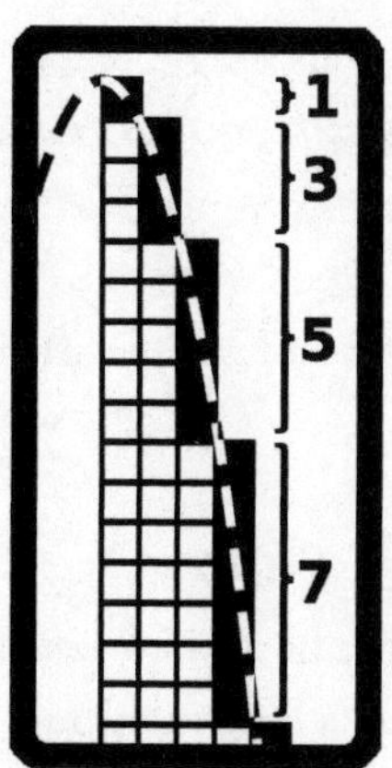

A parabola increasing "as the odd numbers from unity."

> I briefly touch on projectile motion, showing that the shape must be a parabola, supposing the principle of the motion of heavy bodies, that they speed up as the odd numbers from unity. However, I attest to having learned this almost entirely from you.

This was the law of free fall, which Galileo had discovered three decades ago, yet never found the time to publish. The message made him angry.

> Such a warning displays a lack of taste, seeing how I gave this man my studies in utmost confidence. I am now stripped of the fruits of my labor, and passed over by that glory which I crave.
>
> I never doubted his good faith, but I was sorry for my misfortune, which caused my disgust against the will of he who caused it.

Back in Rome, a herd of monks and clergyman were reading Galileo's *Dialogue*. It made them furious.

Benedetto Castelli was present to document their rage. Five years earlier, Pope Urban had heard the monk had skill in hydraulics and promoted him to the University of Rome so that he could design drainage systems for the city. Benedetto enjoyed chatting with his local bookseller there, from whom he learned that Galileo's old enemy Christopher Scheiner had also moved to Rome and frequented the same shop. Christopher had been in just the other day.

> Overhearing a fellow friar sing the praises of your *Dialogue*, celebrating that a major work had come to light, his whole body was affected; his face went pale; a tremor came over his voice and hands. The clerk was amazed. Father Scheiner told him that he would pay twenty times list price for a copy, so that he could reply asap.

"Scheiner has great influence," Galileo had been warned. Yet, confronted two months later, Christopher bowed his head and admitted the book was alright, though byzantine in its construction. Other Jesuits could not even pretend.

From patricians, the praise was fulsome. Here was a scholar, writing in a language they understood, in a format they devoured, featuring protagonists like themselves, who were seemingly right about everything. Never before had natural philosophy so vigorously titillated an audience.

Naturally, there was talk of prohibition; in Italy, this was how citizens knew what was worth reading. Even Cardinal Bellarmino's *Controversies*, the unparalleled defense of contemporary Catholicism, had been briefly banned for denying, in one of its thousand-odd chapters, that the pope was the absolute king of all the world in every respect, no questions asked. Galileo was consoled, by Venetians in particular, that "the subject of prohibition does not diminish the glory of the author. You are read despite malicious envy."

Galileo had barely heard from his friend Pope Urban in six years and heard nothing from him now. Benedetto was running around Rome, always sending optimistic news. "I do not know why you insist on keeping hope alive," Galileo responded, "never hope about anything on Earth." On September 23, his case was passed to the Roman Inquisition.

"I find myself in rank confusion," he wrote two weeks later.

> Rank confusion, because three days ago the Florentine Inquisitor intimated to me that, at the order of the Roman Inquisition, I am to present myself before that same tribunal, where it will be explained to me what I ought to do. Now, recognizing the significance of this business, and out of need for this council, I have resolved to go at once. This will demonstrate to those many parties, more than one of which comes to mind, that I am as I am, an obedient zealot for the Church. I will leave next Sunday.

On that Sunday, Galileo remained in Florence. He was stayed by infirmity.

To be called before the Inquisition, Galileo knew, was an act reserved for the "grossly delinquent"; "It makes me hate all the time I gave to my studies, the fact that I aspired." He wrote this out in a letter begging that he be allowed, as a feeble old man, to take his trial through the post. No, he was told, you really must come at once—even the pope agrees.

Galileo would not go. On December 17, the Roman Inquisition received a doctor's note stating that he was suffering from anxiety, vertigo, indigestion, hernia, arthritis, insomnia, melancholy, and hypochondria. The Inquisition called this "subterfuge." Come, they said, or we will drive into Florence ourselves, tie you up like a hog, and bring you in a cage. Though he could hardly believe it, Galileo knew where he was being forced. The trip to the Holy City ran downhill.

On the way down, Galileo grabbed everything he owned that was relevant to his defense. He took the sixteen-year-old character witness from

Roberto Bellarmino. "It has merely been made known to him the doctrine attributed to Copernicus . . . contrary to scripture . . . not to be held or defended." Galileo had never broken the literal word of the Church. Literal words were what he understood.

In Rome, Galileo stayed at the house of the Tuscan ambassador, which was improper. The law dictated he should have been kept in a cell and charged for it, but fame has always blinded equality. Still, he was not allowed to leave or take visitors. He could not even visit the church next door, though he asked, because there was a garden out back he could wallow in. When the ambassador came back from lengthy discussions with the pope, Galileo finally learned what had caused his dear friend of twenty years to forsake him.

It was alienation and a trifle. There were countless minor agitations, but Galileo had years ago lost contact with Pope Urban VIII, the same man who had once proclaimed they were brothers. The two gave no further chance to soothe their slight differences. The pope did not like that his theological ideas were expressed in Galileo's *Dialogue* by a character whose name sounded a lot like "Simpleton." He did not like the lack of Church dogma from the book generally, did not like that Galileo had insulted the Jesuits, did not like the extralegal influence Galileo had used to get his imprimatur, and did not like the way Galileo buttered his toast on Thursdays.

When servants appealed to the pope on Galileo's behalf, which happened constantly, Urban bellowed, "Enough! Enough!" He revealed that, in the prohibition from 1616, it was his favor which stopped the Inquisition from taking Galileo in for blasphemy then and there. He felt betrayed. Urban stormed, "There is no way out. God save Galilei for having meddled with these subjects."

The week before Galileo was set to testify before the Inquisition, he could not stop screaming in his sleep. His poor daughter Virginia, writing from Florence, believed there would be a happy ending.

On April 12, 1633, Galileo took a private oath before a group of Father Inquisitors to tell the whole truth. They asked him many questions.

Q: And what was decided and made known to you, precisely, in month of February 1616?

A: In the month of February 1616, Lord Cardinal Bellarmino told me that since Copernicus's opinion, taken absolutely, was contrary to Holy Scripture, it could be neither held nor defended,

> but could be taken and used suppositionally. In conformity with this I keep a certificate by Lord Cardinal Bellarmino himself, dated 26 May 1616, in which he says that Copernicus's opinion cannot be held or defended, being against Holy Scripture. I can present a copy of this certificate. Here it is.

This was an extraordinary vindication of Galileo's actions. There was much mumbling. The evidence was taken in, marked with a big letter *B*.

> Q: When you were notified of the above-mentioned matters, were there others present?

Galileo reached into the furthest crevices of his mind, where everything was fuzzy and gray. It was very hard to remember such things.

> A: Yes. There were some Dominican fathers present. But I did not know them, and have not seen them since.

It was unclear to Galileo why this was relevant. The prosecution knew.

> Q: If one were to read to you what you were told and ordered with the injunction, would you remember that?

> A: I do not remember that I was told anything else, nor can I know whether I should recall what was then said to me, even if it were read to me. I say freely what I do recall, because I claim not to have contravened the precept in any way, that is, not to have held or defended the said opinion of the motion of the Earth and stability of the Sun on any account.

At this, the prosecution introduced as evidence a secondary account, unknown to Galileo, apparently obtained from the Vatican archives.

> Friday the 26th [of 1616]. At the palace, the usual residence of the Lord Cardinal Bellarmino, the said Galileo was warned of the error of the aforesaid opinion and admonished to abandon it. Immediately thereafter, the said Galileo was commanded and

> enjoined, in the name of His Holiness the Pope, to relinquish altogether the opinion that the Sun stands still at the center of the world, and the Earth moves. Henceforth, *he was not to hold, teach, or defend it in in any way whatever*, orally or in writing; otherwise the Holy Office would start proceedings against him. The same Galileo acquiesced and promised to obey.

Was it true? Galileo could no longer remember.

> I do not recall that this injunction was given to me any other way than from the mouth of Lord Cardinal Bellarmino. I do remember that the injunction was that I could not hold or defend, and maybe even that I could *not teach*. I do not recall, further, that there was the phrase *in any way whatever*, but maybe there was. In fact, I did not think about it or keep it in mind, having received a few months thereafter Lord Cardinal Bellarmino's certificate dated 26 May which I have presented and in which is explained the order given to me not to hold or defend the said opinion. Regarding the other two phrases in the said injunction now mentioned, namely *not to teach* and *in any way whatever*, I did not retain them in my memory, I think because they are not contained in the said certificate, which I relied upon as a reminder.

Had Galileo broken the compact he had made with Cardinal Bellarmino sixteen years ago? He no longer knew. This was the measure of his guilt in the eyes of the Church. For while Galileo was only suspected of holding or defending the Copernican opinion in his *Dialogue*, there could be no doubt that he had taught it.

Before Galileo arrived in Rome, the pope expressed a belief that "Mr. Galilei spreads troublesome and dangerous opinions under the pretext of running a school for young people." Members of a special panel convened to review Galileo's *Dialogue* and found that "without doubt, he acts with the solicitude of a diligent teacher who desires to produce disciples"; "Under the name of 'The Academician', he acts as a teacher"; "He teaches *ad nauseam*."

On April 27, just as Galileo was finishing his lunch, he was greeted by one of the inquisitors present at his earlier testimony. Something very improper was happening. Galileo was being dealt with extrajudicially. He was informed that the case for his innocence was damaged beyond repair.

Three days later, Galileo appeared before the Inquisition again, with a shocking reversal in tone. ". . . my error was, and I confess it, one of vain ambition, pure ignorance, and inadvertence. This is as much as I need to say on this occasion."

On May 10, he was allowed to present a final defense.

> I am left asking you to consider the pitiable state of ill health to which I am reduced, due to ten months of constant distress, and the discomforts of a long and tiresome journey at the age of seventy in the most awful season. I feel I have lost the greater part of the years which my previous state of health promised me. I am encouraged to ask this by the faith I have in the clemency and kindness of heart of the Most Eminent Lordships, my judges. I hope that if their sense of justice perceives anything lacking among so many ailments as adequate punishment for my crimes, they will, I beg them, condone it out of regard for my declining age.

Galileo's trial was over. At the end of May, he received permission to be driven through the church garden, where he could observe nature through a half-closed window. He was never tortured. He was never shown the instruments of torture. The law was not carried out to its proper extent.

The next month a decision was reached. When Galileo received his sentence, he would be brought down onto his knees; this was the last time he could stand on his own two feet. Before the Inquisition, he proclaimed that he had willfully deceived no one and would never renounce Catholicism. "I do not hold the opinion of Copernicus. I have not held it since being ordered by injunction to abandon it. As for the rest, I am in your hands. Do with me what you wish."

All across the country, there were churchmen and courtiers who applauded that day's events. There were peasants and travelers who did not care. There were monks and scholars who let out little gasps of horror. There were birds in the air and fish in the sea, but it was only, and has only ever been, the Catholic Church that found Galileo Galilei suspect of heresy on June 22, 1633, for teaching that the Earth orbits around the Sun.

Lacunae

Days before Galileo received his verdict, his daughter sent him a letter.

> To give news of everything in the house, I will start with the dovecote. Since Lent the doves have been brooding. The first pair that hatched were devoured in the night by some animal, while the dove tending them was found half eaten, draped over the rafter, completely eviscerated. On this account, the housekeeper believes the culprit to be a bird of prey.

Virginia felt stupid. "I am good for nothing," she repeatedly wrote her father, for whom "the opposite holds true." It was utterly ridiculous to waste his time with her irrelevant life, yet he had asked for a letter once a week. She would have written him every day, if she could.

Like her father, Virginia tended to work late. She usually wrote her letters between eleven P.M. and two A.M. One morning, three weeks after her father's trial, she awoke early to visitors.

Two of Galileo's students were at the convent in a fit. They were demanding the keys to Galileo's house. Virginia knew exactly what had happened. She forfeited them immediately. The students broke into her father's quarters and "did what had to be done." The majority of Galileo's private communications have never been found.

By August, Galileo's official sentencing was circulating through courts and universities as a public warning. The Church had made an example of him. Virginia managed to snag a copy of her own.

Her father had abjured the Copernican opinion, in a form dictated by the Inquisition.

> Furthermore, so that this serious and pernicious error and transgression does not go unpunished, and so that you will be more cautious in the future, and to make an example for others to abstain from similar crimes, we order that the *Dialogue* by Galileo Galilei be prohibited by public edict.
>
> We condemn you to formal imprisonment in this Holy Office at our pleasure. As a salutary penance we impose on

> you to recite the seven penitential Psalms once a week for the next three years.

"I have found in it a small means to do you good, Papa," Virginia responded, "I shall shoulder the burden you have to recite the Psalms. I have already begun to satisfy this obligation with joy." This was not how the law worked, unfortunately, but it was everything she was capable of.

Virginia was waiting for her father to come home. She was pleased to learn that he had eased off the drink, but was devastated when he told her he felt his name "crossed out from the book of the living." She was worried he would die before she was ready.

Two months later, it was not Virginia but Galileo roaming about a convent in an indescribable grief. His daughter had fallen ill to a sudden, violent dysentery. Walking to his house next door, he knew it was he who was not ready.

> By a sinister coincidence, upon returning home I found the vicar of the Inquisition, who informed me that I must stop asking for grace or they would take me back to prison. From this I infer that my present confinement is to be terminated only by that other confinement which is common to all, closed to everything, and lasts forever.

An unknown set of nuns emptied dead Virginia's cell, which smelled irrepressibly of sugar and aloe. They found a collection of hundreds of private letters from her father—her father, that is, Galileo Galilei, known heretic—and put them all in the fire.

Life Inside a Box

A group of church fathers walked past the wasted rural convent, up towards the mountain villa where Galileo lived under house arrest. Their routine was to check up on the convict every season. Today, they did not even need enter the premises; they found him kneeling in the garden, clutching the first bud of spring. He tried to stand up.

"I am ashamed that you see me, looking like a clown. Allow me to go dress up like a philosopher."

The friars were mystified. "Why do you not have someone else do this work?"

"Oh, no no," he said, wiping his tremorous hands on his apron. "I would lose the pleasure."

In February of 1635, Galileo received word that the Catholic Church had extended their censorship of him to include all new material whatsoever. He did not stop writing. As he was so old and frail, most friends expected his mind would deliquesce with his body. There were few left who understood him.

The various foreign intellectuals and Italian diplomats with whom Galileo was acquainted were stunned to receive manuscript copies of his new book. They found a printer abroad. In 1638, the Netherlands became the first country to contain a printed copy of Galileo's *Two New Sciences*. The title, as Galileo complained, was chosen by the publisher to be "popular"; it was attractive, but inaccurate. The new sciences in question, purportedly "resistance to fracture" and "local motion," composed only a fraction of the text. In actuality, *Two New Sciences* was, at long last, Galileo's presentation of every key discovery he had made over a lifetime of research in physics.

In *Two New Sciences*, Salviati, Sagredo, and Simplicius are resurrected to once again converse upon Galileo's insights, but each has been transformed by the years since their last performance. Salviati, who always talked most, is more verbose than ever; he regularly produces pages of theorems without interruption. Sagredo has learned courtesy; both he and Salviati no longer criticize. One writerly friend is an "Archimedes of our age"; another is "read with admiration"; still another has written "commentaries which have illuminated this work." None are recognized as fools except perhaps Simplicius, but he is no longer a straw man for Aristotle. His inquiries are as direct as a brick wall, forcing necessary digressions into unknown territory. If he was modeled on anyone, it was Galileo himself, in his curious youth, finding first interest in the baffling relationship between mathematics and the world at hand.

Early in the book Galileo described elementary toilet plumbing; a hundred pages later he speculated on biology and the natural limits of animal growth. In the third day of dialogue, he finally laid claim to the law of free fall, over thirty years after discovering it. Building upon that, he noted offhand how "any speed once imparted to a moving body on a horizontal plane will be rigidly maintained." This property did not receive a name, although it contradicted the tendency towards rest which Kepler had called inertia. The language of science has a way of smashing opposites together.

For all the frenzy of *Two New Sciences*, ever blushing to stuff itself with more content, its diversity is obviously exceeded by the diversity of what it lacks. Any time the characters approach a topic in theology, they shy away from this "higher science," proclaiming that "we must be satisfied to belong to the less worthy." The implication is that the study of nature had no interconnection with the study of divinity, or more severely, that God is wholly transcendent, and knowledge of his reasoning, the causes of things, cannot be mediated through an earthbound subject. This belief was not so stupefying to Galileo as it would be to proper theologians; much worse for him was the secular abyss of infinity.

In the first day of dialogue, the three friends announce their intention to discuss resistance to fracture, only to immediately fall down a rabbit hole of pure mathematics. The trouble was that the existence of infinitely large and infinitely small quantities were necessary antecedents to Galileo's discoveries. Motion in a vacuum, to Galileo, moved on a continuum from absolute rest to infinite speed; a big force could be composed by an infinite number of infinitely small forces; the resistance to motion provided by an inclined plane grows infinitely small as the plane rotates to horizontal. Galileo had an almost infinite number of encounters with infinity. The only means he had to reckon with them was a medieval sense of proportion. Proportion and fraction are unequipped, by definition, to represent an infinite quantity. Salviati cried, "What a sea we are gradually slipping into without knowing it!"

He went on, "Infinity and indivisibility are by their very nature incomprehensible to us." Galileo had to try, regardless.

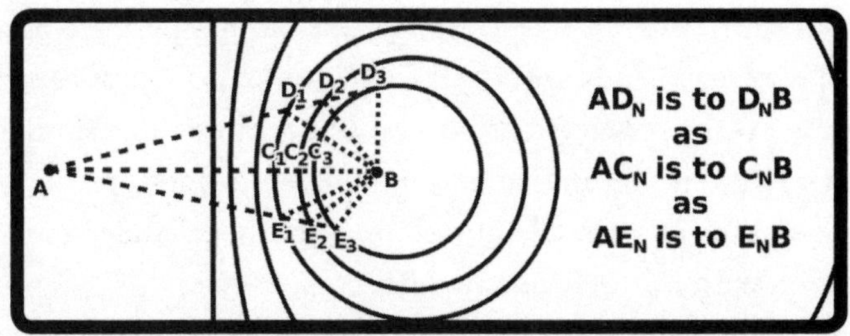

Galileo's first demonstration of a continuum between null and infinity.

To create an intuition for such "marvels which our imagination cannot grasp," he offered a series of geometrical representations of infinity, using only a sense of proportion. One clever example begins with a line *AB*, divided anywhere by point *C*. Galileo described a "remarkable property," that all connected pairs of lines drawn off points *A* and *B* maintaining the proportion of *AC* to *CB* meet at points on the same circle. If dividing point *C* is moved towards the midpoint of *AB*, this circle grows towards a greater and greater size. When *C* is at the midpoint, the circle becomes a straight line! As *C* moves towards the endpoint *B*, the circle dwindles, smaller and smaller, until it vanishes into nothing.

"Now," said Salviati, "what shall we say concerning this metamorphosis from the finite to the infinite?" Was a straight line or a mere point some kind of circle? What would that even mean?

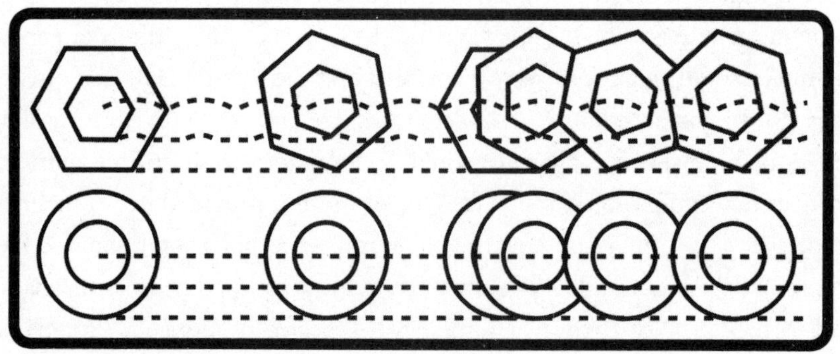

Galileo's second demonstration of a continuum between null and infinity.

In another demonstration, Galileo considered the rotation of different regular polygons, which stamp out a line with their lowest point as they roll along. If a smaller polygon of the same type is inscribed within, it will stamp out a path, which is like a straight line but has periodic bumps. As the inner polygon expands towards the outer one, the bumps contract into nothingness. As the polygon increases its number of sides, a similar reduction occurs. Does this mean that a straight line is not straight but contains an infinite number of infinitely small bumps, or that a circle is not circular but a polygon with infinitely many sides?

Galileo had no answers, but he sensed, through his insufficient sense of proportion, that something much greater than geometry was afoot. He was certain that infinity was exampled in nature; perhaps, even, in the size of the universe, although he was unsure of the specifics. He knew too well his own ignorance, the shadows of what lived outside his outdated mathematics and outside his world of sensible experience.

By July 4, 1637, Galileo's right eye had ceased all functionality. A month before the new year, it finally happened.

> Alas! Your friend and servant Galileo has been blinded beyond hope. This heaven, this world, this universe, which I by my marvelous discoveries and clear demonstrations have enlarged a hundred thousand times beyond what had been previously believed, is for me now shrunk down into a space so small as to be filled by my own person.

The Four Last Things in Cruel Disorder

Galileo had tried to write a fifth chapter for his *Two New Sciences.* There was much he had left to say, but he gave it up when he lost his vision. His struggle was over.

His son Vincenzo lived in Florence and regularly hiked up the mountains to father's house. Galileo told him not to bother, but he kept coming anyway. This was the obligation of family.

Much more intimate than his son was the monk, Benedetto Castelli. After months of fighting against the Church bureaucracy, he attained license to visit his old teacher. At this conclusion to forty years of friendship, Galileo stirred up a single note of positive sentimentality.

> I have been enfeebled by age, all the worse for my failing memory and senses. I pass through my pointless days, so long in their unyielding malaise, yet so short relative to the months and years. No longer am I consoled by the sweet memories of past friendships, of which so few remain, yet I am thankful for this one above all others: I mean that loving correspondence with you, Most Reverend Father.

Outside of the familiars, everyone was pain. His most enduring enemy, Christopher Scheiner, had finally completed his giant book on sunspots, *The Red Bear*. Galileo had no hope of responding at length, but lashed out in a letter. He had lost all will to civil humility.

> This porky asshole dares make a catalog of my errors, but let the beggar behold his fortune. He has realized nothing from his observations, while I have discovered the greatest secrets of nature. For it was my destiny, and mine alone, to observe so many novelties in the heavens.

In Italy, junior mathematicians traded a month's salary on the black market to come away with a copy of the heretical *Dialogue*. Although Galileo could not recognize it, his trial had not ruined but secured his life's reputation with this younger generation. The world was different to these kids. Printing did not appear to them a ragtag, mom-and-pop business but had acquired the character of empire, over and above the limits of government; global trade routes had been split open by technological force, and queens bathed in New World gold; power and prestige were forming shapes, which might preemptively be called industry, a seduction upon which Galileo appeared, stuffed and mounted, as an incidental figurehead. Church dogma did not feel as important. These insolent children, especially from the heathen countries, would not quit molesting Galileo in his last years. Teenagers attacked him, begged to worship him, be his student, write his hagiography, and sculpt an oversized bust of his ambivalent mug.

In 1638, a traveling English boy, with no special credit to his name, imposed himself on Galileo for a full evening. John Milton was a delicate flower, writing timid verse about hillsides and virgins and pure emotion; but during this tour of Europe he radicalized, becoming obsessed with pamphleteering and free speech. Only in midlife did he seriously return to poetry. He contemplated blind Galileo and composed *Paradise Lost*, the great Protestant epic, concerning the devil and the fall of man. The poem is resplendent in its range of Biblical and classical allusion, but among Milton's contemporaries, Galileo was the only one deemed worthy. Every reference to this "Tuscan artist" is grandiose, but not necessarily good; sometimes Galileo's discoveries are used as a metaphor for Satanic magnitude.

. . . His ponderous shield,
Ethereal temper, massy, large, and round
Behind him cast. The broad circumference
Hung on his shoulders like the moon, whose orb
Through the optic glass the Tuscan artist views.

Other verses find them sitting in the parliament of angels.

Of Heaven arrived, the gate self-opened wide
On golden hinges turning, as by work
Divine the sovran Architect had framed.
From hence no cloud, or, to obstruct one's sight,
Star interposed, however small he sees,
Not unconformed to other shining globes,
Earth, and the garden of God, with cedars crowned
Above all hills: as when by night the glass
Of Galileo, less assured, observes
Imagined lands and regions in the moon.

Milton was not blaspheming Galileo, nor idolizing him, nor writing haphazardly. He was trying to pin down the nature of heroism in a fallen world; not, as orthodox Catholics would have it, a virtuous protector descended from on high, but an uncertain mediator between Heaven and Hell, subject to temptation, success, and failure.

Several years earlier, Galileo was pestered by another Englander, much older than Milton and even more invested in politics, by the name of Thomas Hobbes. Hobbes had once served as a tutor in mathematics, when he came to adore the works of Galileo, whose natural philosophy he copied almost verbatim but for amplifying its latent content. He contemplated blind Galileo and composed *Leviathan*, one of the first English works of political theory, where he sought to redefine personhood solely in terms of its material requirements. Men, he wrote, "naturally love liberty and dominion over others," and they therefore require "the introduction of restraint over themselves," which takes the form of the state. This pessimism led to very conservative claims about religion, especially concerning its entanglement with civil government.

> God is the sovereign of all sovereigns, and therefore, when he speaks to any subject, he ought to be obeyed, whatsoever any earthly potentate command to the contrary. But the question is not of obedience to God, but of when and what God has said. To subjects that have no supernatural revelation this cannot be known but by that natural reason which guides them, for the purpose of peace and justice, to obey the authority of their commonwealths; that is to say, of their lawful sovereigns.

Despite this, Hobbes's brand of materialism led many to label him as an atheist. But he was only looking willfully at the present state of politics and science, and attempting to reconcile them with the religious tradition in which he was enmeshed.

Galileo was born too early to read either *Paradise Lost* or *Leviathan*. He was born too early to witness the total collapse of the Papal States and too early to see the lax republican model of Venetian government spread over Europe like jam on toast. He was born too early to know the ultra-enlightened French revolutionaries who declared him "the first who made physics speak the language of truth and reason"; too early to read the centuries of apologia for the Catholic Church arguing he had acted "in bad faith" and "wanted to mock the Inquisition, cheat them with skill"; too early to hear the pacifist Einstein, whose theoretical contributions allowed for the nuclear bomb, label him "the father of modern physics and, in fact, the whole of modern science." And he was born far, far too early to understand that peculiarly modern maxim, "Hell is other people."

The intangible productions of his mind were being rushed offstage, leaving only exploitable biography. "I can truly say," he wrote—or rather, dictated to one of his unseasoned devotees, for he was blind—"I have ended up in Hell on Earth."

There were many onlookers to Galileo's condition in his final years. Most nights his arthritis prevented him from getting even one hour of sleep. He had a double hernia that required him to wear an iron truss and spent so many hours facedown in bed that people thought he was already a corpse. Awake one dawn, he soiled himself. He wrote, "I can hear my little daughter calling me, calling me . . ."

In the last year of his life, he described to an assistant a new sort of clock based upon the pendulum. It was much more consistent than what had come before. With it, there could be scientific agreement, standards, and discipline. Galileo liked discipline. Even blind, bedbound, waiting to die, he had the discipline to attempt yet another dialogue. It was very short. It concerned a sense of proportion and what that meant. He had really figured it out this time. His friends Salviati, Sagredo, and Simplicius are all off in some field, waiting for it to begin. He imagined what Salviati would say: "I feel great consolation today, in seeing our meetings renewed after the lapse of so many years." How joyous that meeting will be, when he finds peace among family and friends. They are waiting for him. He is trying to reach them. He is racing through an infinite black for them.

I strove with none, for none was worth my strife.
Nature I loved, and next to Nature, Art;
I warmed both hands before the fire of life,
It sinks, and I am ready to depart.

—W. S. LANDOR,
Dying Speech of an Old Philosopher (1849)

APPENDIX

Seven Vignettes from the New Astronomy

The Three Roads

Kepler had to appease the Tychonics, whose father had founded "the truest and most accurate" astronomy, according to the opening of the Optics. Such veneration had special cause, as that book's main impetus was Tycho's groundbreaking studies of refraction, but hundreds of pages later Kepler made clear that he sided entirely with Copernican cosmology. The next time around, in *New Astronomy*, Tengnagel found it necessary to add a rambling and convoluted defense of his stepfather before the true preface, but aside from these shenanigans, the imperial mathematician snuck by, for the moment, more or less uncensored.

Kepler's promises would be more affecting. Early in *New Astronomy*, he wrote of Ptolemy, Copernicus, and Tycho:

> I, in the demonstrations that follow, shall link together all three authors' forms. For Tycho, too, whenever I suggested this, answered that he was about to do this on his own initiative even if I had kept silent (and he would have done had he survived), and on his death bed asked me, whom he knew to be of the Copernican persuasion, that I demonstrate everything in his hypothesis.
>
> Furthermore, we shall demonstrate, both right here and throughout the entire book (while doing other things), that these three forms are absolutely, perfectly, geometrically equivalent.

In writing as in life, Kepler was the bridgebuilder. Thus, many proofs are proofs in triplicate, one for each world system; the book is not trivially larger for it. In so doing, he would not only honor Tycho's memory but perchance invite those hoary geocentrists to reconsider their outlook. A difficult trinity had entered his life yet again.

The bearing of this additional burden established Kepler as the last great virtuoso of geometrical astronomy. Just consider, for example, the woodblock prints of his first tripartite proof.

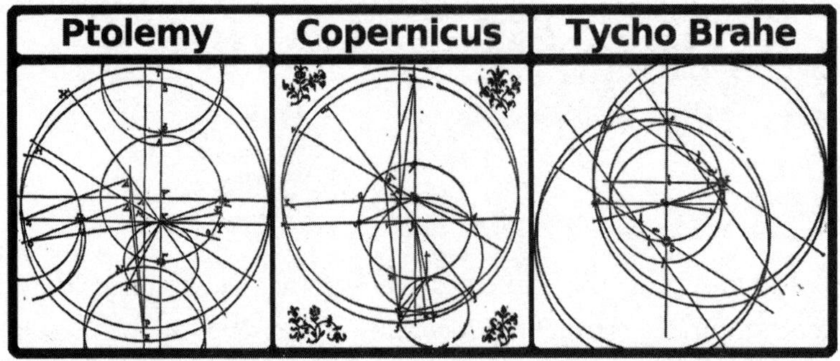

Woodblock prints for each world system from the same proof. Naturally, Kepler has had the Copernican system embellished.

Oh. Oh dear. What a dreadful diagram. What on earth is it for?

The Thing Itself

Each day for exercise, at exactly half past four, the elderly German philosopher Immanuel Kant would exit his modest, unadorned house and walk up and down the street eight times. Housewives would set their clocks by him. He was a man of exactitude.

Today, Kant is recognized as the dominant spirit in traditional Western philosophy. In his *Critique of Pure Reason*, he alluded that his own work would be like a "Copernican revolution," made explicitly metaphysical, changing the very way the world was conceptualized. He divided knowledge into two categories: a *priori*, independent of experience, and a *posteriori*, dependent on experience. He postulated that all humans could know about the material world was bound by their senses, which produced phenomena, as opposed to what was actually being sensed—the things themselves, or noumena. He strove to build a bridge between rational and empirical thought, two tidal forces sweeping across the European imagination. His writing style was abstruse but entertaining; he loved astronomy, called metaphysics a "battlefield," and was born almost precisely 150 years after Johannes Kepler.

A mind which makes its assumptions explicit often repaints its entire surrounding culture. "There can be a very small discrepancy which the senses do not perceive," Kepler noted, when considering how false assumptions in astronomy may yield true conclusions.

His first radical proposal concerned the thing itself, insofar as astronomy could. It is the purpose of the opening act of *New Astronomy*.

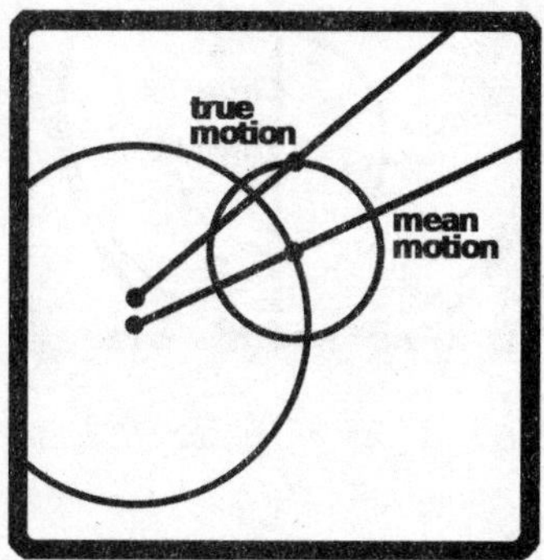

Ptolemaic: true motion runs through both planets.
Mean motion runs through center and epicenter.

Ever since Ptolemy, astronomers had regularly used the mean motion, or the first inequality, where the center of a planet's epicycle appeared in its orbit, rather than the true motion, with its second inequality, where a planet actually was on its epicycle. This was done because it was easier, observations were inexact enough to allow it, and more alarmingly, because astronomy was thought a sort of geometrical game; planets were points like any other and entailed nothing unto themselves.

Kepler felt no such thing. Not only did it now make less sense, with Tycho's meticulous observations, but he had a deep, abiding, religious sense that the planet itself was what mattered, rather than some idealization of it.

Using mean motion and true motion will almost always yield different results, but it was not clear that this difference mattered. How different might these results be?

This is what Kepler's first proof was meant to show. For all three world systems, he meant to prove that using the true motion instead of the mean motion would result in a nontrivial difference in the estimation of Mars's location.

Consider again the Copernican woodblock print. Unlike Ptolemy, who used epicycles, Copernicus had correctly attempted to explain the second inequality as handled entirely by the motion of the Earth. To do this, he measured planets relative to the Earth's orbit about its immaterial center—the mean Sun.

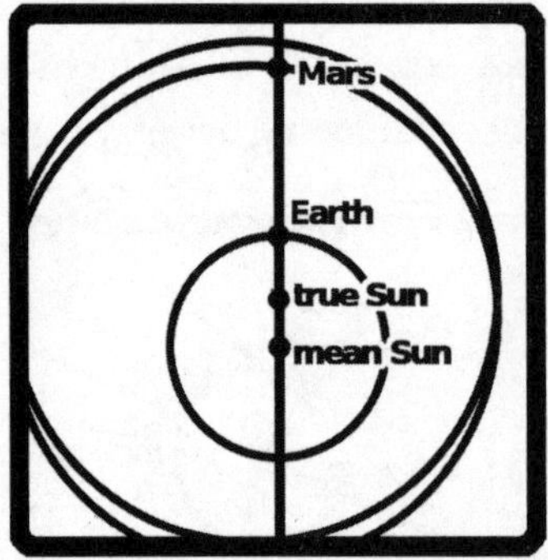

Copernican: when all points align, there is no second inequality.

In the woodblock print, the vertical line furthest right means to demonstrate that, when all three planets are there aligned, Mars appears to be in the same place whether observed from the mean Sun or true Sun; in Kepler's words, "the planet is truly stripped of its second inequality." Copernicus had delighted in oppositions of this type, but they were rare. Kepler then drew a long horizontal line to show where the difference may be substantial.

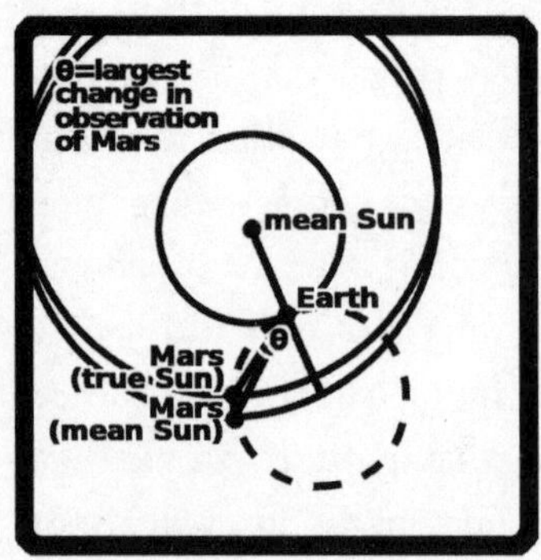

Finding the largest possible difference θ.

For a quantitative example, Kepler created the bottom part of the woodblock. To find how significant the difference could be, he picked the optimal two locations of Mars, where its orbits based on the true Sun and mean Sun would differ most. From this, he assumed the Earth was at a point which, with these two locations, formed a circle tangent to its orbit. Kepler knew from Euclid's *Elements* that all this would yield the largest possible angle.

By means of an extended geometrical calculation, running over multiple pages, Kepler found the radius of this circle, the shortest possible distance from Earth to Mars, almost four tenths the radius of Earth's orbit. He was validating that old Tychonic test: the Copernican model had the Earth draw closer to Mars than the Sun. With this radius, the angles fell into place; true Mars and mean Mars could differ by over one degree, a sizable 1°3'32"!

Kepler's answer was not quite right. At this early stage, he had allowed imprecise measurements and even made a slight error in the calculation. "This is no impediment to us," he wrote, "since we are only performing a 'prelude'"—Tycho's favorite word. These were numbers meant for the mind, to convince the world to once again change its point of reference.

It was his metaphysics of measurement, the very idea of exactness, reconnoitering the practice.

The First Model

Would that Kepler be so simple! If his projection above was that of a slippery theorist, it was blurry and two-dimensional; he now refocused as a tireless practitioner. As the new astronomy took shape, entering into ever thornier thickets and ever richer briars, so too did Kepler, as man and mathematician, become a thornier and richer character.

Kepler's first foray into modeling Mars's motion was under direct inspiration from Ptolemy. He used an equant.

It was a curious choice to return to an equant, which Copernicus had so despised, and Kepler did not immediately state his reasoning. Then, in the middle of a calculation, he lets slip that "physical considerations demand the slowest motion occur where Mars is the greatest distance from the Sun." If he preferred the equant for any reason beyond convenience, it was these "physical considerations." It was a sneaky elision, the shape of astronomy to come.

From Ptolemy's equant, the ever-unsatisfied Kepler demanded a more general approach. Whereas Ptolemy held the equant and Earth (or Sun, for Copernicans) symmetric about the center, Kepler felt this too much to assume; the equant might instead rest anywhere on the entire line of apsides.

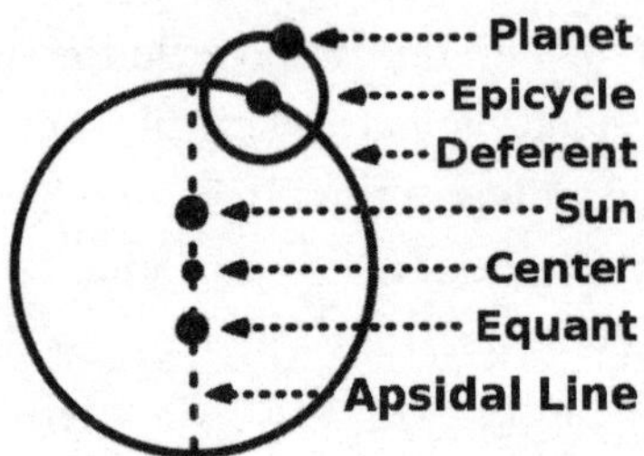

The line of apsides, or apsidal line, is simply the line running through the Sun and equant.

Generality always creates problems. For Ptolemy, a planet's location determined its equant exactly. For Kepler, there were infinitely more possibilities. Only one was optimal. With what steel, what stalwart stuff must a mathematical mind be made to withstand this numerical explosion! "This solution is not geometrical," he despaired, "at least if algebra is not geometrical. Yet algebra too forsakes us here!" To one friend, he wrote a depressed cry for help: "I have sought a proof in vain; I think it is because I am insufficiently skilled." There was no answer.

In his isolation he began. In order to find the optimal location of the equant, he picked three of Tycho's observations of Mars taken at sunset. Three points are enough to uniquely determine a circular orbit, but he added one more "to weaken the sinews of the problem." He drew in the line of apsides and connected all of the dots.

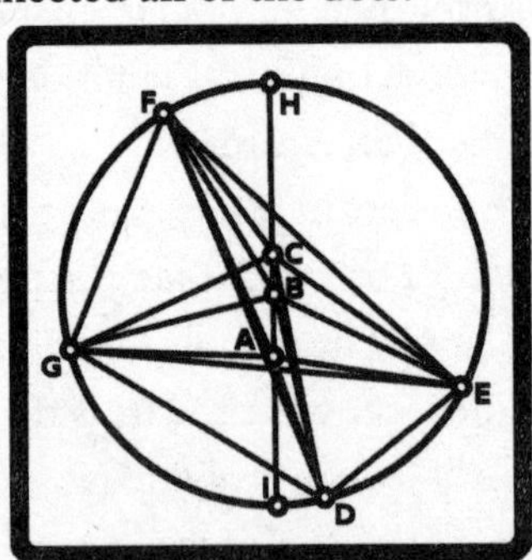

Kepler's circular orbit. He would later name this the vicarious hypothesis.

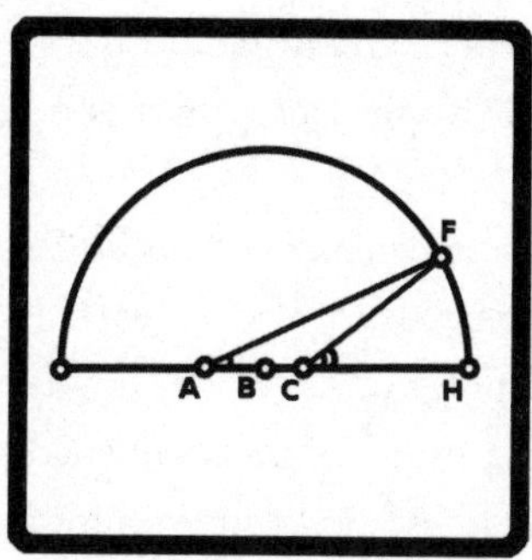

Kepler guessed the values of ∠FAH and ∠FCH to solve the vicarious hypothesis.

His solution was a feat of mathematical endurance. Though Kepler's diagram looks dense, there are only two unknown parameters: the distance from the Sun to the center *AB* and the center to the equant *BC*. Kepler guessed both of these, representing them as angles ∠*FAH* and ∠*FCH*.

Now possessing both unknowns, Kepler had to make one more assumption, which became a point of peculiar deliberation: that the quadrilateral formed by the observations *DEFG* was inscribed within a circular orbit. Because of this, and the fact that opposite angles of a circumscribed quadrilateral must sum to up to 180°, finding angles ∠*GFE* and ∠*EDG* would verify Kepler's guess. Some very extensive trigonometry resolved all the intermediary angles.

When his guess was wrong, there would be nothing for it but to guess ∠*FAH* and ∠*FCH* once more. And once more. And once more. And once more. And . . .

"If this wearisome method has filled you with loathing," he wrote, "it should more properly fill you with compassion for me, as I have gone through it at least seventy times, at the expense of a great deal of time, and you will cease to wonder that the fifth year has now gone by since I took up Mars." The hours unraveled; Kepler, the human computer, performed his iterations.

Finally, it was completed. Kepler stood before Mars victorious, with the most accurate theory that circles and spheres had ever established.

The reader turned the page.

"Who would have thought it possible? This hypothesis, so closely in agreement with the observations during sunset, is nonetheless false . . ."

Kepler's latitudes were imperfect, which was expected. But his longitudes, when Mars was 45° away from the top of its orbit, could vary by up to 8'. Had he done it perfectly, it would have been less than 4' off, an accuracy unheard. Yet for Kepler, it was not good enough.

> Since divine benevolence has vouchsafed us Tycho Brahe, a most diligent observer, from whose observations the 8' error of the Ptolemaic computation is shown in Mars, it is fitting that we with a thankful mind both acknowledge and honor this favor of God. . . . These eight arcminutes alone have led the way to the reformation of all astronomy, and have become the material for a great part of the present work.

Kepler's first solution was no less alarming than his rejection of it. For that he called it his "vicarious hypothesis," because, though he knew it was wrong, it was good enough to use as a substitute when testing all his later theories. By this method did he forge onward, through the thicket, using falsity to cut a path towards truth.

The Great Confusion

Kepler stopped. He gave pause, or, at least, must have done at some prior stage of life once he had contemplated what he now, finally, had experimental evidence for. Here he held the basic Ptolemaic, Copernican, and Tychonic models, each in their most general possible forms, yet each fell short. What could one do, where could one go, but back to first philosophy?

"I shall now prepare myself," he braced, "not through an arbitrary hypothesis but from the very nature of things." Thus, Kepler began to reforge his astronomy, from the ground up.

> My first error was to suppose that the path of the planet is a perfect circle, a supposition that was all the more noxious a thief of time the more it was endowed with the authority of all philosophers, and the more convenient it was for metaphysics in particular. . . .
> . . . An oval is thus substituted for a circular path.

Kepler's oval was curious, but not yet astonishing; once known, it might very well just be some combination of epicycles. But first it had to be made known.

For a mathematical term, "oval" is a frustratingly ambiguous word, because most round noncircular shapes have no geometrical definition. It merely means "egg-shaped." Kepler knew not how to wrestle with it; he could only proceed piecemeal, bit by bit, in awkward steps. He continued to draw the problem as a circle. His language became increasingly opaque.

> I consequently began by dividing up the circle in 360 parts, as if these were least particles, and supposed that within one such part the distance from the Sun does not change.
>
> This procedure is mechanical and tedious.

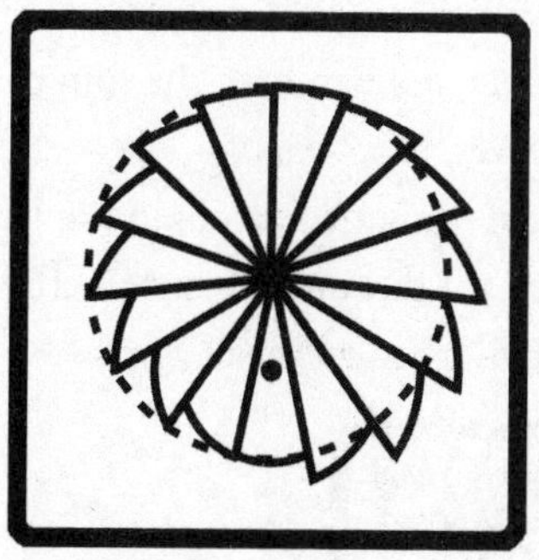

A possible illustration of Kepler's first attempt, in 15 parts rather than 360 (true orbit is dashed).

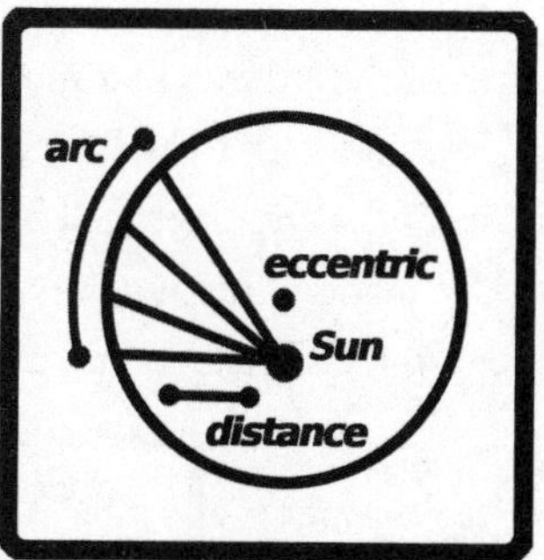

There are an infinite number of possible distances within any arc (four possibilities drawn).

Kepler was approximating the continuous orbit by a discontinuous representation. Even he, the constant computer, could not afford to be so imprecise, impractical, iota by iota. "Unless we can find the sum of all of the distances in an arc," he realized, "(and they are infinite in number) we cannot say how great the time increment is for any one of them."

From this abandonment, he produced an analogue, by one singular, spectacular assumption: that the sum of the distances in an arc of orbit, which was infinite, was represented by the area of the arc.

It was a proofless leap of intuition, an infinitude in a finitude, a continuity from discontinuities. Less than three centuries ago the scholastics had so often written: between the infinite and finite there is no proportion.

Kepler was about to get, he was sadly aware, almost impossibly cryptic, yet his work must be witnessed, if not understood, in order to grasp the simultaneous complexity, swiftness, and turmoil of his thought. Kepler befuddled everyone around him, especially himself.

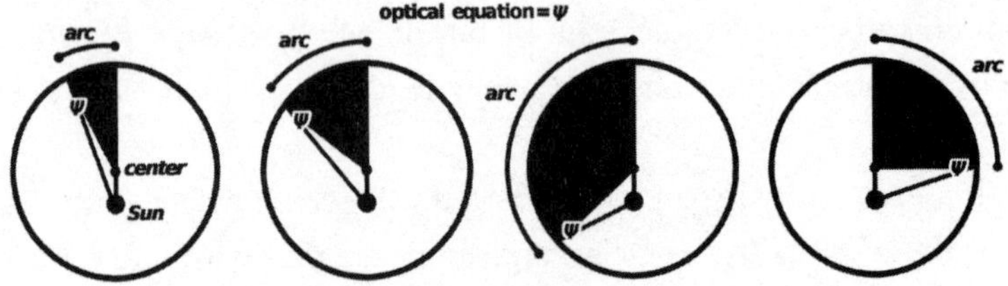

The optical equation, shown for different sizes of arc. The area of the circular part of the arc is blacked out.

The area of the arc was a far more tractable problem than the sum of its distances. Kepler had only to look for the "leftover" bit of triangle after subtracting out the area of the circular arc. This triangle contained an angle Kepler called the "optical equation," because it represented the distance anyone in orbit would see between the visible true Sun and the hypothetical mean Sun.

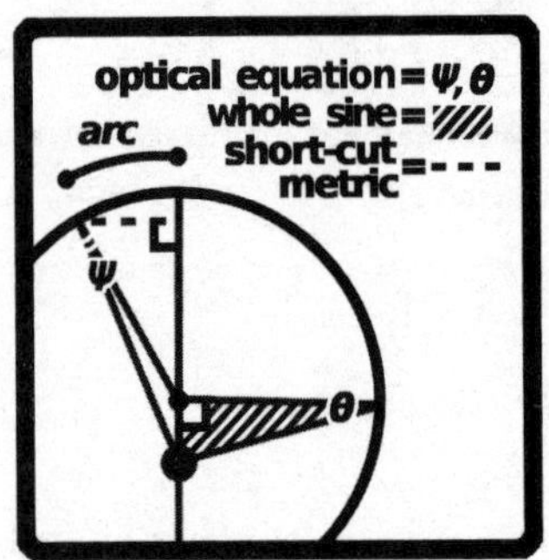

The optical equation.

His preferred way to solve the optical equation was by a short-cut metric. Dropping a perpendicular line from the end of the arc to the line of apsides, he proved this "short-cut" was proportional to the area of that arc's triangle of the optical equation. One such triangle, which contained a right angle, Kepler solved and called the *whole sine*. This whole sine could be easily multiplied by this short-cut to get the area of that short-cut's triangle of the optical equation.

Alas, the arcane solution was imperfect! This method assumed the triangle of the optical equation, like the whole sine, always had a right angle. Kepler caught this error right away and explained it immediately.

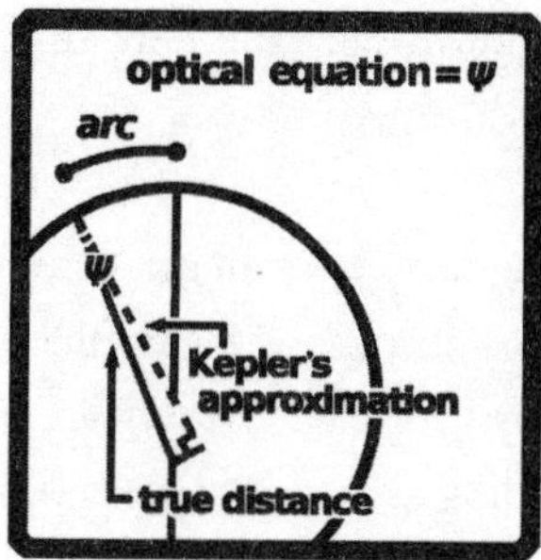

The optical equation extended to a right angle.

Yet as it turned out, this imperfect, area-based answer actually gave the exact correct solution . . . for a true orbit, which was no mere oval, as Kepler would soon reveal. Did the scamp already know, or did he feel it in his bones? How did he always seem to predict the future?

For the moment, he turned back from areas to distances. Taking measurements of the true distance and his approximation for increasing sizes of arc, he lined them up to produce what can only be understood, by the modern mind, as quite possibly the first ever graph of a function.

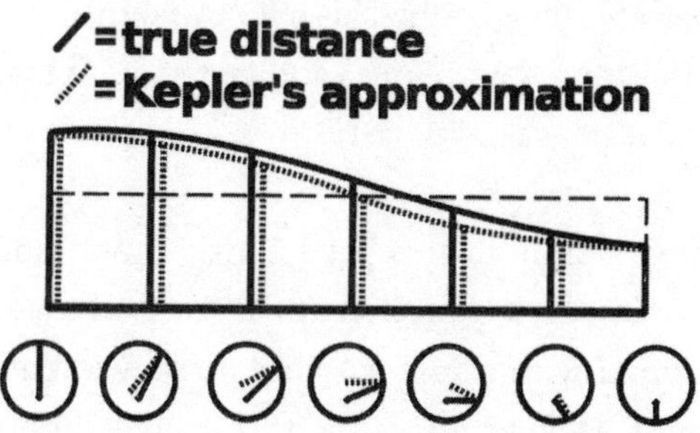

Kepler's approximation always goes through the center and forms a right triangle with the true distance from the Sun to Mars.

Look at that monstrosity! Kepler's approximation was a harmonious, artisanal wave, while the true distances were an uncanny, asymmetric, horrifyingly slight aberration. He had realized from this that areas and distances, though related, could not be treated as the exact same. Finally, he just gave up, deciding it was "not important enough at this time."

> At present there are many first-rate geometers who on occasion labor endlessly on matters whose usefulness is not so clear.

> I call upon one and all to help me here find some plane figure equal to the sum of distances.

He called upon the wrong savior. For all this computation, all this wacky terminology and blatant obscurantism, was the result of his practice butting heads with problems undreamt by geometers. It was an entirely new mathematical language that he was heralding, a calculus far out beyond geometry or algebra.

As a boy, he had yearned for the gift of prophecy.

The Hasty Dog

> What they say in the proverb "A hasty dog bears blind pups" happened to me.

Kepler was a great lover of irony, who delighted in discovering ignorance almost as much as he did in correcting it. Copernicus was "ignorant of his own riches," *New Astronomy* proclaims; Tycho, too, "like most rich men, did not know how to make proper use of his own riches." Kepler by no means escaped this fate.

He had mixed distances with areas and was soon to make a second brilliant conflation. But he was overhasty.

> If I had embarked upon this path a little more thoughtfully, I might have immediately arrived at the truth of the matter. But since I was blind with desire, and did not pay attention to each and every part, staying instead with the first thought to offer itself (a wonderfully probable one, owing to the uniformity of the epicyclic motion) I entered into new labyrinths.

Kepler abandoned his current oval and returned, once again, to his vicarious hypothesis, modified to fit an oval. He could approximate this oval with his earlier method using the optical equation, but this left the same error. "Once again," he cried, "where is the geometer who will show us how to do this?" So, finally, at long last, he changed tack.

For several chapters now, Kepler had been flirting with a certain buxom figure. This oval version of the vicarious hypothesis was "little different from an ellipse," he realized, and again, "it is just as though Mars's path was a perfect ellipse . . . " Oh! Oh. Oh, why not just approximate the approximation, why not just . . .

Let it be a perfect ellipse.

In one sentence, this strange yet elegant child cut the cord from its millenary parents: the ellipse born of the oval born of the circle was adopted as Kepler's theoretical foundation. There were no parties, no parades, just one boy, alone with his theory, laughing in the darkness. "As if," he wrote in a flush, "one were to squeeze a fat-bellied sausage at its middle, they would squeeze and squash the ground meat with which it is stuffed, outwards from the belly towards the two ends."

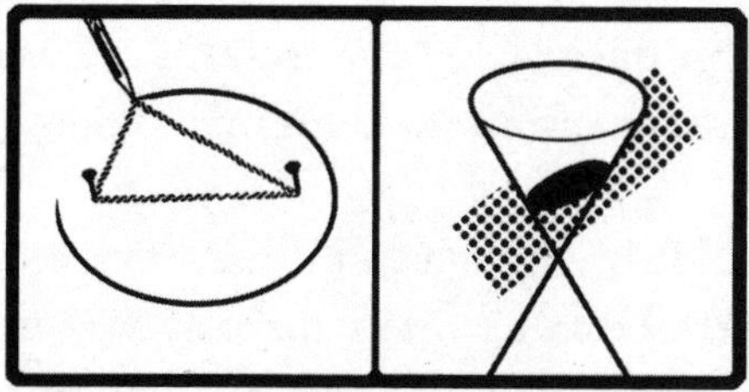

Two ways to generate an ellipse.

The ellipse can be defined simply. It has two foci: Take a loop of string and place it around two distinct points. Now, with a pencil, pulling the string back into a taut triangle, rotate it about the two points. This is the shape. In Keplerland, it was better understood as the closed planar intersection of a cone.

The difference between an ellipse and a circle was a well-known calculation. Kepler could now shift his attentions to a new method, the "little moons," or *lunulae*, which formed between them.

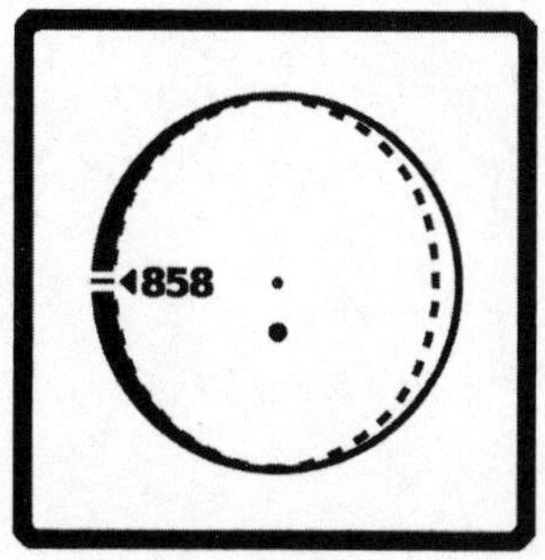

Kepler's lunula in black.

His ellipse was off. The maximum width of the lunula was 858 parts out of 100,000. "By a dogged calculation," Kepler found the correct answer should be 429 parts.

He was thrust so close, yet the proof was not there. From where did this 429 stem?

Let the discoverer speak of revelations.

> I therefore began to think of the causes and the manner by which a lunula of such a breadth might be cut off. While anxiously turning this thought over in my mind, reflecting that absolutely nothing had been articulated, and consequently my triumph over Mars was futile, quite by chance I hit upon the secant of the angle 5°18', which was the measure of the greatest optical equation. And when I saw that this was 100,429, it was as if I were awakened from sleep to see new light. And I began to reason . . .

He had the ellipse. He had its lunula. He had its cause. The jig was up.

Laughably, moments before his victory, the hasty dog stumbled. Like a court jester, a Shakespearean fool, having the truth he mocked it, inflating the ellipse out toward one end, as though it were a true ovum, a fat egg. Still a humorist this far in the thick of it, it became his "puffy-cheeked path": an immediate, uproarious failure. "O ridiculous me!" he chuckled and set it aright.

The Two Laws

> Kepler's First Law: The orbit of a planet is an ellipse with the Sun as one of the foci.
>
> Kepler's Second Law: Line segments from the planet to the Sun sweep out equal areas in equal time.

The New Physic

On a passably clear June day, Kepler sat outside a heavily draped tent clutching a knife, carving notches into a cumbrous, ramshackle, wooden contraption. Several panels, now successfully metered, stood nearly perpendicular a long horizontal post, which swung out upon a socket joint like a jib crane. The whole thing lay low to the ground, resting on a sillily wide rectangular frame. He dragged it back into the makeshift darkroom until only the bronzed tip of the post poked past the curtain. It was just after lunch in the Market Square of Graz. Passing townspeople might ogle the unexplained scene with frank bemusement, but their attentions were soon diverted by a sudden darkening in the noon sky.

Inside the tent, testing the slide and rotation of the post, Kepler brought it to a settle. Rays of outside light shot through its thin pinhole into the darkroom, out onto the leftmost notch. There they projected, mirrored and upside down, till devoured by a glacial moon, one by one, left to right: the slowest, and only, movie in all the world.

A picture of the Sun.

Kepler had made a camera obscura, in order to observe a solar eclipse. He described it in his *Optics*, using the study to resolve many of Tycho's problems with refraction.

He was not the best at this type of operation. In the uncommon case where Brahe's many journals could not provide what was needed, he would jest, "I am going to give you a clown show, in that I use my own observations." Yet he gave them anyway, however reluctantly, in order to verify his theories. Despite the discovery of his two laws, his mighty goal was not to state how something was, except insofar as it got to the why of the matter.

New Astronomy is not only concerned with the shape of planetary orbits but all at once also the reasons for that shape, just as with *The Secret*. Aristotle had one answer, Kepler informed, in which he "introduced to planets separate minds which, it turned out, were gods, as the perpetual administrators of the heavens' motions. He also bestowed a moving soul."

Planets were moved by gods and had a soul: this was the epitome of old-fashioned metaphysics. It gave planets a degree of intentionality, as if they were small-brained, big-bodied, intergalactic actors. But Kepler refused to believe it; it was one leap of faith he could not make.

"With arguments of the greatest certainty," he proclaimed, "Tycho Brahe has demolished the solidity of the orbs, which hitherto was able to serve these moving souls, blind as they were, as walking sticks for finding their appointed road." Traditional Ptolemaic astronomy could have accounted for Kepler's strange elliptical orbit (with enough epicycles, it could account for any orbit!), but Kepler would not allow it. In rejecting his ancient masters, he had to become a master unto himself and propose a replacement: a new "why."

The subtitle of *New Astronomy* is "Based Upon Causes, Or Celestial Physics." The name of its third part was "The Key to a Deeper Astronomy, Wherein There is Much On the Physical Causes of Motions." The nearby cousin of Kepler's optical equation was called the "physical equation."

When confronted with the problem of planetary motion, Kepler invented astrophysics.

This invention was entirely his own, the single idea, which metamorphosed Tychonic observation from Copernican astronomy into Keplerian world-harmony. His planets were not moved by distinct, individual souls but by a single physical force. In its creation, he drew upon the full range of his experience, upon the world of books, from which, by the waning light of day, he read of other forces and upon the world of action, of how he awoke, every dawn, with the foison Sun glittering overhead. He called this physical force the *species*, or motive power.

> There lies hidden in the body of the Sun something divine, which may be compared to our soul, from which flows that species driving the planets around, just as from the soul of a man throwing pebbles a species of motion comes to inhere in the pebbles thrown by him, even when he who threw them removes his hand from them. And to those who proceed soberly, other reflections will soon be provided.

Kepler had such confidence in the species, and the centrality of the Sun, that he almost felt they did not need argument, although he provided it.

> If this very thing which I have just demonstrated a posteriori (from the observations) by a rather long deduction, were demonstrated a priori (from the worthiness and eminence of the Sun), so that the source of the world's life is the same as the source of the light which forms the adornment of the entire world-machine, and which is also the source of the heat by which everything grows, I think I would have deserved an equal hearing.

The species, he thought, would diminish in strength in proportion to its distance traveled. He had shown light to attenuate in almost the same way, with the square of its distance, in his *Optics*.

> If I appear to philosophize with excessive insolence, I shall propose to the reader the clearly authentic example of light, since it also makes its nest in the Sun.

> In all respects and in all its attributes, the motive power from the Sun coincides with light.

It was not quite in all respects. When Kepler finally happened upon the ellipse, he developed a new physical theory of "reciprocation," which caused planets to deviate from a circular path. This reciprocation drew upon a wholly different and barely understood force: magnetism. Each planet was as the opposing and attracting poles of a magnet, pushing or pulling themselves into an ellipse with respect to where they sat on the sides of their orbit. The Sun was a glorious monopole, always attracting. Kepler had learned much about magnetism from a recent book *On the Magnet*, by the Englishman William Gilbert, who created the term "electricity."

> What if all the bodies of the planets are ginormous round magnets? Of the Earth there is no doubt. William Gilbert has proved it. . . .
>
> The reciprocational approach is performed without the action of the mind, by a magnetic force which, though it inheres in the planet and is independent, nevertheless depends for its definition upon the extrinsic body of the Sun. For the force is defined as seeking the Sun or fleeing it.

Electromagnetism was a difficult puzzle. Gilbert had not resolved it, but only begun a long and complex tradition in the English sciences, which were slowly gaining prestige.

Kepler, too, remained unresolved.

> I am satisfied if this magnetic example demonstrates the general possibility of the proposed mechanism. Concerning its details, however, I have doubts.

In the end, the species was, of course, neither light nor electromagnetism. It was simply the species, a wholly distinct physical force, created by wicked speculation, post hoc theorizing, religious surety, and one very lively mind. Its kinks could be resolved in another work, by the next great physicist of the species, who seemed sure to come.

Within those rare reader-scientists who know what is to come, yet celebrate their imaginative and experimental spirits in equal measure, there may rightly lance a pang of regret. Kepler's solution convinced no one. The species is deserted territory, and he was its only practitioner. Yet would history not be wealthier, more spectacular, more phenomenal, had there been more of the species, not one but two, three, four different types of new physic? Alas, history is history. The world is the world, and where it is not can only be read in books.

The end of *New Astronomy* turned toward tighter refinements, as Kepler considered the specifics of planetary latitude. "As long as we are wanting," he wrote, "circumstance compels us to leave this discussion . . . along with many others to posterity, if, indeed, it should please God to indulge humanity enough time on this world to work through such questions thoroughly."

Reader's Bibliography

This extremely abbreviated bibliography is designed for readers who enjoy the material but are primarily seeking a pleasurable reading experience. It focuses on books in English that are currently in print.

Firstly, there is no better mode of access than the source, so I strongly encourage interested readers to try out a translation. These works are admittedly quite difficult, but very worthwhile for those willing to put in the time. Edward Rosen has translated the entire body of work by Copernicus; his most accessible is *Three Copernican Treatises*, recently reprinted by Dover. Tycho Brahe has no English translation in print. Johannes Kepler's work is notoriously difficult to parse, and many of his translations have fallen out of print. The superlative independent Green Lion Press has released some choice *Selections from Kepler's Astronomia Nova*, which I highly recommend. If the reader feels up to it, a second edition of Donahue's complete translation is also available from the same publisher. I think the translation is very fine and retains many of Kepler's strong literary qualities. If read alongside Bruce Stephenson's excellent *Kepler's Physical Astronomy*, the book will be mostly comprehensible to anyone with a basic knowledge of geometry. Galileo Galilei's work is of much less mathematical interest than the others, but he has the advantage of being supremely readable. I would especially recommend Albert Van Helden's translation of *Sidereus Nuncius*, as well as Maurice Finocchiaro's *The Galileo Affair: A Documentary History* and abridged translation of Galileo's *Dialogue*, named *Galileo on the World Systems*. Additionally, if the reader is willing to indulge the niche topic of Christopher Scheiner, I thought Van Helden's collaboration with Eileen Reeves *On Sunspots* was both a tremendous scholarly achievement and an outright gorgeous book.

As to secondhand accounts, there were several books, which I believe are not only good history but have serious literary merit. John Christianson's *On Tycho's Island* is a really wonderful book, with a vivid sense of time and place. The entire second half is a biographical dictionary of people who worked under Tycho, but even those are enjoyable to read. *Galileo's Telescope: A European Story* is a collaboration between several authors (Massimo Bucciantini, Michele Camarota, Franco Giudice, Catherine Bolton) concerning the general impact in Europe of the telescope trade; the book lacks the flow of a single author, but works well if you take each section as a short story. Max Caspar's *Kepler* and Stillman Drake's *Galileo at Work* are two extremely dense but nontechnical biographies that anyone can understand and enjoy, if they don't mind wading through a bit of a bog.

Lastly, a couple of online resources should be noted. The untranslated collected works of Galileo, Kepler, and Tycho Brahe can all be found on the imperious http://archive.org, as well as a great many biographies, which have lapsed copyright. If the reader doesn't mind reading old things on computers, I would recommend Mary Allan-Olney's *Private Life of Galileo*, M. W. Burke-Gaffney's *Kepler and the Jesuits*, and James Brodrick's *Robert Bellarmine: Saint and Scholar*. Poland has exported its national pride online at http://copernicus.torun.pl/en/, which has a wealth of biographical information of Copernicus and some translated minor works. The Royal Library of Denmark has some of Brahe's description of his instruments available under translation by Hans Raeder at http://www.kb.dk/en/nb/tema/webudstillinger/brahe_mechanica/. Finally, the Galileo Project, at http://galileo.rice.edu/, can speak for itself.

A NOTE ON NOTES

This bibliography contains nearly every book I have repeatedly cited; the myriad other sources, mostly articles, are fully cited in place in the endnotes.

Bibliography

PRIMARY SOURCES

Original

Brahe, Tycho. *Tychonis Brahe Dani Opera Omnia*. Comp. J. L. E. Dreyer. Hauniæ: In Libraria Gyldendaliana, 1913. Cited as *TBOO*.

Galilei, Galileo. *Le Opere Di Galileo Galilei*. Comp. Antonio Favaro. Firenze: Barbera, 1890. Cited as *Opere*.

Kepler, Johannes. *Gesammelte Werke*. Comp. Max Caspar and Walther Von Dyck. München: Beck, 1940. Cited as *KGW*.

Kepler, Johannes. *Joannis Kepleri Astronomi Opera Omnia*. Comp. Christian Frisch. Heyder & Zimmer, 1858. Cited as *OO*.

Translated

Brahe, Tycho. *Tycho Brahe's Description of His Instruments and Scientific Work, as given in Astronomiae Instauratae Mechanica (Wandesburgi 1598)*. Trans. Hans Raeder. København: I Kommission Hos E. Munksgaard, 1946. Cited as *Mechanica*.

Copernicus, Nicolaus. *On the Revolutions*. Trans. Edward Rosen. Edited by Jerzy Dobrzycki. Polish Scientific Publishers, 1978. Cited as *De Rev*.

Copernicus, Nicolaus. *Three Copernican Treatises*. Trans. Edward Rosen. New York: Octagon, 1979. Cited as *3CT*.

Copernicus, Nicolaus, Edward Rosen, and Erna Hilfstein. *Minor Works*. Baltimore: Johns Hopkins University Press, 1992. Cited as *Minor*.

Galilei, Galileo. *Operations of the Geometric and Military Compass 1606*. Trans. Stillman Drake. Washington: Smithsonian Institution, 1978. Cited as *Compass*.

Galilei, Galileo. *Dialogue Concerning the Two Chief World Systems, Ptolemaic & Copernican*. Trans. Stillman Drake. Berkeley: University of California, 1967. Cited as *Dialogue*.

Galilei, Galileo, Horatio Grassi, Mario Guiducci, and Johannes Kepler. *The Controversy on the Comets of 1618*. Trans. Stillman Drake and Charles Donald O'Malley. University of Pennsylvania Press, 1961. Cited as *Comets*.

Galilei, Galileo. *Sidereus Nuncius, or, the Sidereal Messenger*. Trans. Albert Van Helden. University of Chicago Press, 2015. Cited as *Sidereus*.

Galilei, Galileo. *Dialogues Concerning Two New Sciences*. Trans. by Henry Crew and Alfonso de Salvio. Dover Publications, 1954. Cited as *Two*.

Galilei, Galileo. *The Galileo Affair: A Documentary History*. Ed. Finocchiaro, Maurice. University of California Press, 1989. Cited as *Affair*.

Galilei, Galileo, and Christoph Scheiner. *On Sunspots*. Trans. Eileen Reeves and Albert Van Helden. University of Chicago Press, 2010. Cited as *Sunspots*.

Galilei, Maria Celeste. *To Father: The Letters of Sister Maria Celeste to Galileo, 1623-1633*. Trans. Dava Sobel. London: Fourth Estate, 2008.

Kepler, Johannes. *The Secret of the Universe (Mysterium Cosmographicum)*. Trans. E. J. Aiton. New York: Abaris, 1981. Cited as *Secret*.

Kepler, Johannes. *Astronomia Nova: New Revised Edition*. Trans. William H. Donahue. Sante Fe: Green Lion, 2015. Cited as *Nova*.

Kepler, Johannes. *Optics: Paralipomena to Witelo & Optical Part of Astronomy*. Trans. William H. Donahue. Green Lion Press, 2000. Cited as *Optics*.

Kepler, Johannes. *Kepler's Somnium; the Dream, or Posthumous Work on Lunar Astronomy*. Trans. Edward Rosen. Madison: University of Wisconsin, 1967. Cited as *Somnium*.

Kepler, Johannes, Aiton E. J., Duncan A. M., and Field J. V. *The Harmony of the World*. Philadelphia: American Philosophical Society, 1997. Cited as *Harmony*.

Kepler, Johannes. *Kepler's Conversation with Galileo's Sidereal Messenger*. Trans. Edward Rosen. 1965. Cited as *Conversation*.

Kepler, Johannes. *The Birth of History and Philosophy of Science: Kepler's A Defense of Tycho against Ursus, with Essays on Its Provenance and Significance*. Trans. Nicholas Jardine. Cambridge University Press, 1988. Cited as *Defense*.

Ptolemy. *Ptolemy's Almagest*. Trans. G. J. Toomer. London: Gerald Duckworth, 1984.

SECONDARY SOURCES

Allan-Olney, Mary. *The Private Life of Galileo: Compiled Primarily from His Correspondence and That of His Eldest Daughter, Sister Maria Celeste, Nun in the Franciscan Convent of St. Matthew, in Arcetri*. Boston: Nichols and Noyes, 1870. Cited as *Private*.

Baumgardt, Carola. *Johannes Kepler: Life and Letters*. New York: Philosophical Library, 1951.

Biagioli, Mario. *Galileo, Courtier: The Practice of Science in the Culture of Absolutism*. University of Chicago, 1993. Cited as *Courtier*.

Biagioli, Mario. *Galileo's Instruments of Credit: Telescopes, Images, Secrecy*. University of Chicago, 2007. Cited as *Credit*.

Bethune, John Elliot Drinkwater. *The Life of Galileo Galilei, with Illustrations of the Advancement of Experimental Philosophy; Life of Kepler*. London: n.p., 1830. Cited as *Life*.

Bindman, Rachel Elisa. *The Accademia dei Lincei: pedagogy and the natural sciences in counter-reformation Rome*. Diss. University of California, Los Angeles, 2000. Cited as *Pedagogy*.

Birkenmayer, Ludwik Antoni. *Nicolas Copernicus: Part One Studies on The Works of Copernicus and Biographical Materials*. Trans. Zofia Potkowska, Zofia Piekarec, Jerzy Dobryzycki, and Michał Rozbicki. Comp. Owen Gingerich. Ann Arbor, Michigan: University Microfilms International, 1981.

Brodrick, James. *The Life and Work of Blessed Robert Francis, Cardinal Bellarmine, S. J.* Burns Oates and Washbourne. 1928. Two volumes. Cited as *Blessed*.

Brodrick, James. *Robert Bellarmine: Saint and Scholar.* Westminster: Newman Press, 1961. Cited as *Saint.*

Bucciantini, Massimo. *Galileo e Keplero: Filosofia, Cosmologia e Teologia nell'Età della Controriforma*. Torino: G. Einaudi, 2007. Cited as *Keplero.*

Bucciantini, Massimo, Michele Camerota, Catherine Bolton, and Franco Giudice. *Galileo's Telescope: A European Story*. Cambridge, MA: Harvard University Press, 2015. Cited as *European.*

Burke-Gaffney, M. W. *Kepler and the Jesuits*. Milwaukee: Bruce, 1944.

Caracciolo, Allì, editor. *Filippo Salviati Filosofo Libero: Atti Del Convegno Nel IV Centenario Della Morte, 18-20 Novembre 2014, Universitaìe Degli Studi, Macerata–Scuola Normale Superiore, Pisa*. Eum, 2016. Cited as *Salviati.*

Caspar, Max. *Kepler.* Trans. C. Doris Hellman. London: Abelard-Schuman, 1959.

Christianson, J. R. *On Tycho's Island: Tycho Brahe and His Assistants, 1570–1601*. Cambridge University Press, 2000. Cited as *Island.*

Danielson, Dennis Richard. *The First Copernican: Georg Joachim Rheticus and the Rise of the Copernican Revolution*. New York: Walker, 2006.

Dobrzycki, Jerzy, editor. *The Reception of Copernicus Heliocentric Theory*. Springer Verlag, 2010.

Drake, Stillman. *Galileo at Work: His Scientific Biography*. Chicago: University of Chicago, 1978. Cited as *Work.*

Drake, Stillman. *Galileo Studies: Personality, Tradition, and Revolution*. Ann Arbor: University of Michigan Press, 1970. Cited as *Studies.*

Drake, Stillman. *Galileo: Pioneer Scientist*. University of Toronto Press, 1994. Cited as *Pioneer.*

Dreyer, J. L. E. *History of Astronomy from Thales to Kepler.* New York: Dover Publications, 1953.

Dreyer, J. L. E. *Tycho Brahe; a Picture of Scientific Life and Work in the Sixteenth Century.* New York: Dover Publications, 1963.

Evans, Richard J. *Rudolf II and His World: A Study in Intellectual History, 1572–1612*. Clarendon Press, 1973. Cited as *Rudolf.*

Favaro, Antonio. *Amici E Corrispondenti Di Galileo Galilei*. Venezia: C. Ferrari, 1906. Cited as *Amici.*

Favaro, Antonio. *Galileo Galilei E Lo Studio Di Padova*. Firenze: Successori Le Monnier, 1883. Cited as *Studio.*

Favaro, Antonio. *Galileo Galilei e Suor Maria Celeste*. G. Barbéra, Editore, 1891. Cited as *Maria.*

Ferguson, Kitty. *Tycho & Kepler: The Unlikely Partnership That Forever Changed Our Understanding of the Heavens*. New York: Walker, 2002.

Field, J. V. *Kepler's Geometrical Cosmology*. London: Bloomsbury Academic, 2013. Cited as *Cosmology.*

Finocchiaro, Maurice A. *Retrying Galileo: 1633–1992*. University of California Press, 2005.

Freely, John. *Celestial Revolutionary: Copernicus, the Man and His Universe*. London; New York: Tauris, 2014.

Friis, F. R. *Sofie Brahe Ottesdatter: En biografisk Skildring*. København: G. E. C. Gad's Universitetsboghandel. 1905.

Gabrieli, Giuseppe. *Contributi Alla Storia Della Accademia Dei Lincei*. Accademia Nazionale Dei Lincei, 1989. Cited as *Contributi.*

Gade, John A. *The Life and Times of Tycho Brahe*. Princeton: Princeton University for the American-Scandinavian Foundation, 1947.

Geymonat, Ludovico. *Galileo Galilei: A Biography and Inquiry into His Philosophy of Science; Foreword by Giorgio De Santillana*. Trans. Stillman Drake. New York: McGraw-Hill, 1965.

Gingerich, Owen. *The Eye of Heaven: Ptolemy, Copernicus, Kepler.* International Society for Science and Religion, 2007.

Gingerich, Owen. *The Book Nobody Read: Chasing the Revolutions of Nicolaus Copernicus*. New York: Walker, 2004.

Gingerich, Owen, and James MacLachlan. *Nicolaus Copernicus: Making the Earth a Planet*. New York: Oxford University Press, 2005.

Goddu, André. *Copernicus and the Aristotelian Tradition: Education, Reading, and Philosophy in Copernicus's Path to Heliocentrism*. Leiden: Brill, 2010.

Górski, Karol. *Łukasz Watzenrode, życie i działalność polityczna (1447–1512)*. Vol. 10. Zakład Narodowy im. Ossolińskich, 1973. Cited as *Lukasz*.

Grindely, Anton. *Rudolf II Und Seine Zeit*. Prague Carl Bellmann's Berlag: 1865. Cited as *Zeit*.

Heilbron, J. L. *Galileo*. Oxford: Oxford University Press, 2010.

Hellman, C. Doris. *The Comet of 1577: Its Place in the History of Astronomy*. New York: Columbia University Press; London: P. S. King & Staples, Ltd., 1944.

Kesten, Hermann. *Copernicus and His World*. Trans. E.B. Ashton, Norbert Guterman New York: Roy, 1945.

Koestler, Arthur. *The Sleepwalkers: The History of Man's Changing Universe*. New York: Macmillan Company, 1959. Cited as *Sleepwalkers*.

Koyré, Alexandre. *The Astronomical Revolution: Copernicus—Kepler—Borelli*. Trans. R.E.W. Maddison. Ithaca, NY: Cornell University Press, 1973.

Koyré, Alexandre. *Galileo Studies*. Trans. John Mepham. New Jersey: Humanities Press, 1978.

Koyré, Alexandre. *To the Infinite Universe: From the Closed World*. The Johns Hopkins Press, 1957.

Kremer, Richard L., and Jarosław Włodarczyk, eds. *Johannes Kepler: From Tübingen to Żagań*. Warsaw: Institut Historii Nauki PAN, 2009.

Kuhn, Thomas S. *The Copernican Revolution: Planetary Astronomy in the Development of Western Thought*. Cambridge, MA: Harvard University Press, 1957.

McMullin, Ernan, comp. *Galileo: Man of Science*. New York: Basic, 1967.

Mosley, Adam. *Bearing the Heavens: Tycho Brahe and the Astronomical Community of the Late Sixteenth Century*. Cambridge: Cambridge University Press, 2007. Cited as *Bearing*.

Needham, Joseph, and Ling Wang. Science and Civilisation in China. Cambridge: Cambridge University Press, 1956

Pedersen, Olaf, and Alexander Jones. *A Survey of the Almagest*. Springer, 2011. Cited as *Survey*.

Neugebauer, O. *A History of Ancient Mathematical Astronomy*. v. 1 & 2. Heidelberg, Berlin: Springer-Verlag, 1975. Cited as *Ancient*.

Prowe, Leopold. *Nicolaus Coppernicus*. v. 1, 2. Berlin, Germany: Weidmannsche Buchhandlung, 1883.

Ricci-Riccardi, Antonio. *Galileo Galilei e Fra Tommaso Caccini. Il Processo Del Galilei Nel 1616 e L'abiura Segreta Revelata Dalle Carte Caccini*. 1902. Cited as *Caccini*.

Rosen, Edward. *Copernicus and the Scientific Revolution*. Malabar, FL: Krieger, 1984.

Rosen, Edward. *Copernicus and His Successors*. London: Hambledon, 1995.

Rosen, Edward. *Three Imperial Mathematicians: Kepler Trapped between Tycho Brahe and Ursus*. New York: Abaris, 1986.

Rothman, Aviva Tova. "Far from Every Strife: Kepler's Search for Harmony in an Age of Discord." (2012): *Princeton Dataspace*. Web. 16 Feb. 2017. http://dataspace.princeton.edu/jspui/bitstream/88435/dsp018623hx767/1/Rothman_princeton_0181D_10092.pdf. Cited as *Strife*.

Rowland, Ingrid D. *Giordano Bruno: Philosopher/Heretic*. New York: Farrar, Straus and Giroux, 2008.

Rublack, Ulinka. *The Astronomer and the Witch: Johannes Kepler's Fight for His Mother*. Oxford University Press, 2015. Cited as *Witch*.

Santillana, Giorgio de, *The Crime of Galileo*. University of Chicago, 1955. Cited as *Crime*.

Shea, William R., and Mariano Artigas. *Galileo in Rome: The Rise and Fall of a Troublesome Genius*. New York: Oxford University Press, 2005. Cited as *Rome*.

Stephenson, Bruce. *Kepler's Physical Astronomy*. New York: Springer-Verlag, 1987. Cited as *Physical*.

Stephenson, Bruce. *The Music of the Heavens: Kepler's Harmonic Astronomy*. Princeton University Press, 2014. Cited as *Heavens*.

Swerdlow, N. M., and O. Neugebauer. *Mathematical Astronomy in Copernicus's De Revolutionibus*. New York: Springer-Verlag, 1984. Cited as *Mathematical*.

Thoren, Victor E., and J. R. Christianson. *The Lord of Uraniborg: A Biography of Tycho Brahe*. Cambridge: Cambridge University Press, 1990. Cited as *Lord*.

Thorndike, Lynn. *A History of Magic and Experimental Science*. Columbia University Press, 1923. Cited as *Magic*.

Voelkel, James R. *The Composition of Kepler's Astronomia Nova*. Princeton: Princeton University Press, 2001. Cited as *Composition*.

Voelkel, James R. *Johannes Kepler and the* New Astronomy. Oxford University Press, 2001.

Westman, Robert S. *The Copernican Achievement*. University of California Press, 1975.

Westman, Robert S. *The Copernican Question: Prognostication, Skepticism, and Celestial Order*. Berkeley: University of California, 2011.

Westfall, Richard S. *Essays on the Trial of Galileo*. Vatican City State: Vatican Observatory, 1989.

Wilding, Nick. *Galileo's Idol: Gianfrancesco Sagredo and the Politics of Knowledge*. University of Chicago, 2014. Cited as *Idol*.

Yates, Frances A. *Giordano Bruno and the Hermetic Tradition*. University of Chicago, 1964. Cited as *Hermetic*.

Zeeberg, Peter. *Tycho Brahes "Urania Titani": Et Digt Om Sophie Brahe*. København: Museum Tusculanums Forlag, 1994.

Zinner, Ernst. *Regiomontanus: His Life and Work*. Trans. Ezra Brown. North Holland, 1990.

Notes

p. 3 **first hint of modernity rising**: These value judgments will irk some medievalists. I am not unsympathetic to their position. Readers who want a more detailed view of the medieval period should refer to the following volumes: *The Autumn of the Medieval Ages* by Johan Huizinga (1924), *A Distant Mirror: The Calamitous 14th Century* by Barbara Tuchman (1978), *The Medieval Machine* by Jean Gimpel (1976), *Medieval Technology and Social Change* by Lynn White (1962), *God's Philosophers* by James Hannam (2009), *Life in the Medieval University* by Robert Rait (1912), and *Le Système du Monde: Histoire des Doctrines Cosmologiques de Platon à Copernic* by Pierre Duhem (1913–59).

p. 4 **"she-goat"**: This and the details of the following paragraph are in Copernicus, *Minor*, pp. 228–34. The farmer who ran away with his wife was Jacob Wayner. The social situation of peasants in Poland, relative to the rest of Europe, except Germany, was especially poor; see chapter 7, "Szlachta," in volume 1 of *God's Playground* by Norman Davies (1981).

p. 4 **Pythagoras**: It seems common practice among historians, and especially scientists, to fetishize Pythagoras far past the point of reason (see, for instance, Bertrand Russell's *History of Western Philosophy*, or Roger Penrose's *Road to Reality*, or the private writings of any of the central characters of this book). There is really no reason to do so. I poke fun instead—though I am nothing, if not a fan of number mysticism and vegans. It is important, I think, to point out what a weirdo he must have been.

p. 6 **University of Krakow**: See Knoll, Paul. *"A Pearl of Powerful Learning": The University of Cracow in the Fifteenth Century*. Brill, 2016, especially p. 43.

p. 6 **"greater pleasure than all else"**: Copernicus, *Minor*, p. 29.

p. 6 **the professor at Krakow**: This was Albert of Brudzewo.

p. 8 **"almost perfection"**: Copernicus, *De Rev*, p. 8. For the mathematically inclined, the basic Ptolemaic system is equivalent to Fourier analysis on the complex plane (truncated, for Ptolemy, to the first two or three terms). The illustration represents the first few terms of the standard Fourier transformation of a square wave for the real part.

p. 8 **"unbelievable pleasures for the mind"**: Copernicus, *De Rev*, p. 7.

p. 9 **classic of Greek geometry, Euclid's *Elements***: Swerdlow, *Mathematical*, p. 93. Also Rosen, *Copernicus and His Successors,* pp. 33–4. Dated as 1492 and 1490, respectively.

p. 9 **baby Jesus impressed upon its cover**: Birkenmajer, p. 53.

p. 9 **"leave incomplete, you must finish"**: This is from the end of the preface to Regiomontanus' *In Ptolemaei Magnam Compositionem, quam Almagestum vocant.* Translated with some sentimental flourishes.

p. 10 **"assistant and witness of observations"**: This was Rheticus. Copernicus, *3CT,* p. 111.

p. 10 **staring at the Moon**: Copernicus, *De Rev*, p. 218, and John Freely, *Revolutionary*, p. 59.

p. 10 **observations taken in Novara's company**: Copernicus, *De Rev*, p. 129. From his personal autograph.

p. 11 **Nicolaus must have learned it**: Goddu, *Copernicus and the . . .*, p. 190. Also Prowe, *Nicolaus Coppernicus*, v. 1, 237.

p. 11 **advise our other members**: Copernicus, *3CT,* p. 327.

p. 11 **perceived failures of the Ptolemaic system**: Swerdlow, *Mathematical*, p. 48.

p. 13 **"very foundation of a friendship"**: Gorski, *Lukasz*, pp. 72–3. Abridged, loosely translated.

p. 13 **"a true portrayal of Nicolaus Copernicus, made out of the self–portrait"**: *"Nicolai Copernici vera efigies ex ipsius autographo depicia."*

p. 14 **"the devil in human shape"**: Koestler, *Sleepwalkers,* p. 127.

p. 14 **"no one has ever seen him laugh"**: Gorski, *Lukasz*, p. 119.

p. 14 **"make pills in the shape of a pea"**: Copernicus, *Minor*, p. 302. Abridged. The recipe was not for Lucas but for the sister of a canon.

p. 14 **peace negotiations with the Teutonic Knights**: Copernicus, *3CT,* p. 332.

p. 14 **first useful independent observation of the skies**: Meaning the earliest quoted in *De Revolutionibus*. Freely, *Revolutionary*, p. 68.

p. 15 **favorite old bookseller back in Krakow**: This was Johann Haller; see Copernicus, *Minor*, p. 22.

p. 15 **"my inspiration stands or falls"**: Copernicus, *Minor*, p. 29. Slight changes.

p. 15 **a small contribution to the new culture of European scholarship**: From Marie Boas Hall, *The Scientific Renaissance 1450–1630,* pp. 24–5: "No one could be considered finished his apprenticeship to humanism unless, as his masterpiece, he produced a creditable Latin translation of a Greek original." Also Goddu, *Copernicus*, p. 195.

p. 15 **intoxicated by the sun's rays**: Copernicus, *Minor*, p. 30.

p. 15 **he declined to follow his uncle's footsteps**: There is much conflict in the literature about this. Rosen (Copernicus, *3CT,* p. 340) holds that Copernicus left Lidzbark in 1510, but multiple older sources refute this (Koyré, *Revolution*, p. 22, Koestler, p. 143, Birkenmajer, p. 129), citing 1512 instead (directly following Lucas's death). Birkenmajer goes so far as to say "All biographers of Copernicus are unanimous in maintaining that his move from Lidzbark to Frauenburg took place in the spring of 1512 soon after the death of bishop Lukas . . ." I contacted Professor Owen Gingerich to resolve this issue, to which they responded that while biographical details are very sparse, 1510 is likely the correct date, sourcing the Regesta Copernicana. Modern sources prefer 1510, though I have no idea what revelation caused the date change. Swerdlow, *Mathematical*, p. 6, is the closest I could find to a discussion of the dating issue. Here, Swerdlow notes (with Rosen) that leaving in 1510 indicates a falling out with Uncle Lucas. This sounds dubious to me, because Copernicus has only ever referred to his uncle with fondness.

p. 15 **Mars leaving occultation**: Copernicus, *De Rev*, p. 268 (book V, chapter 20). And also entering occultation, on January 1, 1512, at 6 A.M.

p. 15 **"blessed memory"**: Copernicus, *Minor*, p. 334.

p. 15 **"We are free."**: Gorski, *Lukasz*, p. 117. This was Erazm Ciolek, Bishop of Płock. The comment is a quotation from Psalm 124.

p. 16 **but he never exercised it**: Copernicus, *3CT*, p. 334.

p. 16 **quadrant . . . triquetrum**: Copernicus, *De Rev*, book 2, chapter 2; book 4, chapter 15.

p. 16 **"respective to its proper center"**: Swerdlow, Noel M. "The derivation and first draft of Copernicus's planetary theory: A translation of the Commentariolus with commentary." Proceedings of the American Philosophical Society 117.6 (1973): 423–51. Cited hereafter as *Derivation*, p. 434; Copernicus, *Minor*, p. 81; Copernicus, *3CT*, p. 57.

p. 16 **aesthetic principles.**: Hardly have I ever agreed with a history of science paper more than Gingerich, Owen. "'Crisis' versus Aesthetic in the Copernican Revolution." *Vistas in Astronomy* 17 (1975): 85–95. It is very much worth reading. This is, I take it, a perfect example of what John Keats meant by the term "negative capability"—"when a man is capable of being in uncertainties, mysteries, doubts, without any irritable reaching after fact."

p. 17 **"the Sun were the center of the universe"**: Swerdlow, *Derivation*, p. 436; Copernicus, *Minor*, p. 81; Copernicus, *3CT*, p. 58. Abridged.

p. 17 ***Little Commentary***: *Commentariolus*. It was published sometime before 1514.

p. 18 **"not to them but to the motion of the Earth."**: This is postulate seven. Swerdlow, *Derivation*, p. 436; Copernicus, *Minor*, p. 81; Copernicus, *3CT*, p. 58.

p. 18 **"ballet of the planets"**: Copernicus, *3CT*, p. 90; Copernicus, *Minor*, p. 90; Swerdlow, *Derivation*, p. 510. Swerdlow uses the phrase "choric dance."

p. 18 **"larger volume."**: Copernicus, *3CT*, p. 59, also Koyré, *Astronomical Revolution*, p. 27.

p. 18 **"common people are Satan."**: Preserved Smith, *Life and Letters of Martin Luther*, xiii. From Luther's *Table Talk*.

p. 18 **"not slaying them."**: From the end of the infamous *The Jews and Their Lies*.

p. 18 **"the fool will overturn the whole of astronomy."**: From Luther's *Table Talk*, again. This quote is cited in nearly every book about Copernicus; hopefully I have given it new context.

p. 19 **"no one is unaware of them . . ."**: Copernicus, *Minor*, p. 176.

p. 19 **"maladminstered state by your ruin, alas!"**: Copernicus, *Minor*, p. 189.

p. 20 **"two months' back salary . . ."**: Scott, Tom, and Robert W. Scribner (eds.). *The German Peasants' War: A History in Document*, pp. 60, 291–301 especially. Humanities Press International, 1991. The town with the housewives was Frankenhausen.

p. 20 **pig entrails and feces**: This is the famous death of Huldrych Zwingli.

p. 20 **please stop killing**: Breaking chronology to allude to St. Batholomew's Day Massacre.

p. 20 **"The chapter has begun to resist"**: Freely, *Revolutionary*, p. 85. Abridged. The town was Elblag. The writer is Tiedemann Giese.

p. 21 **Jupiter and Saturn in opposition.**: Jupiter at 11 A.M. April 30, 1520, and Saturn at 12 P.M. on July 13, 1520.

p. 21 **"They are even ready to die . . ."**: Gingerich, *Making the Earth a Planet*, p. 90; Freely, *Revolutionary*, p. 86.

p. 21 **"second Aesclepius," god of medicine.**: *Prowe*, v. 1, p. 294.

p. 21 **"Love endureth all things"**: Kesten, p. 229.

p. 21 **"justice between brothers"**: This work is the *Antilogikon*, which is reprinted in Franz Hipler, *Spicilegium Copernicanum* (Braunsberg, 1873). Quote from p. 19.

p. 22 **"The whole world is dragged into the fight!"**: Koestler, *Sleepwalkers*, p. 142, quoting Prowe.

p. 23 **"Satan tempts him still"**: Danielson, *The First Copernican*, p. 127. Said by Ambrosius Blarer.

p. 23 **"older men are better."**: Copernicus, *3CT*, p. 196.
p. 23 **"above all an astrologer."**: This was Phillip Melancthon. Freely, *Revolutionary*, p. 117.
p. 23 **"the most remote corner of the earth"**: Copernicus, *3CT*, p. 393
p. 23 **well–received for lecturing on the Copernican worldview.**: Copernicus, *De Rev*, p. 336.
p. 24 **"I want to avoid offending good people."**: Written to Bishop Dantiscus, concerning Anna Schilling. Freely, *Revolutionary*, p. 110.
p. 24 **"to the attention of others"**: This is the letter to Werner. Copernicus, *3CT*, p. 93.
p. 24 **Copernicus liked retirement**: Kesten, p. 308.
p. 24 **"pass judgment on his work."**: Copernicus, *3CT*, pp. 192–3.
p. 24 **preached a unity between Roman and Lutheran creeds.**: This work is the *Antilogikon*, which can be found reprinted in Franz Hipler, *Spicilegium Copernicanum* (Braunsberg, 1873). Quotes are from pp. 25 and 6.
p. 24 **"buried among my papers"**: Rosen, *Copernicus and His Successors*, p. 107.
p. 25 **"begun to conceive the hypotheses of the rest"**: Copernicus, *3CT*, p. 115.
p. 25 **"Regiomontanus . . . highest admiration"**: Copernicus, *3CT*, pp. 109, 121.
p. 25 **"uniform motions of the terrestrial globe."**: *KGW*, v. 1, p. 99; Copernicus, *3CT*, p. 135.
p. 26 **"the Ancients and those who came before him."**: Koyré, *Astronomical Revolution*, p. 30.
p. 26 **"Aristotle would support my teacher"**: Copernicus, *3CT*, p. 142.
p. 26 **and finally, the second coming of Christ.**: Copernicus, *3CT*, pp. 121–2.
p. 26 **"this timeless and perfect number?"**: Copernicus, *3CT*, p. 147; *KGW*, v. 1, p. 105.
p. 27 **"intellectual talents and way of life"**: Freely, *Revolutionary*, quoting Rosen, *Copernicus and the Scientific Revolution*, p. 192. Slight changes.
p. 27 **"pleased me very much"**: Freely, *Revolutionary*. p. 132.
p. 27 **"say something on the matter in a preface."**: Kepler, *Defense*, p. 152.
p. 28 **printshop workers were ironically illiterate**: Gaskell, Philip. *A New Introduction to Bibliography*, pp. 7, 155–6. Oxford: Clarendon Press, 1972.
p. 28 **"let no one expect anything certain from astronomy."**: *De Rev*, p. XVI.
p. 28 **Church conservatives for generations to come.**: It would not be made public until sixty-eight years later, by Johannes Kepler in the preface to *Nova*, p. 4.
p. 28 **"thinness of style reveal them not to be those of Copernicus."**: Written by Michael Maestlin. Westman, *Copernican Question*, pp. 265, 562.
p. 28 **"a little ass wrote this for a big ass!"**: Santillana, *Crime*, p. 101. Quote is by Johannes Kepler. Santillana does not source their quotes, and I could not find the quote in the primary literature. Giordano Bruno also referred to Osiander as an ass.
p. 28 **an eye toward helping his friend's worldview gain support**: Bruce Wrightsman, "Andreas Osiander's Contribution to the Copernican Achievement," *The Copernican Achievement*, pp. 213–42, from Westman.
p. 28 **"called it a 'crime of deceit'"**: Swerdlow, *Mathematical*, p. 29.
p. 28 **"big red *X* in crayon upon the page, like an angry toddler."**: Gingerich, *The Book No One Read*, p. 180.
p. 29 **they now excise him from it.**: Osiander did reenter Rheticus's life *by name*, see Dennis Danielson, *The First Copernican*, p. 119.
p. 29 **"the sun governs the family of planets revolving around it."**: *De Rev*, p. 22 (book 1, chapter 11).
p. 29 **"one revolution in such a short time."**: *Almagest*, p. 44.
p. 30 **"Earth in the same way."**: Copernicus, *De Rev*, p. 11, 16. Abridged.
p. 30 **leaving behind a labyrinth of annotations**: Gingerich, *The Book Nobody Read*, p. 63, 255; Kuhn, *The Copernican Revolution*, p. 134.

p. 30 **"such accurate calculations can be produced."**: Gingerich, *The Book Nobody Read*, p. 87.

p. 30 **so too moves its axis**: Noel Swerdlow has an excellent investigation of Copernicus's theory of axial precession in Westman's *The Copernican Achievement*, pp. 49–89, pp. 70 and 74 are particularly relevant.

p. 31 **"Regiomontanus and Georg Peurbach"**: See Copernicus, *De Rev*, p. 129; Dreyer, *Tycho Brahe*, pp. 354–5. Also Gingerich, *The Eye of Heaven*, p. 26. Birkenmajer has pointed out that this was likely learned from Domenico Da Maria Novara.

p. 31 **Al-Battani, too quickly.**: *De Rev*, p. 383. For details, see Gingerich, *The Eye of Heaven*, p. 22.

p. 32 **same narrative from another's eyes.**: I remember the first time I learned the word "parallax." I was twelve, reading the preface to *Ender's Shadow* by Orson Scott Card: "the parallax is created by Ender and Bean, standing a little ways apart as they move through the same events . . . if I can move that scientific term into literature."

p. 32 **his most confused calculation**: Swerdlow, *Mathematical*, p. 75. For an example of this calculation, see Copernicus, *De Rev*, p. 252 (book V, chapter 9).

p. 32 **"parallax was given as p"**: This is a naive description. Parallax could be identified with respect to latitude, longitude, right ascension, etc. On Copernicus and the tangent function, Swerdlow, *Mathematical*, pp. 234, 240, 257. This definition is of geocentric parallax, see Neugebaeur's *Ancient*, p. 100.

p. 33 **"perplexing the mind with an almost infinite multitude of spheres . . ."**: Copernicus, *De Rev*, p. 20 (book I, chapter 10). Emphasis added. Translation changed for clarity.

p. 33 **he just called it "immeasurable"**: Our conception of immeasurable is, today, rather different than Copernicus's. Cf. Dennis Danielson, *The First Copernican*, p. 62. Also, Koyré, *From the Closed World*, p. 34.

p. 34 **"only on his last breath on his last day."**: Rosen, *Copernicus and the Scientific Revolution*, p. 168.

p. 35 **"more purely Copernican"**: This does not mean more accurate; see Gingerich, Owen, *Early Copernican Ephemerides*, p. 406 Studio Copernicana XVI, Wydawnictwo, Wroclaw, 1978. The same article is in Gingerich's *Eye of Heaven*, p. 205. Kepler also thought ephemerides were the essential tool to disseminate the Copernican worldview; Rothman, *Strife*, p. 156.

p. 35 **"has understood the mind of Copernicus better than he."**: This was Caspar Peucer; Danielson, *First Copernican*, p. 185.

p. 35 **Literature was turning into conversation**: I am following Elizabeth Eisenstein, *The Printing Press as an Agent of Change.* If the printing press had been convenient fifty years earlier, we would all probably write about Regiomontanus instead.

p. 36 **"sodomitical misdeeds."**: Danielson, *The First Copernican*, pp. 143–8.

p. 36 **"none of his works would have seen the light."**: This was said to Valentine Otho. Westman, Robert S. "The Melanchthon Circle, Rheticus, and the Wittenberg Interpretation of the Copernican Theory," *Isis* 66.2 (1975): p. 183. Also Copernicus, *3CT*, p. 357.

p. 36 **"not only my teacher . . . my father."**: Letter to Emperor Ferdinand, preface to Johannes Werner, *De Triangulis Sphaericis* (Krakow, 1557); Danielson, *First Copernican*, p. 222.

p. 39 **a space dragon**: This line owes some inspiration to a similar description in James Joyce's *Ulysses.*

p. 39 **"light streaming out in all directions."**: Clark, David H., and F. Richard Stephenson. *The Historical Supernovae.* Elsevier, p. 175. 2016.; cf. Needham, Joseph, *Science and Civilization in China*, v. 3, p. 428.

p. 39 **changeless and eternal being**: Aristotle, *Metaphysics*, 1073a23.

p. 40 **"I leave that to others."**: *TBOO*, v. 3, pp. 59, 62. I have simplified Maestlin's method; he made two lines using four stars and did a little trigonometry.

p. 40 **"I agree with Copernicus"**: Maestlin's annotations appear to be more extensive than any other copy; see Gingerich, Owen. *An Annotated Census of Copernicus' De revolutionibus (Nuremberg, 1543 and Basel, 1566)*, pp. 219–27. Leiden; Boston, MA: Brill, 2002.

p. 40 **"without assumptions"**: *TBOO*, v. 6, p. 88, in reference to Peter Ramus. Cf. Thoren, *Lord*, p. 35.

p. 40 **"Phoenix of Astronomers."**: Said by Johannes Kepler, *KGW* v. 2, p. 48. Six years later, John Donne wrote in his *Anatomy of the World*, "Every man alone thinks he has to be a Phoenix."

p. 40 **"will be greatly amplified in its new splendor"**: Thorndike, *Magic*, p. 82.

p. 40 ***Prelude to a Restored Astronomy:*** *Astronomiae Instauratæ Progymnasmata.* There is no good English translation for *progymnasmata*, which were rhetorical exercises for young Greek and Roman patricians to learn the art of persuasion. It implies that Tycho's scientific work was part of a rhetorical, civil, and political endeavor, rather than just scientific. I find the standard translation, "preliminary exercises," to be grossly verbose. I have focused on its preliminary character as a "prelude," exclusively because it better fits my themes.

p. 41 **"readying my tool, I attacked."**: *TBOO*, v. 2, p. 308. (*Progymnasmata*, chapter 3). See Clark, David H., and F. Richard Stephenson. *The Historical Supernovae*, p. 174. Elsevier, 2016, for a more literal and less abridged translation.

p. 42 **"just as this fake star did."**: *TBOO*, v. 3, p. 310 (Progymnasmata, Conclusions). J. L. E. Dreyer goes into great detail about this in *Tycho Brahe*, p. 194.

p. 42 **"over the course of spring"**: *TBOO*, v. 3, p. 194. More precisely, "after the vernal equinox."

p. 42 **they said, let us ready for the second coming of Christ.**: Theodore Beza, for example. The Landgrave of Hesse, as a metaphor. Hellman, *Comet of 1577*, pp. 115, 117, 398.

p. 42 **"consume all with fire."**: Doris C. Hellman, *The Comet of 1577*, p. 287. From Thomas Twyne, updated to modern English. The "group" he was in contact with was the Elizabethan court, particularly John Dee.

p. 42 **some hilltop one night to watch**: Caspar, *Kepler*, p. 37; also Koestler, *Sleepwalkers*, p. 232.

p. 42 **Tycho smiled**: Dining with Charles de Dancey and Johannes Pratensis; Dreyer, *Tycho Brahe*, pp. 42–3, Thoren, *Lord*, p. 62. This introduction unabashedly riffs on the one in Gingerich's *Tycho Brahe and the Nova of 1572, O. Gingerich*, ASP Conference Series, v. 342, 2005.

p. 42 **"*On the New Star.*"**: *Tychonis Brahe, Dani De Nova Et Nullius Aevi Memoria Prius Visa Stella, iam pridem Anna a nato Christo 1572, mense Novembri primum Conspecta, Contemplatio Mathematica* ("The Mathematical Contemplations of the Dane Tycho Brahe *On the New Star*, which None Recall Having Ever Seen Before, which was first seen now 1572 years after the birth of Christ in the month of November").

p. 43 **"without the knowledge of my parents"**: Thoren, *Lord*, p. 4; *TBOO*, v. 5, p. 106; Brahe, *Mechanica*, 106.

p. 43 **the gall to try improve themselves.**: Christianson, *Isis* 58 (2): 199, *Tycho Brahe at the University of Copenhagen*.

p. 44 **"unacquainted with art and good order."**: Tycho's guardian was Anders Sorensen Vedel. Quote from S. Vedel, Om den Danske Krønike at beskrive (Copenhagen, 1581, reprinted 1787), p. 18. Translation by Leon Jespersen, "Court and nobility in early modern Denmark." *Scandinavian Journal of History* 27.3 (2002): 131.

p. 44 **"without mentioning it to anybody."**: Brahe, *Mechanica,* p. 107. Abridged.

p. 44 **"childish and of doubtful value"**: Brahe, *Mechanica,* p. 110.

p. 44 **"I would prefer that one really be a master of arts."**: Thoren, *Lord*, p. 137.

pp. 44–5 **the good brother**: F. R. Friis, *Sofie Brahe Ottesdatter. En biografisk Skildring.* pp. 6, 22. The phrase is more of an honorific, although it does not seem shared across the family.

p. 45 **a Danish drinking buddy**: This was Manderup Parsberg, a third cousin.

p. 45 **"selection of the humble life–companion."**: Gade, *The* Life, p. 35.

p. 45 **partaking in "an evil life"**: Dreyer, *Tycho Brahe,* p. 71; Thoren, *Lord*, p. 373; Gade, *The Life*, p. 49.

p. 45 **"They always want more from me"**: *TBOO*, v. 7, p. 26; Christianson, *Island,* p. 8. Freely translated. Abridged.

p. 46 **"to her husband, satisfying fecundity."**: Gade, *The Life*, p. 48.

p. 46 **"offered up his backyard to host it."**: *TBOO*, v. 2, p. 343. The alderman was Paul Hainzel.

p. 47 **"Forty grown housemen"**: Thoren, *Lord*, p. 33, *TBOO*, v. 2, p. 344. An off–the–cuff approximation from the illustration, at 50 pounds per cubit of dried oak, indicates to me that this is a lower bound.

p. 47 **a lone older relative, another uncle**: This was his mother's brother, Steen Bille, living at Herrevad Abbey.

p. 47 **"whole of creation . . . the study of celestial phenomena."**: *TBOO*, v. 1, p. 18; v. 5, p. 108.

p. 47 **"my lovely sister Sophia Brahe, then a maiden of about fourteen"**: Dreyer, *Tycho Brahe*, p. 73, communicates her age as seventeen years old, which Thoren repeats in *Lord of Uraniborg*, p. 75. But this is contradicted by Tycho in *TBOO*, v. 1, p. 131. I think Sophie was born in 1559, and the new star both texts refer to was observed in 1573.

p. 47 **"venture into this sublime science."**: *TBOO*, v. 1, p. 14.

p. 48 **"who delights in the heavens before the earth."**: *TBOO*, v. 1, pp. 69, 70. A Danish translation is provided in *Tycho Brahe Uraniaelegi. Nyoversættelse, tekst og kommentarer*, Renæssanceforum v. 3, 2007, by Peter Zeeberg.

p. 48 **a pen pal Tycho much enjoyed hearing from.**: In *TBOO*, v. 8, p. 52, Tycho writes to Maestlin with glowing praise. Tycho also has a great deal of praise for Maestlin in his *Prelude*, in *TBOO*, v. 4 pp. 207, five stories tall 238.

p. 48 **"as a body agree with Copernicus's demonstrations."**: Rosen, *Copernicus and his Successors*, p. 218. An obvious lie; specialists did not agree, and even Copernicus's mathematical demonstrations contained occasional errors. This type of pedagogy is fascinating and still very much in practice. Newtonian mechanics will likely always will be taught in schools, and many chemistry textbooks still teach a simplified version of the law of conservation of mass.

p. 48 **"the appendix of his textbook explaining Copernican ideas."**: Hellman, *The Comet of 1572,* p. 138.

p. 48 ***Theater of Astronomy***: Though never completed, Brahe detailed his plan outline quite quickly in a letter to Casper Peucer, *TBOO,* v. 7, p. 132.

p. 48 **"but because I am in harmony with the truth."**: Brahe, *Mechanica,* p. 123.

p. 49 **"I saw the little island of Hven . . ."**: Christianson, *Island,* p. 22.

p. 49 **"family which has always been dear to me."**: Christianson, *Island*, p. 22; *TBOO*, v 7, p. 27. Abridged.

p. 49 **"Liquid and light," Galileo would call it**: Heilbron, *Galileo,* p. 83; Drake, *Work*, p. 421.

p. 49 **"tucked into the corner of the east dining hall."**: The winter dining room, see Brahe, *Mechanica,* p. 130.

p. 50 **"all the ale they can hold."**: Gade, *The Life*, p. 63; Christianson, *Island*, p. 296; Thoren, *Lord*, p. 194.

p. 50 **literal ton of gold.**: I could not find the quote for this. It may be a lovely exaggeration, although for perspective, this would "only" be about $20 million in 2017 (with a grain of salt; gold was trading, loosely, at $525 an ounce, as the market had just been flooded from the New World). It is very commonly stated, see Gingerich, Owen. "Tycho and the ton of gold." *Nature* 403.6767 (2000): 251. Thoren, *Lord*, p. 188, projects that 1 percent of the total revenues for the crown went to Tycho every year.

pp. 50–51 **"five stories tall"**: Sophie Brahe married Otte Thott in 1579, and they lived in Eriksholm Castle; Christianson, *Island*, p. 163. Her child was Tage Thott, born May 27, 1580.

p. 51 **in her fat purse**: This line is stolen directly from the Sylvia Plath poem "Metaphors" concerning her feelings on pregnancy. I think it is very apposite for a sixteenth-century Danish noblewoman.

p. 51 **a simple "Soph."**: See, for instance, *TBOO*, v. 9, pp. 47, 74, 76. For ease of use, here is a listing of all the places I found where Sophie's name occurs in Tycho's meteorological journals: *TBOO*, v. 9, pp. 47, 74, 76, 79, 81, 84–6, 88, 89, 91–3, 95, 97, 99, 101, 107, 109, 111–2, 115, 117, 119, 122–4, 128, 133, 136, 139, 141, 146. I suspect I have missed some; Zeeberg, *Urania*, p. 20 reports that she stayed five times in 1589, thirteen times in 1590, seven times in 1591, and seven times in 1592; I counted further one time in 1586, three times in 1593, five times in 1594, two times in 1595, three times in 1596, and one time in 1597. This does not get at the length of her stays, which were sometimes for a day and sometimes for several weeks.

p. 51 **He also invented its modern title: *Little Commentary***: Koyré, *Astronomical Revolution*, p. 76; Rosen, *Copernicus and His Successors*, p. 73.

p. 51 **a friend of Tycho's, who spread its secret wisdom about.**: The friend was Thaddeus Hagecius.

p. 52 **than contact with any of the arts.**: This line is ripped directly from the end of Virginia Woolf's *Orlando*, and again, I think it is more literally fitting here than it was in its original text.

p. 52 **"Thinking the planets do them harm."**: Full poem in Zeeberg, *Urania*, pp. 182–3, lns. 11–2, 19, 30. These lines have been stitched together to remove the specificities of names and characters.

p. 53 **a cheery drunken conversation.**: *TBOO*, v. 7, p. 321. A full quotation will appear in the endnotes after a couple thousand more words.

p. 53 **a penchant for unwise monetary investments.**: He may have already funded a rather silly trading company for expeditions to the East; see Zeeberg, *Urania*, p. 21.

p. 53 **"your nighttime love."**: From Tycho's poem to Erik, *TBOO*, v. 15, pp. 3–5, lns. 20, 32.

p. 53 **He signed it as a "dear friend"**: Rosen, *Three Imperial Mathematicians*, p. 224.

p. 53 **for two whole weeks**: It is not clear when she left. I say she left on the August 23, so that I can spin the tale. But her departure is not mentioned in the books, only her arrival. Given that she seems to have been absent from Hven for the past few years, and the queen's arrival was very important, I think this interpretation also makes the most sense. *TBOO*, v. 9, p. 47.

p. 53 **sat for portraits**: Descriptions of Tycho and Sophie stem from their pictures in Christianson, *Island*, pp. 117, 259, respectively.

p. 54 **"which they knew crammed the house"**: *TBOO*, v. 6. p. 64.

p. 54 **"Copernicus's self-portrait peaceably stared out."**: Westman, Lindberg, *Reappraisals of the Scientific Revolution*, p. 185. Readers further interested in the very heavy Copernican influence on Tycho Brahe may enjoy the succinctly titled article

"Copernican Influence on Tycho Brahe" by Kristian Moesgaard in *The Reception of Copernicus's Heliocentric Theory,* edited by Jerzy Dobrzycki, pp. 31–55.

p. 54 **"immediately, a heroic poem.":** Brahe, *Mechanica,* p. 46.

p. 54 **"What can genius not overcome?":** *TBOO,* v. 6, p. 266–7. Readers will note I only use the term "genius" when it appears in a quote from someone else. I do not like what the word has become. From Latin, I usually translate it as "wit" or "talent," except in cases like this, where it would not make sense.

p. 54 **"illustrious astronomer of our age.":** Dreyer, *Tycho Brahe,* p. 83; Thorndike, *Magic,* v. 6, p. 4; *TBOO,* v. 3 p. 146.

p. 55 **that day's humidity gave for a brief:** Details like this are possible because of the specificity of Tycho's meteorological journal; *TBOO,* v. 9, p. 47.

p. 55 **"different means of softening them.":** *TBOO,* v. 9, pp. 324–6. Translated below.

p. 56 **"that I hide the stars of the heavens . . . ":** *TBOO,* v. 6, p. 273.

p. 57 **inscription below read.:** *TBOO,* v. 6, p. 275. Abridged.

p. 57 **"benefited to no small degree.":** *TBOO,* v. 9, pp. 324–6. Translated below.

p. 58 **"The song of the skylark has returned.":** Actually, "We heard the song of the skylark," but *TBOO,* v. 9, p. 70, has the line. From *TBOO,* v. 9, p. 81.

p. 58 **garden at Sophie's castle for inspiration:** Because I will later describe Galileo's garden, I have not gone into detail on the garden here—such descriptions are so often lists, and historically speaking, very much an extrapolation, so I think once is enough. But interested readers should check out Christianson, John R. "Tycho Brahe in Scandinavian Scholarship." *History of Science* 36.4 (1998): 473, and Ørum–Larsen, Asger. "Uraniborg—the most extraordinary castle and garden design in Scandinavia." *The Journal of Garden History* 10.2 (1990): 97–105. Sophie's garden almost certainly consisted of what Tycho's almost certainly consisted of: the tree standards apple, pear, plum, and cherry; the exotics figs, walnuts, quince, and apricots (introduced to Denmark by Tycho's kinsman); and the medicaments, angelica, blessed thistle, bloodwort, elecampane, gentian, juniper, rhubarb, prunella, saffron crocus, sweet flag, wormwood, and valerian.

p. 58 **"no end of strife and diligence and hard work":** This is from an important letter in *TBOO,* v. 9, pp. 324–6. Translated below.

p. 58 **"not without a little superstition.":** *TBOO,* v. 9, pp. 324–6. Translated below.

p. 58 **"she sent me a rather long letter":** From an important letter in *TBOO,* v. 9, pp. 324–6. Christianson, p. 260 has quoted the best part of the full letter, which is featured (in Danish) in Friis's *Sofie Brahe,* pp. 41–6.

I have translated Tycho's full letter (rather loosely) from Friis's Danish below, for readers like myself further interested in this early case study for women in the science

For context, Tycho was considering publishing some of her letters in a book of his scientific correspondence. (He did not do so.) This was, of course, virtually unheard of for a woman, to the point where he wrote the following forward, warning the conservative reading public.

"Tycho Brahe's Report on two [now unknown] astrological letters

About the reason and cause of the following Danish letters, so that the good reader may understand why they are introduced here.

I have a sister called Sophie; a good six years ago, she became a widow after the passing of a brave nobleman, but she remains quite young and has only one child, a son. She has in these years, as can easily happen to a widow, had many sorrows and worries, so she has sought different means, from time to time, of softening them. Firstly, she has at her farm in Eriksholm, located in Scania and built like a castle, laid out a tremendously beautiful garden such that can hardly exist in these northern climes; it has, to keep a long story short, cost her no end of strife and diligence and hard work, and has been done in a very proper manner, taking into account the planting of trees and sundry herbs. All else is done so rightly that there has hardly been such a place before. When she finally finished, she had not yet been sufficiently liberated from her oppressive sorrow, and therefore also learned to prepare spagyric medicaments, and has been so successful that she has not merely distributed them to her friends and well-off people, as needed, but has also given them to the poor, free of charge, such that both parties benefited to no small degree. Finally, as yet unable to satisfy the urgings of the mind which settled ever more, she zealously threw herself into the making of horoscopes, either due to her sharp wit which continually demands harder and harder tasks, or because those of her gender have a tendency for contemplation of things to come, and not without a little superstition. I gave her help with directions and guidance, as she wanted it, more with pyronomics than with gardening, in which she had enough insight of her own, but I seriously warned her to desist from astrological speculations, because she should not strive after subjects too abstract and complicated for a woman's talents. But she, who has an unbendable will and such great self-confidence that she will never yield to men in intellectual matters, cast herself all the more energetically into her studies and in a short time learned the basic principles of astrology, partly from Latin authors whom she had translated into Danish at her own expense, and partly from German authors on the subject (for she has an excellent grasp of that language). When I saw the clear signs of this, I quit opposing it and simply advised her to moderation in her ongoing studies. And I could do this with some happiness, for I knew that she was feeling some comfort in her widowhood and all what follows; her mind was thus led to ever more excellent things. Nonetheless, I thought that she ought not go too deeply into these investigations, and that she herself would grow weary when she realized how difficult they were. When she guessed my thoughts—clever, she is!—she sent me a rather long letter, in which among other things she clearly described her progress in astrology and drew upon her pyronomic studies to show she could acquire this science. Although I reminded her it contains not but a few difficulties, she seeks to overcome them, and has made enough progress in her preparations (which she mentions with more than enough clarity) to show that it presents no obstacles she cannot surmount to reach the depths of this science. She states she has no regret for the time and money lost in study, but that she actually attains a greater and greater love for it. She asks me to promote this business, such that she may learn more and more secret things. Finally, she presents me with three astrological problems, which are not irrelevant and quite difficult, and asked me to solve them for her. This is the reason I would include her letter and its content, as stated. But since this letter contains several *Secret*s of the astrological science, and since there are some things therein that are considerably more learned than one would expect of a woman, I thought perhaps I should not include it. I know it is with her applause that I would, otherwise she would not have written the letter so finely and carefully. I would include it first in Danish, as written by her own hand and as I have received it, from nearby Scania. I would leave several of her artistic spoonerisms in place, as she is familiar with them and uses them routinely. In addition, I would add a Latin translation, such that those ignorant of our language could understand it; of my Danish answer I would do the same. It contains things that would almost certainly help astrologers. No one might be upset that I so blend Danish and Latin letters, for the German landgrave did something similar. And as it has been done in German, it may be done in Danish. It is quite possible some would doubt such a letter was written by a Danish noblewoman, but they must know it was written by my aforementioned sister, sent from her garden, resting in the Skane, and what more, that she understands the questions it treats. Yes, and she tirelessly cultivates the study of ever more difficult questions, inasmuch as her household duties allow. It really should not surprise anyone, not after Fulvia Olympia Morata [an early female classical scholar from Italy] journeyed to the German mountains, and published writings of a very similar style and sensibility. Perhaps it is more admirable, as my sister takes on a far more difficult science than the Italian, whose language is almost as Cicero's. It is not so difficult for an Italian mind to write and speak with a certain eloquence [because Italian is close to Latin]. In and of themselves, women (if they are of good sense) are in possession of certain natural talents, due to the influence on the Moon and Mercury, such that many (practically all) are quite chatty. But as far as I know, no woman has yet existed who has scientifically and knowledgeably treated of astrology; there are very few even among men, who desire to be called scholars. I do not doubt that if my sister had knowledge of the Latin language, she would pass the muster of this Fulvia Morata and any other woman, because of her insight into heavenly things which she has acquired with such energy. She would deserve to be called Olympia Fulvia and the consequent Morata, for she understands Olympic and heavenly things alike. She is more radiant in gifts than Fulvia, and has a character at least as fine, if not better; she denotes Urania, which we have begun to call her because of her rare gifts, as a short and easy reference—not to mention, she has been named Sophie [goddess of wisdom] from the cradle. I do not mention all this to impugn the name of Fulvia Morata, and I even wish such names rose ever higher—to publicize her and other women of good character, even if they were less learned than my sister, for the sake that it would be easier to believe the following letter is written by her and that she understands its content, and learns each day even harder things. I cannot believe that all this should really arise disgust or discomfort in someone else because, without regard to their judgment, I have borne witness to these truths of my sister, whom I know in an especial way. The reason why, as noted above, she has not learned Latin, and what has prevented her from doing so, is that she is young, and has a great interest in this language and has even begun to achieve something therein. This she indicates, somewhat unclearly, in her letter in question. Now, we would like to publish the letter itself, which will explain this all in a comprehensive way, and afterwards, the promised Latin translation will follow. It is our hope that it will be received frankly by the reader, as honestly written and communicated.

T.B."

The Friis collection has several late-period letters from Sophie as well. Zeeburg, Urania, pp. 254–96, contains a great deal of relevant source material. Sophie is also mentioned, enjoyably enough, in 1610, to Galileo by Martin Hasdale; *Opere* v. 10, Nr. 375, p. 417 ("a very honorable old lady, who writes well of mathematics, and translates Latin books to German for her pleasure.").

p. 59 **"I caught a sudden star."**: *TBOO*, v. 4, p. 5.

p. 59 **a "curly lock of hair on its head"**: *TBOO*, v. 4, p. 6.

p. 59 **"proclamations have leaked through"**: *TBOO*, v. 4, p. 180.

p. 59 ***Of the Aetherial World***: *De Mundi Aetherei Recentioribus Phaenomenis Liber Secundus*, Book Two of the Recent Phenomena of the Aetherial World, printed in 1588. "Book Two" came before book one, because Tycho had a preplanned sequence for his *Progymnasmata* (*Prelude*), a detail I have deliberately left unmentioned to avoid confusions.

p. 59 **dear friend Michael Maestlin was his favorite**: *TBOO*, v. 4, pp. 207–38. At the end, Tycho congratulates Maestlin for avoiding astrology: "This is weighted with little truth in the writings of Maestlin, and with [the opinions of] ours being found together, led to moderation, out of this I was detecting due diligence and industry, which was appearing to me dignified, in which a fuller and more guarded consideration was given. In others, some are not so shrewd and sedulous in content."

p. 60 **"adapted to the immobility of the Earth."**: Thoren, *Lord*, p. 86, *TBOO*, v. 1, pp. 172–3, Ferguson, Tycho, and Kepler, p. 62.

p. 60 **"Most Suitable to the Apparent Motions."**: *TBOO*, v. 4, p. 158. A campy translation.

p. 60 **Tychonic system**: See account from Kuhn, *Copernican Revolution*, p. 202.

p. 61 **"re-centering the camera on the moving Earth"**: If the Copernican and Tychonic systems were both modeled mathematically they would be related by a simple isomorphism.

p. 61 **"my own discovery, was suspect to me."**: Westman, "Three Responses to Copernican Theory," *The Copernican* Achievement, p. 329.

p. 61 **"They are not really in the heavens"**: *TBOO*, v. 4, p. 159.

p. 61 **an orbit to be an oval!**: *TBOO*, v. 4, p. 162; see also Dreyer, *A History of Astronomy*, p. 366.

p. 61 **"contrary to physical principles."**: *TBOO*, v. 1, p. 149. This has been translated in a forgotten paper by Raymond Coon, "Tycho Brahe: Translation of De disciplinis mathematicis." *Popular Astronomy* 37 (1929): 311.

pp. 61–2 **"a consequence of the incorrect observations of the ancients."**: Brahe, *Mechanica*, p. 113.

p. 62 **but of course, found none**: Blair, Ann. "Tycho Brahe's Critique of Copernicus and the Copernican System." p. 364, *TBOO*, v. 8, p. 209. Cited hereafter as *Critique*.

p. 62 **"no evidence for the annual motion of the Earth"**: Gingerich, Owen, and James R. Voelkel. "Tycho Brahe's Copernican Campaign." *Journal for the History of Astronomy* 29 (1998): 9. Cited hereafter as *Campaign*.

p. 62 **"sin against first principles"**: Blair, *Critique*, p. 359.

p. 63 **"long-accepted Ptolemaic hypothesis cannot stand."**: *Gingerich & Voelkel, Campaign*, p. 24.

p. 63 **while contemplating a mistake by Copernicus**: Thoren, *Lord*, p. 223.

p. 64 **"Is this it?"**: Christianson, *Island*, p. 109. Untranslated: *Quid si sic?*

p. 64 **Black velvet was draped over his fat coffin**: Jespersen, Leon. "Court and nobility in early modern Denmark." *Scandinavian Journal of History* 27.3 (2002): 132. I can find no evidence that Tycho attended the funeral, although one assumes he did.

p. 64 **by a council stacked with Tycho's relatives and supporters.**: Thoren, *Lord*, p. 341.

p. 64 **"heavy drinking, extravagance, and immoderation."**: *TBOO*, v. 1, p. 199.

p. 65 **"dumb mouth hung open."**: *TBOO*, v.7, pp. 321–2. A very similar quotation is provided by Tycho and translated in Christianson, *Island*, p. 90 and Rosen, *Three Imperial Mathematicians*, p. 40; I prefer this one for narrative.

p. 66 **"like a raving maniac."**: Rosen, *Three Imperial Mathematicians*, pp. 251–2.

p. 66 **a fool, a visionary, a jackass, the master of evil.**: All used descriptions, from Rosen, *Three Imperial Astronomers*, pp. 247, 251.

p. 66 **had translated Copernicus's *On the Revolutions* into German.**: Granada, Miguel A. "Essay Review: Early Translations of De Revolutionibus De Revolutionibus: Die erste deutsche Übersetzung in der Grazer Handschrift," (2008): 265–71. Reymers was also the first to publish the prosthaphaeretic formulas, but he did not discover them (a fact which strengthens claims of plagiarism). I read it more as a case of, let us say, "extreme influence." Though little is known about his life, Reymers was far from the charlatan most biographies paint him as. Thanks are due to Nick Jardine's excellent research in this area.

p. 66 **"merry band of men."**: Gade, *The Life*, p. 178, Ferguson, Tycho & Kepler, p. 222. He also claimed to have had sex with Tycho's eleven-year-old daughter.

p. 66 **"Learn to safeguard your possessions better."**: Rosen, *Three Imperial Astronomers*, p. 195; Thoren, *Lord*, p. 393.

p. 67 **"admitting in the preface . . ."**: Rosen, *Three Imperial Astronomers*, p. 223; *TBOO*, v. 8, p. 204.

p. 67 **"counsel to our king."**: *TBOO*, v. 14, pp. 100–101. Abridged. This was Christian Friis.

p. 67 **"absolutely no interest."**: *TBOO*, v. 14, p. 101. Abridged.

p. 68 **another, the following month, in the winter room**: Christianson, *Island*, p. 197, *TBOO*, v.9 p. 146.

p. 68 **hoopoes still alit upon its cliffs.**: Thoren, *Lord*, p. 214, *TBOO*, v. 9, p. 82.

p. 69 **"as if you were our equal."**: Gade, *The Life*, pp. 152–7.

p. 69 **beyond the judgment of man.**: Cf. Christianson, *Island*, p. 196 on absolutism and divine right in Christian IV's reign.

p. 70 **"So fare thee well!"**: TBOO, v. 9, pp. 208–11. *Abridged.* Christianson translates more than enough to English, *Island*, pp. 216–7.

p. 70 **He dedicated to Rudolf a new book.**: Dreyer, *Tycho Brahe*, p. 261; *TBOO*, v. 5, pp. 5–10.

p. 70 ***Instruments for a Restored Astronomy***: Astronomiae Instauratae *Mechanica*. This is presently the only work of Tycho's translated into English: Brahe, *Mechanica*, trans. Raeder. I would recommend his *Epistolae*, for any willing translators.

p. 71 **giant armillary sphere**: This armillary sphere was equatorial. Tycho's began to use the equatorial coordinate system rather than the ecliptic because of his documentation of the fixed stars. The equatorial system was preferable because it corresponded to the earth's daily rotation, and the ecliptic system has since never come back in vogue. Mentioned repeatedly in Needham, because the Chinese used the same, cf. *Science and Civilization in China*, v. 3, pp. 266, 270, 340, 366, 372, etc. Also Woolard, Edgar W. "The Historical Development of Celestial Co-ordinate Systems." Publications of the Astronomical Society of the Pacific 54.318 (1942): 82.

p. 71 **"any region is his fatherland."**: Brahe, *Mechanica*, p. 63.

p. 71 **"You will always find a home in your sister"**: This whole paragraph is from *TBOO*, v. 14, pp. 157–9 (mostly the second and last paragraphs); Zeeburg, *Urania*, pp. 266–9. Tycho responds in *TBOO*, v. 14, pp. 178–82. This is, apparently, the only letter Sophie ever sent to Tycho. Her son, Tage Thott, also wrote to Tycho, *TBOO*, v. 14, pp. 170–2.

p. 71 **"that several refused"**: Thoren, *Lord*, pp. 374–5, 398, 408.

p. 73 **"Foremost of Mathematicians, by his own hand."**: In *KGW*, v. 13, n. 82; pp. 154–5. A nearly complete translation is in Rosen, *Three Imperial Mathematicians*, pp. 83–4.

p. 77 **"'I might find the cause,' he wrote, 'in vain.'"**: *Kepler*, *Optics*, p. 216. This is in reference to his myopia, but he was also polyopic, see Kepler, *Optics* p. 216 again, also Koestler, p. 230.

p. 77 **"convex bifocals"**: Kepler, *Optics*, p. 217; Kepler, *Conversation*, p. 72; Burke-Gaffney, *Kepler and The Jesuits*, p. 17. Note that bifocals were already invented centuries ago (though were a rarity), but had largely lacked a theory explaining them.

p. 77 **religious art was diverging into a macrabe, peculiarly German strain."**: . . . the appearance of distinctly national tendencies in art added their bit to the same unsettlement [away from traditional religion]" E.A. Burtt, *The Metaphysical Foundations of Modern Physical Science*, p. 28.

p. 77 **such horrors were borne first witness**: I allude to Matthias Grunewald's *Isenheim Altarpiece* (even though he was Catholic), but see no reason to mention it by name. Kepler may have learned of it, as Emperor Rudolf would spend the years 1597–1601, when Kepler was in his service, trying to bring it into his possession. But I use it only to demonstrate the German Gothic tradition, of which Durer, though more cosmopolitan, was also a part. Kepler references Albrecht Durer in his *Harmonies of the World,* Cf. *Harmony*, p. 75.; Grunewald worked on pieces by Durer, and the two have historically been conflated. Rudolf collected Durer paintings in a frenzy.

p. 77 **a weakness further justifying his poor health.**: *OO,* v. 8–2, p. 672. *KGW,* v. 19, p. 320.

p. 78 **his conception was the cause of the marriage rather than the other way around.**: A speculation I have seen mentioned nowhere else except for *Johannes Kepler and the New Astronomy*, Voelkel, p. 12. I think it is very probable.

p. 78 **"a luckless spirit."**: *OO,* v. 8–2, p. 672. Also translated in Koestler, *Sleepwalkers*, p. 231.

p. 78 **"the mother imparts much upon the fetus."**: *OO, v.* 5, p. 262; *KGW,* v. 6, p. 281. From Aiton, Duncan, Field, *Harmony of the World,* pp. 376, 379, with slight edits.

p. 78 **"could not overcome the inhumanity of her husband."**: *OO,* v. 8–2, p. 672.

p. 78 **"he ruined everything."**: *OO,* v. 8–2, p. 672. Also translated in Koestler, *Sleepwalkers*, p. 231.

p. 78 **"he wrote dithyrambs"**: And a wordsmith he remained, for it is to Kepler that we owe the proper definition of *orbit* ("Orbita." Before it had really just meant "path." There was no real cause to consider it, because Ptolemaic models only concerned the movements and angles of epicycles, and not the previous position of planets. In fact, the illustration of the Ptolemaic epicycles given in the prologue to this section is a sensational modernization; Ptolemy would never have thought of his model as drawing out an "orbit" as such. Kremer, *Zagan*, p. 101, E. L. Davis, "Astronomia *Nova*: The Classification of Planetary Eggs"), the first use of *inertia* (Unart. First, naturally, because his conception of motion was not modern. Kremer, *Zagan*, p. 59, William H. Donahue, "Kepler as a Reader of Aristotle." Also Koyré, *Galileo Studies*, p. 144. Kepler's inertia will be mentioned a bit later, when we reach his *Epitome of Copernican Astronomy*), the term *camera obscura* (Kepler, *Optics,* p. 67), and the optical notion of *focus* as the fiery center to which rays of light tend (Rosen proposes this in p. 123 of Kepler's *Somnium*). Kepler was the first person to use the term *satellite* in astronomy (Kepler, *Conversation*, pp. 76–7). Apparently, he was also the first to use the word *pencil* ("penicillum"; *KGW,* v. 4, p. 368).

p. 79 **"but questioned his Physics."**: *KGW,* v. 19, pp. 328–9, *OO*, v. 5, p. 476; Koestler, *Sleepwalkers*, pp. 239–40. Kepler's Latin is very cryptic here. Both translators made mistakes or omissions. I may have as well. Here are, for my purposes, other highlights:

"In boyhood, he was must given to play; in adolescence, other things delighted his mind . . . because he was cautious with money, he was driven from play [gambling], and often played with himself [wanking]. And let it be known, this caution was not for wealth, but to raise above the fear of poverty . . . but the love of coin captures many."

"His speaking, writing, was forever running between new thoughts or words or matters or ways of speaking or argument, or of some new plan."

[Kepler goes on a long digression about luckiness, unluckiness, and its causes.] "Let it be noted, if it seems above that he desires a life of leisure, this is not strictly true. He would die without work."

"If by chance he should be dragged into military service, it will have been wholly by chance. For he is no more a soldier, than he who descends into that camp wholly out of despair. Here is present anger, intent to deceive, vigilance against constant, sudden attacks, and not a chance of happiness."

p. 79 **was said to have lived in the south tower.**: Mentioned in Baumgardt, *Johannes Kepler*, p. 23. His first name was technically Johann, and his presence at Maulbronn is actually extremely dubious. See *Faust. Und Faust*, p. 128, Günter Mahal, Attempto–Verlag,

1997. Consider, from Northrup Frye's *Anatomy of Criticism*, p. 39: "the *alazon* may be one aspect of the tragic hero as well: the touch of *miles gloriosus* in Tamburlaine, even in Othello, is unmistakable, as is the touch of the obsessed philosopher in Faustus and Hamlet."

p. 79 **he recited his favorite sermons.**: From Caspar, Kepler, p. 40; *KGW*, v. 19, p. 337.

p. 79 **for both his personality and build**: Kepler describes himself as "agile, wiry, and well-proportioned" in Koestler, 236. Paraphrased also in Caspar, 369, from Frisch's *OO*, v. 5, p. 47.

p. 79 **comely, penitent Mary Magdalene.**: *OO*, v. 8, pt. 2, p. 676; *KGW*, v. 14, Nr. 226, p. 275, ln. 479.

p. 80 **his first astronomy class freshmen year**: Kepler entered Tubinger Stift on September 17, 1589. Kepler wrote he studied with Maestlin six years ago when he was twenty-four (in 1596), see *KGW*, v. 1, p. 9, or Coelho, Victor, *Music and Science in the Age of Galileo*, v. 51, p. 46. Springer Science & Business Media, 1992. ("Kepler, Galileo, and the Harmony of the World" by Gingerich).

p. 80 **with a style lucid and direct**: This is obvious from his appendix to Kepler's *The Secret,* containing a detailed (though basic) mathematical exposition of the Copernican system which is in fact the clearest I have read, by anyone, in any language. For a translation, see Grafton, Anthony. "Michael Maestlin's account of Copernican planetary theory." *Proceedings of the American Philosophical Society* 117.6 (1973): 523–50.

p. 80 **"arguing it comes about by the Earth's revolution."**: *Mysterium Cosmigraphicum*, p. 63, or *KGW*, v. 1, p. 9.

p. 80 **yet now he was a math teacher.**: Westman, Dobryzcki, *The Reception of Copernicus's Heliocentric Theory*, p. 8.

p. 80 **"ought to receive the highest mention."**: Westman, Dobryzcki, *The Reception of Copernicus's Heliocentric Theory*, p. 8.

p. 80 **an A- in astronomy.**: Rosen, Edward. "7.1. Kepler and the Lutheran attitude towards Copernicanism in the context of the struggle between science and religion." *Vistas in Astronomy* 18 (1975): 317. Also, Gingerich, Owen. "Johannes Kepler and the new astronomy." *Quarterly Journal of the Royal Astronomical Society* 13 (1972): 346, which points out that Kepler's excellent grades were rather uniform across all students. Some of Kepler's grades can be seen in *KGW*, v. 19, Nr. 7.9, p. 317.

p. 80 **"The lighter the weight"**: Jarrell, Richard A. "The Life and Scientific Work of the Tübingen Astronomer Michael Maestlin." PhD diss., University of Toronto (1971), pp. 41, 170.

p. 81 **Kepler's kooky and endearing single mom**: Jarrell, Richard A. "The Life and Scientific Work of the Tübingen Astronomer Michael Maestlin." PhD diss., University of Toronto (1971), p. 160; *KGW*, v. 13, Nr. 119, p. 328.

p. 81 **The Bible apparently claimed the Earth stood still.**: In many famous verses, which Kepler contested. See Donahue's translation of *Astronomia Nova*, p. 30.

p. 81 **a distinction of kind.**: Scholars of every stripe have long recognized that neither religion nor science has ever been necessarily internally coherent (and thank goodness!). It was Thomas Aquinas who made this distinction as "*sui generis*," which is all over the *Summa*. Interested readers can find quotations in Clark, Ralph W. "Aquinas on the Relationship between Difference in Kind and Difference in Degree." *Thomist: A Speculative Quarterly Review* 39.1 (1975): 116.

p. 81 **"carry over into common usage."**: Kepler, *Nova*, p. 30. Page 33 features a lovely quote, which I have sadly not found a place for in my narrative: "Advice for idiots: whoever is too stupid to understand astronomy, or too pathetic to believe Copernicus without it affecting his faith, I advise them to mind their own business. Betake themself home to scratch their own patch of dirt. They can be sure that they worship God no less than the astronomer."

p. 81 **"while they are jokes."**: Rothman, *Strife*, p. 34, or *KGW*, v. 13, Nr. 80, p. 151. I have added the ellipsis for dramatic effect.

p. 81 **"The person cannot be divided in two."**: Rothman, *Strife*, p. 42, citing D. Martin Luther Werke, Weimarer Ausgabe (Weimar, 1883–1929), Schriften, 26, 332b. Slight changes.

p. 81 **"a union of these two kinds of life"**: *KGW*, v. 13, Nr. 8, p. 10.

p. 81 **"my freedom and soundness of mind."**: *KGW*, v. 13, Nr. 8, p. 10.

p. 82 **"might find greater enjoyment."**: This is from *Harmony of the World*, *KGW*, v. 6, p. 93. I believe the Aiton, Duncan, and Field translation, p. 129, has made a mistake (which I make all too commonly!) of accidentally giving the line the opposite of its intended meaning. Kepler explains the "Natural Method" on p. 283.

p. 82 **"in such a short span of time"**: Bruno, Giordano. *Jordani Bruni Nolani Opera Latine Conscripta*, v. 1, pp. 324–5. Niapoli: Frommann, 1879; Rowland, *Giordano Bruno*, p. 219.

p. 82 **pantheistic creator–God, who so endowed creation.**: See Yates, *Giordano Bruno and the Hermetic Tradition*, p. 143: "The infinite universe and the innumerable worlds are for him new revelations, intense accentuations of his overpowering sense of the divine."

p. 83 **"But it may turn out to be of public interest."**: *KGW*, v. 13, Nr. 23, p. 33.

p. 84 **"another indicated a smaller circle."**: Field, *Cosmology*, p. 47, with edits. Also, *KGW*, v. 1, pp. 11–2.

p. 84 **"same as the orbits of Saturn to Jupiter!"**: Kepler, *Secret*, p. 67.

p. 84 **"the egg of Christopher Columbus."**: Rothman, *Strife*, p. 150. They also explain the rather strange "egg" metaphor.

p. 84 **"I shied from no computation, however difficult."**: Koestler, *Sleepwalkers*, p. 251.

p. 84 **"I could never have given birth."**: *KGW*, v. 12, Nr. 60, p. 105. For the elephant, see *KGW*, v. 12, Nr. 64, p. 113. Maestlin's "fire" is Vulcanus, the god of fire. He is making a metaphor with the Greek gods. In the case of Kepler, I have given special attention to metaphors where he refers to himself as a woman.

p. 84 **The Secret of the Universe:** *"Prodromus dissertationum cosmographicarum, continens mysterium cosmographicum, de admirabili proportione orbium coelestium, de que causis coelorum numeri, magnitudinis, motuumque periodicorum genuinis & proprijs, demonstratum, per quinque regularia corpora geometrica"*, forerunner of the *Cosmological Essays*, which contains the *Secret of the Universe*; on the Marvelous Proportion of the Celestial Spheres, and on the True and Particular Causes of the Number, Magnitude, and Periodic Motions of the Heavens; Established by Means of the Five Regular Geometric Solids. I have favored the title of Aiton's translation. Although it is hardly the most accurate, it is certainly the most appealing.

p. 84 **"the number, shape, and motion of the circles"**: Kepler, *Secret*, p. 63.

p. 85 **"looked to those five regular, same-faced solids."**: Kepler, *Secret*. p. 63. "Same-faced" is an vain attempt to work in the definition of regular solids without being verbose. They also require all angles to be equal.

p. 85 **"see Euclid, Book XIII, scholium after Proposition 18."**: Kepler, *Secret*, p. 97 (picture derived from p. 153. Kepler uses "semi–diameter," which maintains the same ratio to the orbit, circumference. Drawing from *KGW*, v. 13, p. 32, Nr. 22.

p. 85 **marvelously close to the actual ratios as measured in Kepler's era**: Kepler "cheats" on Mercury, using the circle inscribed in the square formed by the middle of the octahedron rather than the sphere inscribed in octahedron itself. And, of course, the numerical relationships are not exact.

p. 86 **too many crimes**: *KGW*, v. 19, p. 336, and Caspar, Kepler, p. 40.

p. 86 **a most irreverent, quasireligious method**: Kepler, *Secret*, p. 149.

p. 86 **"in measure and number and weight."**: From King James, Wisdom of Solomon, chapter 11, verse 20. Thanks to J. V. Field's excellent piece on the *Harmony of the Worlds*, which heads Kremer, *Zagan*, for establishing this reference.

p. 86 **"God is praised through my astronomy!"**: Kepler, *Secret*, p. 19. See also Kepler, *Somnium*, p. 239. With simplifying edits.

p. 86 **great lay theologian, Dante Alighieri**: Also John Milton. See the opening of Book III of *Paradise Lost*. Kepler would believe in the coeternality of number all his life, and repeated it in Kepler, *Harmony*, pp. 146, 367.

p. 86 **"Justice moved my maker on high."**: From Canto III, *Inferno*.

p. 87 **"the physical properties of the planets from immaterial things"**: A fear which would make some quantum physicists laugh. *Secret*, p. 123.

p. 87 **"the brilliance of its light."**: Kepler, *Secret*, p. 201. Simplified.

p. 87 **"mine were physical, or rather metaphysical."**: Kepler, *Secret*, p. 63. Italics added.

p. 87 **even chose to add Rheticus's First Account as an appendix**: See Rothman, *Strife*, p. 51. Also, Voelkel, *Composition*, p. 64. Also *KGW*, v. 13, Nr. 52, p. 97. Maestlin added a very professional supplement as well. Actually, Maestlin's contributions to Kepler's first book are so substantial that I would suggest that without Maestlin, there would be no Kepler. Kepler acknowledged this. For a fuller breakdown of Maestlin's contributions to *The Secret*, with a translation of his appendix, see Grafton, Anthony. "Michael Maestlin's account of Copernican planetary theory." Proceedings of the American Philosophical Society 117.6 (1973): 523–50.

p. 87 **"set the type for the printer himself."**: See Kremer, *Zagan*, p. 113, "Kepler versus Lansbergen: On Computing Ephemerides, 1632–1662", O. Gingerich. Also *KGW*, v. 13, Nr. 52, p. 95.

p. 87 **"if you do not get back to me promptly."**: *KGW*, v. 13, p. 36.

p. 88 **"follow after the creation of the universe."**: Kepler, *Secret*, p. 65.

p. 88 **"the five–fold pattern of the starry spheres."**: Kepler, *Secret*, p. 225. Thanks to the footnote, p. 253. Aiton's added commentary has been invaluable.

p. 88 **disdain for the Church.**: Details drawn from Rowland, *Giordano Bruno*, p. 10. I step into histrionics here, naturally.

p. 88 **he did send a personal copy of his work to Tycho Brahe**: That publication was *Acrotismus camoeracensis*. See Kremer, *Zagan*, Michael A. Granada, "Kepler and Bruno on the Infinity of Solar Systems," p. 143. Also Evans, *Rudolf II*, p. 232.

p. 88 **"quite rather, exile to his infinite space"**: This was, of course, written to Galileo; since I have not made clear Kepler and Galileo knew each other yet I have not specified this. *KGW*, vol 4, p. 304. Also Kepler, *Conversation*, pp. 36–7.

p. 88 **"his vexation flipped God and Man about in circles."**: *KGW*, v. 16, p. 142, Nr. 488. Also Kremer, *Zagan*, "Kepler and Bruno on the Infinity of the Universe and of Solar Systems," Miguel A. Granada, p. 135.

p. 89 **working under an essentially feudal system of labor.**: The peasantry made up about 85 percent, or 13.6 million of the German population in 1600.

p. 89 **"course and suspicious peasants."**: Caspar, *Kepler*, p. 237. Strangely enough, Kepler only seems to talk about the lower classes in his broken German dialect.

p. 89 **"I might well call them my teachers."**: Slight changes, from Caspar, *Kepler*, p. 172.

p. 89 **"woman is ever a fickle and changeable thing."**: *"Varium et mutabile semper foemina,"* a famous quote. *KGW*, v. 1, p. 36, or Aiton's translation, p. 117.

p. 89 **"I was led to my wife."**: *KGW*, v. 13, Nr. 60, p. 104.

p. 89 **"private misfortune arose."**: From Caspar, *Kepler*, p. 75. The full letter is in *KGW*, v. 13, Nr. 64, pp. 113–9.

p. 89 **"which they openly condemn the state."**: *KGW*, v. 13, Nr. 64, p. 115.

p. 90 **"blood red to yellow."**: From Caspar, *Kepler*, p. 77; *KGW*, v. 13, Nr. 99, p. 228; Voelkel's *Johannes Kepler and the New Astronomy*, p. 38.

p. 90 **"the Copernican heresy for many years now."**: *KGW*, v. 13, Nr. 75, p. 143. Cf. Nr. 73, 76.

p. 90 **"I love your system of the world."**: Rosen, *Three Imperial Mathematicians*, p. 86. Koestler, p. 298, claims that Kepler requested Reymers send a copy of the work to Tycho, but this is (as far as I can tell) untrue.

p. 90 **"capable of helping and harming me."**: Rosen, *Three Imperial Mathematicians*, pp. 90–1.

p. 90 **with which their story began**: A good deal of the literature (Christianson, *Island*, p. 299, Caspar, *Kepler*, p. 70) suggests that Kepler sent a copy of *The Secret* to Tycho. Rosen (*Three Imperial Mathematicians*, p. 106) disputes this, because of the earlier letter to Reymers, claiming only the letter. I agree with Rosen.

p. 91 **"And his dearly beloved friend."**: *KGW*, v. 13, Nr. 92, pp. 197–202. The letter has been abridged slightly elsewhere. Partial translation also available in *Three Imperial Mathematicians*, p. 110.

p. 91 **"Everyone loves themself!"**: Koestler, *Sleepwalkers*, p. 298.

p. 91 **"You seem to have written such things rashly."**: Kepler, *Defense*, p. 15.

p. 92 **attributed to one "Mr. Johannes Repler"**: Not once, but twice. Kepler, *Defense*, pp. 53–4.

p. 92 **"uttered in a poetic spirit"**: From Rosen, *Three Imperial Mathematicians*, p. 145.

p. 92 **"Now a daughter is born to me . . ."**: *KGW*, v. 14, Nr. 132, p. 43, ln. 1–4. Most of the quotes in this section are from this very important letter.

p. 92 **"honorably adorn astrology . . ."** : *KGW*, v. 14, Nr. 132, pp. 44, ln. 36, 50, 78, 81–3, 110, 112–3. I have abridged to avoid unneeded names. Concerning lns. 112–3, there are some relevant comments in Grafton, Anthony. "Kepler as a Reader." *Journal of the History of Ideas* 53.4 (1992): 561. There is a partial translation concerning Ursus in Kepler, *Defense*, p. 66.

p. 93 **"Nor am I any more respected in setting about my work . . ."**: *KGW*, v. 14, Nr. 132, p. 56, ln. 523.

p. 93 **"violence would have the embroidery of justice . . ."**: *KGW*, v. 14, Nr. 132, pp. 56–7, ln. 560–9. Also Caspar, Kepler, p. 98.

p. 93 **She was best at salads and cooked tortoise.**: Kepler somewhat famously remarks on these two dishes; see Caspar, *Kepler*, p. 76, or *KGW*, v. 13, p. 185, Nr. 89, ln. 227 for tortoise, and Ferguson, *Tycho & Kepler*, p. 296 or *KGW*, v. 1, p. 285 for salad.

p. 93 **"never salvation at home with a riot outside . . ."**: *KGW*, v. 14, Nr. 132, p. 57, ln. 597–9. Kepler goes on to say that he wouldn't mind guards for his wife.

p. 94 **"ears—eureka."**: *KGW*, v. 14, Nr. 132, p. 46, ln. 135–40.

p. 94 **"I do not yet wish my words to be unpopular with great men."**: *KGW*, v. 14, Nr. 132, p. 54, ln. 469–70. I have, naturally, skipped over most of Kepler's mathematical work on world-harmony here; it will be touched on later.

p. 95 **"who knows what fate has in store?"**: *KGW*, v. 14, Nr. 133, p. 59. This is the next letter. In Nr. 142, p. 86, Kepler tells Maestlin flat out he is leaving, but the letter is not so interesting. This official was Herwart von Hohenburg, who was the chancellor of Bavaria, not Graz, although he was a devote Catholic.

p. 95 **"friend to both Kepler and Tycho."**: This was Baron Hoffman.

p. 95 **"I can offer you nothing."**: *KGW*, v. 14. Nr. 178. "I can offer you nothing" is specifically referring to a professorship at Tubingen.

p. 95 **"he would start an altercation."**: *KGW*, v. 20.1, p. 18. See also Kepler, *Defense*, p. 135.

p. 95 **"judge in this quibble over words."**: Ibid.

p. 96 **"the Bear himself, to his face."**: Ibid.

p. 96 **had fled the city**: Reymers may have left to escape Tycho, to avoid the plague, or because he had fallen out with the emperor. Tycho, naturally, thought it was because of him.

p. 96 ***Tycho Brahe's Triangle Practice:*** *Triangulorum Planorum et Sphaericorum Praxis Arithmetica Qua Maximus Eorum Praesertim in Astronomicus Usus Compendiose Explicatur* (Arithmetic Practice of Planar and Spherical Triangles Which Have Been Specially and Concisely Explained to Be of Greatest Use in Astronomy). This book is not often mentioned, but the technicalities of Brahe's research program are pretty fascinating. Cf. J. L. E. Dreyer, "On Tycho Brahe's Manual of Trigonometry," *The Observatory*, v. 39, pp. 127–31 (1916).

p. 96 **a list of "dogmas" for just such use cases.**: The above account is based on Voelkel, *The Composition of Astronomia Nova*, pp. 100–101. Useful information about Longomontanus can be found in *On Tycho's Island*, pp. 314–6.

p. 96 **to size up his future coworkers.**: The ride is about 100 kilometers. See *TBOO*, v. 8, p. 246. Thoren says six hours, *Lord of Uraniborg*, p. 414.

p. 97 **a tryst with one of Tycho's daughters**: *KGW*, v. 14, Nr. 187. Also Thoren, *Lord*, p. 463.

p. 97 **"loving and advancing all the best stuff"**: *KGW*, v. 14, p. 109, Nr. 157.

p. 97 **"several difficulties insert themselves."**: Ibid.

p. 97 **Tycho Brahe was living a posthumous life.**: From the final letter of John Keats: "I have an habitual feeling of my real life having passed, and that I am leading a posthumous existence. God knows how it would have been—but it appears to me—however, I will not speak of that subject."

p. 97 **"it has been difficult to bring my family together . . ."**: This is pieced together from *KGW*, v. 19, 2.1, pp. 37–40, ln. 1, 7–8, 11–3, 42, 54–5, 122. I believe I have counted line numbers correctly (fifty per page), but this section is not numbered. For the sake of narrative, I have written about documents 2.1 and 2.2 as if they are more or less the same.

p. 97 **"In me, he must have faith."**: *KGW*, v. 19, Nr. 2.1, p. 40, ln. 148–52.

p. 98 **"concede to me my philosophical freedom."**: *KGW*, v. 19, Nr. 2.2, p. 42.

p. 98 **"which I do with vigor and zeal?"**: Ibid.

p. 98 **"to sustain this convenience of yours."**: *KGW*, v. 19, Nr. 2.4, p. 45.

p. 98 **"I am able to establish nothing certain."**: *KGW*, v. 19, Nr. 2.4, p. 46.

p. 98 **"Old age creeps in upon him"**: *KGW*, v. 19, Nr. 2.1, p. 40. This is a teeny bit of an anachronism (Kepler would have noted this right before the contract).

p. 98 **"begin to cry?"**: Caspar, *Kepler*, p. 374, notes that Kepler broke into tears upon realizing his *Secret of the Universe*.

p. 98 **"when no man can work?"**: I realized this quote from Irving Howe's "Reflection On the Death of My Father," which also ends on this relevant quote: "To make a myth of the man I should have mourned as a father, to cast him at the center of the only story I had to tell, was to reach a kind of peace between generations."

p. 98 **"excessive arrogance."**: Indeed, Brahe said a good deal more than this. The narrative here is irreparably damaged (though for flow the story does not concede this) by the loss of one or more letters Kepler sent to Brahe from Prague, which were, apparently, very rude. Tycho had this to say, writing to his employee Jessenius (who was also the scrivener at the earlier meeting):

"With the others you were listener and spectator of that which was said and done by Kepler, around the very moment of departure, and how immediately after, about to bid adieu near the carriage, he was suffering a certain regret of what led before, and that everyone, whose moods he had not been able to control, he should excuse, and as I am no stranger to gift-giving, I then whispered into your ear: that in Prague you should cause an offering be drawn up for me, to testify that he himself, has conducted himself badly, has derailed and desires to return to me in friendship, and that with you having affirmed that this would be done, and telling me this freely and separately beforehand, that on the journey he would be seriously admonished by you, firmly rebuked, that he had been powerlessly carried by bile, against what befits men of learning and honor: on that account I send you hence specially written letters by his own hand, in which unrestrained petulance and excessive arrogance and mordancy, which not my wine, nor my little song, nor any other method which may be pretended, will be able to excuse, if he was not then deranged, (that the strength of his diseases may be seen by him to stand together as the principal cause, he secretly cherishes), what he was writing with a sound mind, and by his own pen, which he ignorantly (though to him I have now begun to forgive, should he recover his senses) commanded."

This is in *TBOO*, v. 8, p. 298; *KGW*, v. 14, Nr. 161, p. 113. Unfortunately, story and space have not fully permitted me to express Tycho's extraordinarily abstruse, pronoun-laden, Latinate style. I have indulged in a bit of "translationese" above to try and convey it. Unlike previous biographers, I do not believe Brahe's account above deserves much weight.

p. 99 **Kepler was an agent of Ursus**: A possibility suggested in Thoren's *Lord of Uraniborg*, p. 441, sourcing *TBOO*, v. 8, p. 299. It is a bit of stretch, but is not totally unreasonable and can be played well for effect.

It is extremely rewarding to compare Kepler's experience with Brahe with that of another German from Christianson's *On Tycho's Island*, p. 151.

p. 99 **"(methinks it his head or heart!)"**: *KGW*, v. 14, Nr. 161, p. 13 (also above quote). Also Koestler, *Sleepwalkers*, p. 306.

p. 99 **"And he hates baths."**: Pieced together from *KGW*, v. 19, p. 336, Nr. 30. Koestler has a much longer translation in *Sleepwalkers*, p. 236.

p. 100 **"the intemperance and sickness of my mind . . ."**: The beginning of Nr. 162 from *KGW*, v. 14, p. 114. Abridged. Koestler has translated more of this in *Sleepwalkers*, pp. 306–7. Caspar notes that this may never have been sent to Tycho, as it is missing the usual salutations.

p. 100 **"a beggarly and contemptible man"**: *KGW*, v. 13, Nr. 117, p. 311 (Kepler thus thought this before he encountered Tycho, although the affair could not have helped.).

p. 100 **it was no longer with desire but pity.**: Baumgardt, *Johannes Kepler*, pp. 62–3.

p. 100 **she could not decide which.**: Caspar, *Kepler*, p. 76, *KGW*, v. 14, Nr. 188. Caspar indicates this is Barbara's only letter to him, saved because Kepler wrote astronomical sketches in its margins (Caspar, p. 119). See also Dreyer, *Tycho Brahe*, p. 300.

p. 100 **Stray, worldly instruments loomed over.**: Rublack, *Astronomer and Witch*, p. 147, *KGW*, v. 14, Nr. 188.

p. 100 **"be a big bride."**: *KGW*, v. 14, Nr. 187, also p. 480.

p. 100 **"To merit this fuss in the future, you want to use others."**: *KGW*, v. 14, Nr. 191. That Secretary was Johannes Eriksen; Christianson, *Island*, p. 273.

p. 101 **"but it is very cold."**: *KGW*, v. 14 Nr. 186.

p. 101 **as Kepler called the Brahe manor.**: Ferguson, p. 254.

p. 101 **"the apogee of this, the nodes of that."**: *KGW*, v. 14, Nr. 168, ln 39, 105, 110–3.

p. 101 **persuade against any further theorizing.**: *KGW*, v. 14, Nr. 170, also p. 475. Notably, Longomontanus's work with lunar theory did involve some Copernican methods.

p. 101 **hawking a method for squaring a circle**: An impossible task, as is well known. Longomontanus did exactly this. See Christianson, *Island*, p. 318. I do give

Longomontanus short shrift here; cf. Swerdlow, Noel M. "Tycho, Longomontanus, and Kepler on Ptolemy's solar observations and theory, precession of the equinoxes, and obliquity of the ecliptic." *Ptolemy in Perspective*, 151–202. Springer Netherlands, 2010. Even in Swerdlow's depiction, Longomontanus seems to miss the big picture.

p. 101 **"he becomes childish."**: *KGW*, v. 14, Nr. 203, ln. 21–3.

p. 101 **"He always resembles a lost man."**: Caspar, *Kepler*, p. 120.

p. 102 **"but the culprit died."**: *TBOO*, v. 8, p. 371. Also Thoren, *Lord*, p. 454. Abridged.

p. 102 **barely ten copies of Reymers's work remain.**: Of his *Astronomicis Hypothesibus*, specifically. *The Book Nobody Read*, p. 118.

p. 102 **"win or lose, I am always defiled."**: This friend was George Rollenhagen. *Three Imperial Astronomers*, p. 126; *TBOO*, v. 8, p. 51. One is reminded of the famous Sayre's Law: "Academic politics is the most vicious and bitter form of politics, because the stakes are so low."

p. 102 **for publication, rather than trial.**: Thoren, *Lord*, p. 459, *Three Imperial Mathematicians*, p. 302. This book is in translation as *The Birth of History and Philosophy of Science: Kepler's A Defense of Tycho Against Ursus* by Nick Jardine.

p. 102 **"I write of Ursus, and nothing else"**: *Kepler, Defense*, p. 27.

p. 103 **"Let me not seem to have lived in vain."**: *TBOO*, v. 13, p. 283; Rosen, *Three Imperial Methematicians,* p. 313. "Drinking" is "victus," which Rosen more accurately translates as "way of eating," but this doesn't really make sense, given that Tycho's problem is uremia.

p. 103 **"My grief is too great to get the name across my lips."**: Zeeberg, *Urania*, p. 61.

p. 103 **"acknowledged as his wedded wife"**: Dreyer, *Brahe,* p. 367. Slight changes. Sophie seems to be the only Brahe to warmly acknowledge Kirsten throughout her life, see Friis, *Sofie Brahe Ottesdatter*, p. 25.

p. 103 **"in the peace of a simple country home."**: Thoren, *Lord*, pp. 45–6, 461.

p. 104 **"my marriage bed empty because of it!"**: Zeeberg, *Urania*, pp. 94, 136–7; *TBOO*, v. 9, p. 193, lns. 23–4.

p. 104 **believing it to come an hour later.**: Thoren, *Lord*, pp. 324–5. Thoren is a particularly specialized in Tycho's lunar theory (particularly his nutation, or "variation of the inclination"). Cf. An Early Instance of Deductive Discovery: Tycho Brahe's Lunar Theory, *Isis*, v. 58, No. 1 (Spring, 1967), pp. 19–36; *Tycho Brahe's Discovery of Variation*, Centaurus, 1967: v. 13: no. 3: pp. 151–66.

p. 104 **"it is necessary to redo the examination."**: *Tycho Brahe's Discovery of Variation*, p. 16.

p. 105 **"which I now believe I have achieved"**: *Tycho Brahe's Discovery of Variation*, p. 165.

p. 105 **"much faster in the syzygies as it is slower in the quadratures."**: *Tycho Brahe's Discovery of Variation*, p. 163. I have changed the punctuation. Kepler uses this fact in his *Harmony*, p. 426. Emphasis added.

p. 105 **The analogy of husband and wife to Sun and Moon**: The husband-wife/sun-moon analogy was a ubiquitous reference in medieval European, especially German, poetry; cf. Wunder, Heide. *He is the sun, she is the moon: Women in early modern Germany*, esp. pp. 205–6. Harvard University Press, 1998.

p. 105 **"sunless, starless . . . I accept it."**: *TBOO*, v. 9, p. 193, ln. 25; p. 202, lns. 10–4. Simplified.

p. 105 **to call Tycho quixotic would be a mistake**: A mistake committed by Arthur Koestler, *The Act of Creation*, p. 428.

p. 106 **"Not to seem, but to be?"**: Christianson, *Island*, p. 114. There are several noteworthy astronomical references in Hamlet: "It is most retrograde to our desire"; "I could be bounded in a nutshell and count myself a king of infinite space"; "She is so conjunctive to my life and soul / That as the star moves not but in his sphere." Hamlet stabs King Claudius; Claudius is Ptolemy's first name. And, surely, Tycho's *Theater of Astronomy.*

All the world is a stage, no? I was pleased to discover I was not the only one who enjoyed such foolhardy speculation: cf. Usher, Peter. "Hamlet's Transformation." *Elizabethan Review* 7.1 (1999): 48–64. The Tycho–Hamlet connection is not literal, of course, but Shakespeare was naturally reacting to astronomy of his time, as Tycho did poetry.

p. 106 **"parts of its brickwork were stripped for a lesser house."**: Dreyer, *Brahe*, p. 376.

p. 106 **Grave robbers broke into his tomb and stole his nose.**: The thieves ought to have been disappointed, as his "death nose" was not his formalwear and had a good deal of copper; Johns Hopkins University Press, 2002. Hellman, C. Doris. "Tycho Brahe," *Dictionary of Scientific Biography*, v. 10, pp. 473–5.

p. 106 **"I don't want to die like Tycho Brahe"**: "Nechci umřít jako Tycho Brahe," *A Rough Guide to Prague*, Rob Humphreys, p. 92. Also Hven's Gate, J.L. Heilbron, *London Review of Books*, v. 22 No. 21, 2 November 2000, p. 24.

p. 106 **"live well, and drink to my health!"**: A rather deliberate mistranslation, interpolating Christianson, *Island*, p. 164 ("drain a tankard") and Ferguson, *Tycho & Kepler*, p., 112, 376 ("to the very dregs"). In *TBOO*, v. 7, p. 327, the "artem" is simply "ebibo" (drinking) but is reported with such style that I feel no guilt for this poetical flourish. Today there is a distillery on Hven called "The Spirit of Hven," which produces a specialty single malt whiskey called "Tycho's Star."

p. 106 **"Am I rejoicing, or am I grieving?"**: *KGW*, v. 12, p. 234, ln. 10. This is from Kepler's *Elegy Upon the Death of Tycho Brahe*, written the day after his funeral. There is a German translation on *KGW*, v. 12, p. 405.

p. 106 **"a brilliant hope for mathematics is at hand"**: That message was from Matthias Hafenreffer. *KGW*, v. 14, Nr. 198, ln. 12, p. 194. This is a bit of a mistranslation, to compress the lines that follow: "I heard of Tycho's death not without great sorrow: to me, because of my great hopes in mathematics, I am embittered . . . but God arranges all. A different hope lessens my grief, that we survivors, by the grace of God, are able to complete Tycho's imperfect work; which thus repairs some amount of the grief I've carried from Tycho's premature death, to such a degree that there is joy and congratulations . . ."

p. 107 **"if you love me still, testify by writing a letter."**: *KGW*, v. 14, Nr. 203, ln. 5–13, 26–9, 37–40, 245, pp. 202–8. This is a rather stitched together abridgment.

p. 107 **"and took especial care of the observations."**: *KGW*, v. 15, Nr. 357, ln. 22–7, p. 232; Voelkel, *Composition*, p. 144.

p. 107 **There were twenty-four books in total**: Kepler, *Optics*, p. 13.

p. 107 **four of her spinster sisters**: Christianson, *Island*, p. 368. Christianson has compiled some of the only good English-language information on Tengnagel. This happened in 1604, but it is clear that Tengnagel was providing for them before then. Tycho's four other daughters were Kirstine, Sophie, Sidsel, and Magdalene. Only one, Sidsel, would get married in 1608, see *Danmarks Adal Aalsborg*, v. 5, p. 105. Tengnagel's daughter was named Ida Catherine.

pp. 107–8 **after Emperor Rudolf**: Rudolf Tycho Gansneb genannt Tengnagel. While there are mentions of him dating back centuries, I could not find in any of them, nor in *Danmarks Adel Aalborg*, his exact birthdate. It was sometime in 1604.

p. 108 **"which I would undertake to earn my supper."**: *KGW*, v. 15, Nr. 323, ln. 229–30.

p. 108 **"you will witness the power of my industry"**: *KGW*, v. 15, Nr. 323, ln. 14. Kepler is here talking about lunar theory, but he goes on to explain most of his pursuits over the past several years.

p. 108 **"time and experience will expose as worthless"**: Field, J. V. "A Lutheran Astrologer: Johannes Kepler." *Archive for History of Exact Sciences* 31.3 (1984): 232. I will skip over Kepler's *On the New Star*, preferring to mention it later.

p. 108 ***"Optical Part of Astronomy"***: *Ad Vitellionem Paralipomena Quibus Astronomiae Pars Optica Traditur*, "Paralipomena to Witelo whereby The Optical Part of Astronomy is Treated." An awful title.

p. 108 **"whatever else the name may be . . ."**: Voelkel, *Composition*, p. 147, or *KGW*, v. 15, Nr. 232.

p. 108 **"unwilling relinquishes the good parts of the work."**: *KGW*, v. 15, Nr. 357, ln. 22–7, p. 232. This directly follows the previous quotation. Again, see Voelkel, *Composition*, pp. 150–1. The contract is in *KGW*, v. 19, Nr. 5.1, 5.2, pp. 189–91.

p. 108 **"but offers it to no other."**: *KGW*, v. 15, Nr. 281, p. 23. Koestler, *Sleepwalkers*, p. 346, Caspar, *Kepler*, p. 140. Kepler's words here are not obscure, but dogs do not (or should not) eat hay. His dog metaphor is really something!

p. 109 **"This is not my profession."**: Voelkel, *Composition*, 148, *KGW*, v. 15, Nr. 281, p. 23.

p. 109 **asking for help with observation.**: *KGW*, v. 15, Nr. 296, p. 58. Tengnagel had earlier needed help from Kepler to understand the dispute between Ursus and Tycho, see Kepler, *Defense*, pp. 67–71.

p. 109 **Kepler obliged.**: He observed the new star with Tengnagel (and Jost Burgi) on November 21, 1604. *KGW*, v. 1, p. 159.

p. 109 **handwritten copies of his findings.**: see Voelkel, *Composition*, p. 153, *KGW*, v. 15, Nr. 325, ln. 69–72. Voelkel's book is excellent and filled with quotations.

p. 109 **"may justly be considered new."**: *KGW*, v. 15, Nr. 304, p. 71, see also Voelkel, *Composition*, p. 152.

p. 109 **"For what it is I dare to disturb?"**: *KGW*, v. 15, Nr. 305, p. 73, ln. 59. Abridged. See T. S. Eliot, "The Love Song of J. Alfred Prufrock": "And indeed there will be time / To wonder, 'Do I dare?' and, 'Do I dare?' / . . . / Do I dare / Disturb the universe? / In a minute there is time / For decisions and revisions which a minute will reverse." Kepler, of course, dares.

p. 109 ***"New Astronomy"***: *Astronomia Nova ΑΙΤΙΟΛΟΓΗΤΟΣ, seu Physica Coelestia, tradita commentariis de Motibus Stellae Martis, Ex observationibus, G. V. Tychonis Brahe* (New Astronomy Based Upon Causes or Celestial Physics treated by means of commentaries on the motions of the star Mars from the Observations of Tycho Brahe, Gent). What a tongue for titles young Kepler had!

p. 110 **This was his war.**: Interested readers may enjoy Barreca, Francesco. "The Allegory of War in Johannes Kepler's Astronomia *Nova*." *Nuncius* 26.2 (2011): 312–33.

p. 110 **"A vicious rumor"**: What an astonishing portrait of Copernicus this gives us, unlike any standard English biography! *KGW*, v. 3, p. 8, with some abridgments.

p. 110 **"fathered by you, body I will be."**: *Kepler*, *Nova*, p. 15. Abridged.

p. 110 **"difficult and strenuous war."**: *Kepler*, *Nova*, p. 5.

p. 110 **mythic poet of war.**: Kepler sprinkles the work with literary references, but none so clear as Virgil. For Ovid, see Kepler, *Nova*, p. 316. For Horace, p. 376. On Virgil, concerning the "puffy–cheeked path," he quotes the *Eclogues* in chapter 58, Kepler, *Nova*, p. 428. Also several times in the introduction, Kepler, *Nova*, pp. 28–9, and p. 83, amongst others. Granted, Virgil is about as standard a poetic reference as they come, but it still seems clear to me that Kepler loved Virgil in the similar way that, say, Galileo loved Ariosto, but this has been noted by none of his biographers I have encountered. Virgil was to him the epitome of poetic style.

p. 110 **"Cease, O Keplerus, to do battle in a war against Mars."**: Kepler, *Nova*, p. 10.

p. 110 **"war is breaking out again in full force."**: Kepler, *Nova*, p. 379.

p. 111 **"fifth year has now gone by since I took up Mars."**: Kepler, *Nova*, p. 190.

p. 111 **named the "physical equation."**: The angle formed from the center of the orbit to the end of the arc to the equant. Named as much, I assume, because the work initially began by assuming it was equant distance, which provided a physical explanation to planetary movement. See Koyré, *Astronomical Revolution*, Appendix I, p. 365.

p. 111 **Kepler invented astrophysics.**: A claim I assumed early, but do not see stated so often. It appears in Gingerich, *The Eye of Heaven*, p. 321. In any case, Kepler surely invented an astrophysic. I believe it is the first one with scientific validity.

p. 112 **"who threw them removes his hand from them."**: Kepler, *Nova*, p. 285.

p. 112 **"one and all to help me here!"**: Kepler, *Nova*, p. 314. More accurately, Kepler asked geometers to use geometry to solve a problem that could only be solved with calculus.

p. 113 **"as though Mars's path was a perfect ellipse . . ."**: From Whiteside, Derek T. "Keplerian Planetary Eggs, Laid and Unlaid, 1600–1605." *Journal for the History of Astronomy* 5.1 (1974): 11. Every one of Kepler's essential devices in *New Astronomy* seems to have exactly one wonderful mathematical article on it. Whiteside's is exemplary.

p. 113 **"Let it be a perfect ellipse."**: Kepler, *Nova*, p. 354. "It" is the symbols representing the orbit, which I have edited out. It has been commonly said that Kepler discovered the ellipse by way of a triangulation procedure which only occurs in the later chapters of *New Astronomy*; I consider that triangulation so insignificant that my account does not mention it. Cf. Wilson, Curtis. "Kepler's Derivation of the Elliptical Path." *Isis* 59.1 (1968): 4–25. A similar triangulation procedure was used at the beginning of Kepler's researches on Mars, but this only gave Kepler justification for his equant; see Wilson again or Voelkel, *Composition*, p. 106. I take it for granted that the final bridge which Kepler crossed from oval to ellipse was, in essence, an aesthetical one.

p. 113 **"from the belly towards the two ends."**: Kepler, *Nova*, p. 338. Kepler had written this regarding his discovery of a noncircular first inequality, but I have merged this fact with his discovery of the elliptical first inequality a few chapters later, because the distinction is simply too subtle for this brief summary.

p. 113 **"see new light."**: Kepler, *Nova*, p. 407.

p. 114 **the pacifist, master of Mars, the god of war.**: Kepler, *Nova*, p. 437. Donahue wisely chose this as the cover of their translation; it is a beautiful symbol. Drinkwater has on "updated" version in their *Life of Kepler*, p. 34.

p. 114 **it is a poem about astronomy.**: Dante's astronomy is, of course, not particularly scientific, but it is fascinating and worth an extended study. Cornish's *Reading Dante's Stars* is a recent one which is clearly written (but quite fairly assumes one to have good knowledge of the poems). Chapter 1 is particularly relevant, especially p. 15.

p. 114 **"To be such that I smiled, so mean did it appear."**: *Paradiso*, Hollander translation, Canto 22, p. 601. Dante travels through nine spheres in total, one for each of his planets, and then the Sphere of the Fixed Stars, which he has just reached. The ascent is completed with a vision of God in the Realm of Pure Light. I think the idea of smiling at meanness is an excellent encapsulation of Kepler's entire worldview.

p. 114 **"the beginnings of a thought before it comes."**: *Paradiso*, Hollander translation, Canto 10, p. 250.

p. 115 **"to whom I gave myself for my salvation."**: *Purgatorio*, Hollander translation, Canto 30, p. 669.

p. 116 **"His Friend in unyielding faith."**: *KGW*, v. 15, Nr. 322. Also Koestler, *Sleepwalkers*, p. 347. This letter is strongly abridged.

p. 116 **"she supposed to refuse to eat?"**: *KGW*, v. 17, Nr. 643. Kepler quotes them to contest such claims, but they do not seem to seem to entirely lack truth. See also Caspar, *Kepler*, pp. 173–6, 206–8; Rublack, *Witch*, pp. 278–80.

p. 117 **love for her children.**: Caspar, *Kepler*, p. 176.

p. 117 **three dead in infancy.**: I include Kepler's stepdaughter, Regina, whom he loved like his own. His son, Friedrich, died of smallpox at the age of six on February 19, 1611, which is close enough to infancy for me. This triggered Barbara's depression. Kepler's other two were Susanna (named after his first daughter, who died), born in 1603, and Ludwig, born 1608.

p. 117 **"Is this the dress of salvation?"**: Caspar, *Kepler*, p. 207. I have presented this anachronistically, not fixing her death with a date. Barbara Kepler died on July 3, 1611. I did this for three main reasons: I wanted to write her out before talking about Galileo, who, for obvious reasons, must be presented before 1610; I wanted to give her death the veneer of timelessness; and I wanted to use it as an opportunity to parallel her childbearing with Kepler's authorship.

p. 117 **"To lose all the fear, to reach something better."**: Caspar, *Kepler*, p. 207. Also, *KGW*, v. 12, p. 214; v. 19, Nr. 7.14, p. 322. The entire pamphlet is unspeakably touching; it includes poems for his dead infant son, which I did not mention. Here is a more literal translation: "Thus now you discern hollow images / Since, having suffered, from the aevum your very own light / you will discern: why do you fear to let go of these / O eye, and to pursue the good?"

p. 117 **"married the stars."**: *KGW*, v. 12, p. 213. The next line states "It was bigamy." Kepler was, by virtue of his irenicism, extremely guilty about his own cruelty.

p. 117 **"should it befall me to die."**: *KGW*, v. 15, Nr. 304, p. 71. Partial translation in Voelkel, *Composition*, p. 152.

p. 117 **"works and deeds."**: *KGW*, v. 15, Nr. 324, p. 144. It should be noted they agreed to do so only as it accorded with the emperor's wishes.

p. 118 **"People like me are short-lived"**: *KGW*, v. 15, Nr. 358, p. 247; Koestler, *Sleepwalkers*, p. 350.

p. 118 **"I think it to have happened by divine arrangement"**: Kepler, *Nova*, p. 134. Astute readers will notice me referencing the book (as with *The Secret of the Universe)* as if it were a child.

p. 121 **"nigh impossible to escape."**: *Opere*, v. 9, p. 31. Heilbron has a detailed exposition in *Galileo*, pp. 28–33.

p. 121 **"other men in as many years!"**: From Alessandro Marsili, Olney, *Private Life*, p. 292. Marsili was not a student, per se—this opening takes place in mock–1605, but largely ignores chronology in favor of scene-setting. The chronology of almost all events before 1609 should be understood as very loose, not in just this biography, but in all discussions of Galileo.

p. 121 **"through having been your disciple."**: From Benedetto Castelli; Heilbron, *Galileo*, p. 195. Viviani, who is rather too euphoric for my tastes, said that Galileo was "sent by God as an example of what an artist could be."

p. 121 **jostled between one another for a view.**: Bethune, *Life of Galileo*, p. 16, Drake, *Work*, p. 105. Favaro, *Studio*, v. 1, p. 140, disputes that the crowd was quite this large.

Not much is known about Galileo's public teaching, but as this author expects this career was formative, some brief dry description is here desirable. For the years 1593–4, 1599–1600, 1603–4, Galileo taught Sacrobosco's *Sphere* and Euclid's *Elements*. In 1594–5, he taught *Almagest*; in 1597–8 he taught *Elements* and Aristotle's *Mechanics*. This is from Favaro, *Studio*, v. 1, p. 142. In the last year provided, 1604–5, Galileo taught on "theory of the planets," which may have offered him a chance to discuss, in a cool and unbiased manner, his pet project of Copernican astronomy.

Padua, about 1600, had about fifteen hundred students and about fifty-five faculty. Given that Castelli, teaching at Pisa (about five hundred students, about forty-five faculty) attracted about thirty students, it seems reasonable to suggest that Galileo publicly taught about one hundred students, maybe a little less; mathematics was, apparently, unnecessary enough that Padua had left the chair open for four years after

Galileo's predecessor, Giuseppe Moletti. His public lessons were almost certainly not so fun as his private lessons, because he was constrained to classical texts, though he was, apparently, quite popular.

Interested readers should confer with Favaro, *Studio*, v. 1, pp. 137–77; Biagioli, Mario. "The Social Status of Italian Mathematicians, 1450–1600." *History of Science* 27.1 (1989): 41–95; Grendler, Paul F. *The Universities of the Italian Renaissance*. JHU Press, 2002, and a rather fun case study on Galileo's first university, Schmitt, C. B. "The University of Pisa in the Renaissance." *History of Education* 3.1 (1974): 3–17.

p. 121 **"Then I saw where Signor was heading"**: From Guidobaldo del Monte; *Opere*, v. 10, Nr. 33, p. 45.

p. 122 **"a somewhat philosophical mind."**: From Virginio Cesarini, Drake, Work, pp. 263–4; *Opere*, v. 12, Nr. 1,349, p. 414.

p. 122 **"chatting with young mathematicians after class"**: This is Benedetto Castelli, see *Opere*, v. 11, p. 605 or Drake, *Work*, p. 222. Comically, in the paragraph above, Castelli rejects a charge against him of anti-Aristotelianism.

p. 122 **"the poems of Ariosto, who was his favorite."**: From Vincenzo Viviani, Santillana, *Crime*, pp. 145–146.

p. 122 **never took sick leave.**: Favaro, *Studio*, v. 1, p. 142.

p. 122 **"It is unique in the world."**: Favaro, *Amice*, book VIII, p. 17; *Opere*, v. 11, p. 170. I have turned this into dialogue; it was actually said after Sagredo returned from Syria. It was a very common view; Jean Bodin noted a few decades earlier, touring through the city, that "the civil wars, the fear of tyranny, the acrimonious politics which ruin study—this is the only city which seems immune to these kinds of servitude"; *Galileo e Keplero*, Bucciantini, p. 24.

p. 122 **Venetian "casino"**: Nothing is known about Galileo's lover but her name and that she was Venetian and died in 1612. She was called a prostitute by contemporaries; see Heilbron, *Galileo*, p. 104. She was almost definitely a prostitute (read: "courtesan"). There is no shame in it.

p. 122 **adjacent plot of land for a larger backyard.**: The house was moved into in 1599, garden was added in 1603, see Favaro, *Studio*, v. 2, p. 55.

p. 122 **divine the personalities of the girls.**: Heilbron also provides a translation of both daughter's horoscopes on pp. 92–3.

p. 122 **"best eighteen years of my life."**: This is very often quoted, see, for instance, Heilbron, *Galileo*, p. 63. From *Opere*, v. 18, p. 209.

p. 123 **trench beds of rhubarb, spinach, aloe, and rose.**: These are all foods and ingredients Galileo sent his daughter Virginia in the convent, see Galilei, *Father*, pp. 9, 37, 41, 53, 67,73, 153, 171, 279. I make a creative leap by saying first, what is quite certain, that Galileo did not buy these at market, and second, what is almost certain, that he was growing them at this time. Galileo may also have grown cinnamon and pomegranates.

p. 123 **he was beginning to memorize the works of Ariosto**: Galileo put together serious notes for his disputation against Tasso and for Ariosto in about 1620, after which the numerous references to Ariosto begin to appear in his work. But there is a reference to Orlando, from Ariosto's *Orlando Furiosa*, in the *Dialogue of Cecco*, (*Opere*, v. 2, p. 332), and a couple of other nebulous references from his early period. Ariosto "coined" (such things are never clear) the word humanism in the sixth satire, Campana, Augusto. "The Origin of the Word 'Humanist'." *Journal of the Warburg and Courtauld Institutes* (1946): 60–73.

p. 123 **"everything remained a verdant green."**: Ludovico Ariosto, *Orlando Furiosa*, Canto X, st. 63. I have used the prose translation by Guido Waldman, p. 100, but have added line breaks in keeping with the "prose poetry" style used throughout this book. Galileo recommends this passage over Tasso in *Opere*, v. 9, p. 137. See also Panofsky, Erwin. "Galileo as a Critic of the Arts: Aesthetic Attitude and Scientific Thought." *Isis* 47.1 (1956): 6.

p. 123 **"Agriculture provided him with philosophy."**: From Vincenzo Viviani, Favaro, *Studio*, v. 2, p. 55.

p. 123 **"come into the hands of others."**: From Shea, W. "Galileo and the Super*Nova* of 1604." *1604–2004: Supernovae as Cosmological Lighthouses*. v. 342. 2005, p. 17. This is concerning publishing his lectures on the star, but one imagines it applies just as well to the book, *Dialogo de Cecco di Ronchitti da Bruzene in perpuosito de la stella Nuova.*

p. 124 **"more than once for accuracy."**: Galilei, *Two*, pp. 178–9. All quotes relating to this experiment are taken from this section except where otherwise noted. I have changed the quote somewhat because Galileo's lengthy description is somewhat cumbersome.

This is not entirely how Galileo discovered the law of free fall, but further exposition is extremely dry, very uncertain, and cannot quote from Galileo directly, as it depends on his obscure working notes. For readers who want an interpretation drawn from Galileo's working notes, see Drake, *Pioneer*, pp. 9–31. No interpretation, however, is correct beyond doubt; Galileo's working notes are undated.

p. 124 **"for each time of descent."**: Galilei, *Two*, pp. 178–9.

p. 124 **"natural magic."**: This is a reference to Giambattista della Porta's *Natural Magic*, published in 1558, which sold very well indeed, and described the camera obscura and telescope. Galileo is often credited with having removed from science the need to deal with "causes," in the Aristotelian sense, but this is insignificant. Court magicians and medicine men had been doing so for centuries.

p. 124 **"as the squares of the times."**: Galilei, *Two*, pp. 178–9.

p. 124 **here was something new.**: Sort of new. Galileo was not the first to postulate the law of fall. I am convinced that Nicole Oresme did and Domingo de Soto did as well. It was very difficult to discern the relation between acceleration, velocity, distance, and time before calculus, but the time-squared relation was simple enough for people to discover, and rediscover, although virtually all natural philosophers (Michael Varro, Leonardo da Vinci, Albert of Saxony) got the underlying relationship that velocity is linearly proportional with time wrong, as did Galileo at first. After correcting this mistake, what Galileo did was popularize the law, by taking the principle of uniform acceleration as a starting point for a new physics, and so the law is "his." The task of building a system of physical thought upon this law was what took him thirty years to publish it.

Cf. Wallace, William A. *Prelude to Galileo: Essays on Medieval and Sixteenth-Century Sources of Galileo's Thought*. v. 62, pp. 43, 84. Springer Science & Business Media, 2012; Drake, *Pioneer*, chapter 2, Drake's *History of Free Fall*, or Koyré's *Galileo Studies*. There are many books and articles on the topic, none of which successfully resolve the confusion surrounding this issue. I have considered this discussion too tangential for this kind of story.

p. 124 **"indubitable principle to take for an axiom"**: Sent to the infamous Paolo Sarpi, *Opere*, v. 10, Nr. 105, p. 115; Koyré, *Galileo Studies*, p. 67; Renn, Jürgen, et al. "Hunting the white elephant: When and how did Galileo discover the law of fall?" *Science in Context* 14.1 (2001): 137, for translations. The latter paper contains some very questionable assertions, but goes far in demonstrating how much guesswork scholars have imbued into Galileo's early career.

p. 125 **after reading about magnetism.**: The instruments of Galileo are worth reading about independently in McMullin, *Man of Science*, pp. 256–92, "The instruments of Galileo Galilei" by Silvio Bedini.

Galileo was reading *De Magnete* by William Gilbert. This author regrets that the biography of William Gilbert will never be written, because virtually all of his biographical materials perished in the Great London Fire of 1666. He was a scientist of the first order, and his importance cannot be overstated.

p. 125 **"I am at his beck and call."**: Translated freely, *Opere*, v. 10, Nr. 97, p. 106.

p. 125 **"pardon me; excuse my weakness"**: *Opere*, v. 10, Nr. 97, p. 107. This was Duke Vincenzo Gonzaga of Mantua.

p. 126 **"appearance similar to a parabola."**: From Guidobaldo del Monte, in Naylor, Ronald H. "Galileo's Theory of Projectile Motion." *Isis* 71.4 (1980): p. 551.

p. 126 **"I believe he is a believer."**: This quote is from Silvestro Pagnoni. Poppi, Antonino. *Cremonini, Galilei e gli inquisitori del Santo a Padova*. Centro, 1993.; pp. 54–5. These documents were discovered extremely recently, given the age and popularity of the Galileo affair, and are more accurately recounted in Heilbron, *Galileo*, pp. 104–5.

p. 126 **ought to pay him more.**: Drake, *Work*, p. 47. This story relates to his reappointment in 1599, but applies just as well here.

p. 126 **personal lessons in cosmography, optics, arithmetic**: Drake has done the dirty work of sifting through Galileo's accounts and compiling a convenient table of his private tutoring; *Work*, p. 51.

p. 127 **"full of virtue, religion, and holiness."**: Favaro, *Amice*, book XXI, p. 2. It is not clear whether Castelli's family was noble (probably not), but Benedetto relinquished all claim to noble titles by joining his religious order.

p. 127 **Tycho Brahe's nephew**: This was Otto Brahe, see *Opere*, v. 14, p. 150, Biagioli, *Credit*, p. 8, Favaro, *Studio*, v. 1, p. 185.

p. 127 **"will publish nothing before he dies."**: *Opere*, v. 10, Nr. 95, pp. 104–5, translated in Wilding, *Idol*, p. 30. Tycho sent Galileo a letter himself, to which Galileo did not respond, for unclear reasons; Heilbron, *Galileo*, pp. 118–9 for translation; Drake, *Studies*, p. 130 for exposition.

p. 128 **"I refuse the pain."**: *Opere*, v. 10, Nr. 57, pp. 67–8, *KGW*, v. 13, Nr. 73, pp. 130–1. This letter has been frequently translated, see McMullin, *Man of Science*, "Galileo's Contribution to Astronomy," Willy Hartner, p. 181; Koestler, *Sleepwalkers*, p. 356; Drinkwater, *Life*, p. 15.

p. 128 **magnitude and proportion.**: This is the entire fifth book of Euclid's *Elements*.

For my treatment of a "sense of proportion" in Galileo's life, I am indebted to Stillman Drake, whose *Scientific Biography* thoroughly convinced me that this was how Galileo viewed the world; cf. *Work*, pp. 422–35. Proportionality was how Galileo came to terms with mathematical problems that would only be rightly resolved by the calculus.

In equal measure to Drake, however, I am indebted to Virginia Woolf's depiction of Sir William and Septimus Smith in *Mrs. Dalloway*: "Proportion, divine proportion, Sir William's goddess . . . Worshiping proportion, Sir William not only prospered himself but made England prosper, secluded her lunatics, forbade childbirth, penalized despair, made it impossible for the unfit to propagate their views until they, too, shared his sense of proportion . . . "; "To us, they protested, life has given no such bounty. He acquiesced. They lacked a sense of proportion. And perhaps, after all, there is no God? He shrugged his shoulders. In short, this living or not living is an affair of our own?"

I do not entirely follow Woolf's judgment on the matter. I am different; my characters are different; the politics are different; the seventeenth century is different; the twenty-first, too.

p. 129 **"those with little schooling."**: Drake, *Compass*, p. 52.

p. 129 **"Be happy!"**: *Opere*, v. 2, p. 371; Drake, *Compass*, p. 41.

p. 129 **apamphlet on the new star**: This was *Astronomical Considerations Concerning the New Star of the Year 1604*. It may well have been written by his tutor, Simon Mayr, in Capra's name, but I have excluded Mayr from my narrative; cf. Westman, *Copernican Question*, p. 582.

p. 129 **"witty teachings . . . care on mathematical matters."**: *Opere*, v. 2, pp. 291, 293.

p. 129 **"beasts" . . . "my ox."**: *Opere*, v. 2, pp. 289, 292.

p. 129 **"called your devoted servant!"**: *Opere*, v. 10, Nr. 131, pp. 153–4. Galileo tutored Cosimo II.

p. 129 **"slanders and impostures against me"**: *Opere*, v. 2, p. 519. As before, this translation may very well have been done by Capra's tutor, Simon Mayr.

p. 129 **"despised, evacuated, abused"**: *Opere*, v. 2, p. 518. Very freely translated.

p. 130 **"poisoning all the sense and more with his tongue."**: *Opere*, v. 2, p. 532.

p. 130 **their brief candle held aloft in the dark.**: Brodrick, *Saint*, pp. 254–5.

p. 130 **the Catholic Order . . . refuse work forever.**: The Jesuits, specifically, who will be introduced when the time is right. On Sagredo, Favaro notes that "on the whole for a Venetian patrician the status of his service was quite wretched." I find him to be winsomely repulsive.

p. 130 **"a report reached my ears"**: Drake, *Studies*, p. 140; Drake, *Work*, p. 139; Van Helden, *Messenger*, pp. 36–37.

p. 131 **"come forward!"**: From *KGW*, v. 13, Nr. 76, pp. 144–5; *Opere*, v. 10, Nr. 59, pp. 69–71. Partial translation in Koestler, *Sleepwalkers*, p. 359. I have a little license with the excerpt; here is the full letter.

"Your letter, kind fellow, written August 4, received September 1, brought me twin joys: first, that I begin friendship with you, the Italian; next, that we accord on Copernicanism. Thus, since at the end of your humane letter to me you invited further letters, I would not willingly miss such a chance, even though I write to you presently as a noble youth. I suppose by this time, if you have had the leisure, you have become familiar with my book; I would simply love to hear how your thoughts have progressed: for I am in the habit of writing anyone whatsoever, insisting upon their unadulterated opinion. But how truly I wish that you, so blessed with a keen intellect, felt differently! Although you are wise and hide, your example of, as you say, withdrawing from general ignorance, not rashly inveighing or opposing the madness of common teachings, following the way of Plato and Pythagoras, our original teachers, in this age there is Copernicus, then many others, even the smartest mathematicians, setting about this immense work, no longer so new, of moving the earth. Perhaps it will succeed, with the support of community, urging this speeding chariot towards its turning point, so that, because small crowds may be swung by reason, with more and more support we begin to overwhelm with authority, and perhaps through this trickery we may be able to induce knowledge of the truth; for which reason your laboring may assist all sharing the disadvantaged opinion, for they take comfort in your agreement, and are aided by your support. Perhaps your land of Italy is not stirred, they do not perceive, are not able to believe; here in Germany we do not enter into great thanks for this doctrine. Still, there are reasons which protect us from trouble. Firstly, there are divisions among a great many people, so that from no one deed could I drain out the din of so much shouting. Then, those close to me are commonfolk, for whom this is very abstruse, so that they say they do not understand, though they wonder, and do not wish to believe, should they think on it. It would be wiser to teach indifference, so that they mix caution with these mathematical struggles. In fact, they can be obsessed, I speak from experience, with the decrees of mathematical astrology, when they hear that we have ephemerides constructed out of the Copernican hypothesis.

Whoever writes ephemerides today may follow Copernicus, but it is asked that we leave out what the mathematics would demonstrate, that it is not possible to have such phenomena without the motion of the earth. For though this is not claimed or declared as self-evident [αὐτόπιστα], they do not concede to the mathematics; and since that is what is true, why is it not thrust forward as irrefutable? They stand, therefore, devoid of mathematics, which must do the better part of the work. They, because they have the same name [Ii, cum nomen idem habeant . . . ?], do not relinquish claims without demonstration, in which they are unskilled, so it takes more effort to perform. Nonetheless, there is a remedy held out: solitude. A mathematician can be anywhere; wherever it is, is best. Then if another has the same opinion elsewhere, they procure letters from them; for which reason, your demonstrations are an auspice (in that they help me), are able to awaken the imaginations in the minds of teachers, as if all professors everywhere agreed. But why deceive with this accomplishment? Have faith, Galileo, and come forward. If my guess is right, there are but a few mathematics teachers in Europe who would wish to rebel against us: such is the power of truth. If your Italy is unsuiting to publication, and if there are any impediments, perhaps Germany will grant more freedom. But enough of this. At least you write to me privately, if you dislike the public, for you have found the boon of Copernicus.

"Now it would be agreeable to ask a few observations from you: quite naturally, lacking instruments, I take refuge in others. Do you have a quadrant with which you are able to note the certainty of single scrupules and quarter degrees? If so, when it is about December 19, deduce the altitude of the tail of Ursa Major and Minor that same night. Then around December 26 observe both again and the altitude of the polar star. First observe the star and then around March 19 year 98, the altitude at night, hour 12; the other, around September 28, and also hour 12. For if, in what I have chosen, the difference in both observations comes between one and the other scrupule, by greater than ten or fifteen, for the sake of all astronomy the argument must be spread very widely; if not, there is no comprehensible difference to be captured. However, victory may be demonstrated in these most noble puzzles, which thus far none have taken a hold of, which we will report back to the community. It would be wise to say as much.

"I send two further copies, because [the messenger] Hamberg has said to me that you desire more. And whomever you will give them to, that you make an offering of the book is payment enough. Farewell, you bright fellow, and return to me with a very long letter. 13 October year 97, with thanks.

"Your courteous and loving

"Prof. Johannes Kepler

"To the brightest sir Galileo Galilei,

Paduan Mathematician, entrusted. Padua."

Kepler's letter is endearingly oblivious, in that he asks Galileo to check for parallax in the stars (as if no one had thought of that before!) and refers (as much as I can tell) to "scrupules," an old Chaldean unit of eighty arc-seconds, rather than degrees or minutes. Though he is a delight, he is also a vague annoyance, and it is easy to see why Galileo did not respond; it is unclear to me whether Galileo ever took a single naked-eye astronomical observation in his life, especially before 1598.

p. 131 **very tall with giant ears.**: I have read that Galileo was nearly two meters tall (six feet, three inches). I can find no scholarly backing for this, although Drake notes in *Pioneer Scientist* that six feet was "a very convenient height" for Galileo.

p. 131 **His right eye was off-kilter.**: This is just attributable to a convex lens, though in both drawings of Galileo by Ottavio Leoni his eyes are seriously and intentionally askew due to a fluxion. It is not known when this started. His eyes look hazel in a young portrait, but appear to have browned with age. They were always striking. Portraits are reproduced in the color plates of Heilbron, *Galileo.*

p. 132 **"to study medicine and mathematics."**: *KGW*, v. 16, Nr. 547, p. 268; also Bucciantini et al., *European*, p. 95. This chapter, "Peregrinations," is some of the best on Horky in English; also Rothman, *Strife*, pp. 168–88. The best overview of the affair is in Favaro's biography of Magini in Favaro, Antonio, and Giovanni Antonio Magini, *Carteggio Inedito Di Ticone Brahe, Giovanni Keplero, E Di Altri Celebri Astronomi E Matematici Dei Secoli Xvi. E Xvii. Con G.A. Magini, Tratto Dall' Archivio Malvezzi De' Medici in Bologna, Pubblicato Ed Illustrato Da A.F. Bologna: Nicolò Zanichelli*, 1886, pp. 118–37. There is also a decent account in Westman, *Question*, pp. 460–83. All that is known about Horky is, in effect, the letters he sent. His story is regularly shafted by Galileo biographers, but I can think of no better person to represent the impact of Galileo's astronomy on the public.

p. 132 **defeated the immature Galileo Galilei**: Drake, *Work*, p. 14.

p. 132 **the astrological findings of Domenico Novara.**: Birkenmajer, Ludwik Antoni. *Mikołaj Kopernik. Część Pierwsza: Studya Nad Pracami Kopernika Oraz Materyaly Biograficzne. Krakow: Skład Główny W Ksifgarni Spółki Wydawniczej Polskiej*, 1900, p. 424.

p. 132 **to write his biography?**: This was also asked of Galileo, as suggested earlier; Heilbron, *Galileo*, p. 119; Drake, *Studies*, p. 130. Magini declined as well as Galileo; no one wanted the task, and as far as I can tell no one, to the detriment of history of science, would attempt to write Tycho's biography until Pierre Gassendi in 1653.

p. 132 **brought Magini his copy of Kepler's first book**: *KGW*, v. 16, Nr. 548, p. 271. This letter mentions the book was given by a *nobili germano*, which Drake, *Studies*, p. 131, takes to mean Horky. It is not actually clear to me that this is the case, but I have followed Drake's assumption because I like it so much.

p. 132 **"light and leader of my country."**: *KGW*, v. 16, Nr. 547, pp. 268–9.

p. 132 **"highest haven."**: *KGW*, v. 16, Nr. 552, p. 280, ln. 11.

p. 132 **ephemerides in the Tychonic style.**: This would not precisely continue, cf. Voelkel, James R., and Owen Gingerich. "Giovanni Antonio Magini's 'Keplerian' Tables of 1614 and Their Implications for the Reception of Keplerian Astronomy in the Seventeenth Century." *Journal for the History of Astronomy* 32.3 (2001): 237–62. On the whole, Magini's scientific practice seems to open with doubt but remain amenable to novelty.

p. 132 **"Wenceslaus Horky"**: *KGW*, v. 16, Nr. 560, pp. 295, 449. Kepler apologized in *KGW*, v. 16, Nr. 574, p. 310, which Horky accepted, with the face-saving reply, "Nearly everyone from Bohemia has the name Wenceslaus" (Nr. 577). The mistake was actually a pretty serious offense and is mentioned in several other letters.

p. 132 **University of Bologna had a vast library.**: I could not find any exact figures, but as the oldest European university in existence, this must be true. A lovely work on the topic is Blair, Ann M. *Too Much to Know: Managing Scholarly Information before the Modern Age*, especially pp. 37, 161.Yale University Press, 2010.

p. 133 **all 550 copies**: Bucciantini et al., *European*, p. 84. Blair, *Too Much*, p. 52 mentions 1,000 was a common size for most works, but this does not apply to specialist astronomical texts. For context, Gingerich, *No One*, pp. 126–7 calls 1,000 "a huge number" and suggests about 550 copies were printed of Kepler's *Rudolphine Tables*, 500 copies for Copernicus's *On the Revolutions*, and 400 and 750 copies of Isaac Newton's *Principia* in the first and second printings, respectively.

p. 133 **when he got his hands on the thing**: I assume from *KGW*, v. 16, Nr. 562, p. 298, that Horky actually read the book, which makes sense, given that he was working with

Magini in Italy, but the note does not actually specifically say this. As Bucciantini et al., *European*, p. 83 quite wonderfully states, "One did not even need to know Latin to understand it, as, paradoxically, it didn't matter if one actually read it or not."

p. 133 **"named the Medicean stars."**: Galilei, *Sidereus*, pp. 26–7. One cannot approach scholarly discussion of the *Sidereal Message* without hearing about one of the most glorious displays of pedantry in the field, the extended arguments over how to translate its title. As this is a narrative, I have incorporated the change into the story, beginning with Galileo's trip to Rome. For curious readers, I simply direct them to Van Helden, *Sidereus*, p. ix–xi.

p. 134 **"face of the Earth itself"**: Galilei, *Sidereus*, p. 40.

p. 135 **"it is hardly believable."**: Ibid., p. 59; *Opere*, v. 3 pt. 1, p. 76.

p. 135 **"right up to our day."**: Ibid., p. 64. Slight changes.

p. 135 **"universe is impossible."**: Galilei, *Sidereus*, p. 84. Sharply abridged.

p. 135 **"True or false, I do not know . . ."**: *KGW*, v. 16, Nr. 562, p. 298.

p. 135 **"the reflection of the Moon"**: Bucciantini et al., *European*, pp. 90, 264; *Opere*, v. 10, Nr. 303, p. 345.

p. 135 **"howling wolves"**: *KGW*, v. 16, Nr. 564, pp. 299–300; *Opere*, v. 10, Nr. 288, p. 311. Magini was bringing Horky into a very trivial quarrel with an astronomer named Origanus over the quality of his ephemerides.

p. 135 **"four fictitious planets"**: *KGW*, v. 16, Nr. 564, pp. 299–300; *Opere*, v. 10, Nr. 288, p. 311.

p. 135 **"I love it, I kiss it"**: *KGW*, v. 16, Nr. 570, p. 306. This letter has since been lost. "To kiss" was (and, as much as I am aware, still is) an extremely common expression of joy and respect in Italy; it concludes many letters to and from Galileo. Horky the Bohemian was probably starting to pick up the Italian style.

p. 136 **"feeble messenger back to health."**: This glorious and borderline incoherent description has been translated often; *KGW*, v. 16, Nr. 570, p. 306; *Opere*, v. 10, Nr. 301, pp. 342–3; Heilbron, *Galileo*, p. 161; Wilding, *Idol*, p. 119; Bucciantini et al., *European*, p. 92.

p. 136 **"but in the heavens it deceives."**: *KGW*, v. 16, Nr. 570. p. 308.

p. 136 **"no thanks for the favors."**: *Ibid.*

p. 137 **"his golden fame."**: *Opere*, v. 3, p. 145, abridged.

p. 137 **slow, methodical style**: *Sidereal Message* was written at the rate of about 250 words a day, with at least a month of unnecessary observations. This was, for Galileo, a discomfiting speed, and he continued to work on it right up to publication. Galileo was extremely lucky he was able to maintain his priority, though the fact that he did does demonstrate the degree to which his craftsmanship exceeded other astronomers of his day.

p. 137 **"I read your *Peregrination*"**: *KGW*, v. 16, Nr. 585, p. 323, *Opere*, v. 10, Nr. 376, p. 419. Abridged to remove names.

p. 137 **whatever he could and leave town.**: *Opere*, v. 10, Nr. 334, p. 376.

p. 137 **"His opinions have been greatly damaged."**: *KGW*, v. 16, Nr. 597, p. 342, lns. 20–7, 34; *Opere*, v. 10, Nr. 419, p. 457. Abridged.

p. 137 **after a "little beauty."**: *KGW*, v. 16, Nr. 599, p. 345. This is Horky's last letter.

p. 138 **"had begun twelve years earlier."**: *KGW*, v. 4, p. 288; Kepler, *Conversation*, p. 9.

p. 138 **"most unusual thesis!"**: *KGW*, v. 4, p. 288; *Opere*, v. 3 pt. 1, p. 105; Kepler, *Conversation*, p. 9.

p. 138 **"He loves the truth"**: Kepler, *Conversation*, pp. 12–3. Abridged. I would be very much unsurprised if Kepler referenced Galileo's old letter to him in writing this.

p. 138 **It was his most read work.**: I do not think there can be actual proof of this, but consider: it was very short, and Kepler's only astronomical work which could be read by a layperson. As a fun case study, it is the only work by Kepler which Robert Burton cites in his *Anatomy of Melancholy*.

p. 138 **"humor as more pleasant."**: Kepler, *Conversation*, p. 5.

p. 138 **"hyperbolic nipple."**: Kepler, *Conversation*, p. 20, slight changes.

p. 139 **"anticipate the senses attain greatness"**: *KGW*, v. 4, p. 304; Kepler, *Conversation*, p. 37.

p. 139 **"lest that '*spicillum* burn his eye"**: *KGW*, v. 4, pp. 295–6; Kepler, *Conversation*, p. 22.

p. 139 **"proffering complete faith"**: *KGW*, v. 16, Nr. 587, p. 327; *Opere*, Nr. 379, p. 421. I have offered a largely positive picture of the Galileo-Kepler relationship, so it is worth noting that Kepler's *Conversation* was not a wholly positive work. What it was, was an unbiased and mostly favorable account from the top living astronomer. But there were criticisms cogent enough that Horky misread Kepler as opposing Galileo, and Maestlin thought Kepler had "plucked Galileo's feathers." As is often the case with good-natured, honest opinions, Kepler's words were immediately distorted by everyone to make them say whatever they wanted.

p. 139 **"Tnisoidohversesoagarowtivtonovmbullse"**: *KGW*, v. 4, p. 344; *Opere*, v. 19, p. 229. I have taken the extreme pleasure of translating these anagrams. The original was "smaismrmilmepoetaleumibunenugttaurias."

p. 139 **ciphers for his alchemy.**: Not specifically anagrams, necessarily. See Mosley, *Bearing*, p. 91.

p. 139 **"these priests one everything!"**: Reported by Martin Hasdale. Bucciantini et al., *European*, p. 117; *Opere*, v. 10, Nr. 378, p. 420.

p. 140 **"the king could own people, too!"**: This entire description is taken (with some poetical flourish) from E. Neumann, "Das Inventar der Rudolfinischen Kunstkammer von 1607/11" in *Queen Christina of Sweden, Documents and Studies* (*Analecta Reginensia*, I, Stockholm 1966), pp. 262–5, as well as Evans, *Rudolf*, pp. 181, 198. Thanks to Bucciantini et al., *European*, p. 116, for pointing out the important relation between the telescope and Rudolf's cabinet of wonders.

p. 140 **"has not been abortive."**: *Opere*, v. 3 pt. 1, p. 183; *KGW*, v. 4, p. 317. Heilbron, *Galileo*, p. 219, reports that gynecology was considered obscene, so this is quite a shocking introduction.

p. 140 **"Saith too soon: view Mars, revolving doublets."**: Kepler's attempt is *Salve umbistineum geminatum Martia proles [Hail, fiery twins, children of Mars]*; *KGW*, v. 4, p. 344. The fact that both of Kepler's guesses for both of Galileo's anagrams were actually proven true hundreds of years later is absolutely astonishing.

p. 140 **"revolted by jokes"**: Galilei, *Comets [Assayer]*, p. 185.

p. 140 **"cannot still his soul."**: Bucciantini, et al., *European*, p. 126; *Opere*, v. 10, Nr. 384, p. 426.

p. 141 **his *perspicillum* to the Venetian senate.**: *Opere*, v. 10, Nr. 231, p. 253. See Drake, *Work*, p. 141; Drake, *Studies*, p. 150, Bucciantini et. al, *European*, p. 39.

p. 141 **"exposed to the open air."**: *Opere*, v. 10, Nr. 277, p. 302; Thiene, G., and C. Basso. "Galileo as a Patient." *The Inspiration of Astronomical Phenomena VI*, v. 441. 2011, for a helpful listing of Galileo's medical conditions.

p. 141 **"all was wet with shadow."**: *Opere*, v. 10, Nr. 190, p. 213.

p. 141 **"weakness of their servant."**: *Opere*, v. 10, Nr. 277, p. 298. Much of the surrounding information comes from this very important letter.

p. 141 **"soon you will see it."**: *Opere*, v. 10, Nr. 313, p. 358. This was written to Matteo Carosio, who appears to have been attached the Marie de Medici and the French court; he is the subject of the last book (Nr. XLI) of Favoro's *Amici*, almost entirely because of this letter.

p. 141 **"arrive in Florence to serve us."**: *Opere*, v. 10, Nr. 359, p. 400.

p. 142 **"I shall always lust after."**: *Opere*, Nr. 209, p. 233, abridged; Biagioli, *Courtier*, p. 29.

p. 142 **fell to the floor like a common wench**: *Opere*, v. 11, Nr. 570, p. 173; Allan-Olney, *Private*, p. 66.

p. 142 **The very last personal letter . . . only he found time to write directly to Kepler**: Galileo wrote personally to Kepler again in 1627, in a rather anticlimactic way, recommending one of his students for employment. See *KGW*, v. 18, Nr. 1054, p. 308; *Opere*, v. 13, Nr. 1838, p. 374.

p. 142 **"laugh with you forever?"**: *Opere*, v. 10, Nr. 379, p. 423. This quote has been abridged and reordered.

p. 142 **"I have observed two moons, lo, vigils to Saturn."**: Galileo's actual solution was *altissum planetum tergenum observavi* (I have observed the highest planet as a triplet). *KGW*, v. 4, p. 345.

p. 142 **two whole months to set up house**: *Opere*, v. 10, Nr. 402, p. 440, Galileo says to Giuliano De Medici he will not be accommodated before All Saints' Day. *Opere*, v. 10, Nr. 382, p. 424, he tells Belisario Vinta he cannot travel all the way on a horse because of his "indisposition," a smart euphemism.

p. 143 **a sort of natural television.**: This description is based on the engraving of the Porto di Mezzo by Giuseppe Zocchi, which is, granted, from about 1750.

p. 143 **"live in the mountains."**: *Opere*, v. 11, Nr. 476, p. 46.

p. 143 **"Le Selve, the villa of Signor Filippo Salviati."**: *Opere*, v. 11, Nr. 461, p. 27.

p. 143 **first friend Galileo had made in court.**: Galileo's first biographer, Vincenzo Viviani, claimed Galileo had taught Salviati in Padua. This is almost certainly wrong. Caracciolo, *Salviati* ("Homo Novus"), pp. 100–103, speculates the relationship began on Galileo's earlier trips to Florence, which may be true, but it was only made concrete after Galileo moved there.

p. 143 **guiding his young bride about the floor**: Filippo married Ortensia Francesco Guadagni in 1602, with a banquet, a ball (attended by Grand Duchess Christiana), and a large breakfast next morning. See Biagioli, Mario. "Filippo Salviati: A baroque virtuoso." *Nuncius* 7.2 (1992): p. 85, hereafter cited as *Baroque*.

p. 143 **Next year he was a father**: His daughter's name was Alexandra. She was born in 1603 and died in 1610.

p. 143 **all for a household of two.**: Biagioli, *Baroque*, p. 90, ft. 1.

p. 143 **turned twenty-four**: See Salviati's funeral oration, Arrighetti, Niccolò. *Delle lodi del Sig. Filippo Salviati*, Florence, Giunti, 1614, p. 28; hereafter cited as O*razione*. The reason why this happened should be understood as the central question of Salviati's life. No one has any clue, but the answer contains, in microcosm, the entire scientific culture of baroque Italy.

p. 143 **Salviati had lately read . . . Simplicius**: Arrighetti, *Orazione*, p. 30.

p. 143 **"divine" and "superhuman"**: Favaro has compiled a listing of Galileo's references to Archimedes (over one hundred!) in *Opere*, v. 20, pp. 69–70.

p. 143 **"the liberal philosophy."**: *Opere*, v. 11, Nr. 878, p. 510. See the very good essay in *Salviati*, "Filippo Salviati filosofo libero, un homo novus accanto a Galileo" by Allì Caracciolo, pp. 77–127.

p. 144 **They invited women to speak**: I am alluding here to Margherita Sarrocchi, who is the subject of the first book in Favaro's *Amici*. See, in particular,r the elegant Ray, Meredith K. *Margherita Sarrocchi's Letters to Galileo: Astronomy, Astrology, and Poetics in Seventeenth-Century Italy*, p. 17–8 Springer, 2016.

p. 144 **the word "new," *novus, nuovo***: This hardly needs evidence, but the word *nuovo* appears in Salviati's brief funeral oration seventeen times! Novelty is quite the opposite of the usual purpose for a panegyric. Cf. Thorndike, Lynn. "Newness and Craving for Novelty in Seventeenth-Century Science and Medicine." *Journal of the History of Ideas* (1951): 584–98.

p. 144 **Sagredo would reminisce of his foreign excursion**: He mentioned it repeatedly up to a year before his death. See *Opere*, v. 12, pp. 258, 273, 335, 342, 377, 452.

p. 144 **"father but greetings and stories."**: *Opere*, v. 11, Nr. 915, p. 553.

p. 144 **"loaded up like a right jackass."**: *Opere*, v. 10, Nr. 219, pp. 242–3.

p. 144 **"made to serve me, and not me it."**: *Opere*, v. 11, Nr. 915, p. 554.

p. 144 **"the true philosophy."**: Favaro, *Amici*, Bk. 8, p. 16, Opere, v. 12, Nr. 1096, p. 157.

p. 145 **"having to spend a penny."**: *Opere*, v. 10, Nr. 287, p. 310. Abridged.

p. 145 **"had noticed any such thing."**: *Opere*, v. 10, Nr. 434, p. 481.

p. 145 **the Moon was quite brilliant.**: The above diagram is taken almost precisely from the lovely article "Copernicus Didn't Predict the Phases of Venus" by Neil Thomason in *1543 and All that: Image and Word, Change and Continuity in the Proto-scientific Revolution*, v. 13, p. 293, edited by Guy Freeland and Anthony Corones. Springer Science & Business Media, 2013. It is a frequently used picture; Heilbron, *Galileo*, p. 168.

p. 145 **Galileo had requested a copy of Kepler's Optics**: *Opere*, v. 10, Nr. 402, p. 441; Bucciantini et al., *European*, p. 64. He also requested Kepler's *On the New Star*. It is not clear that he received them.

p. 146 **"the smallest part of what you describe."**: This was Johannes Papius. From Görlich, Paul. "14.5. Kepler's Optical achievements." *Vistas in Astronomy* 18 (1975): pp. 843–4; *KGW*, v. 15, Nr. 375, p. 314. Abridged.

p. 146 **"As I oft collect these darn immaturities vainly, roughing men . . . oh no!"**: Galileo's actual anagram was *Haec immatura a me iam frustra leguntur o.y.* (These immaturities are now brought together by me in vain o.y.), *Opere*, v. 10, Nr. 435, p. 483.

p. 146 **"I look forward to what Kepler has to say"**: *Opere*, v. 10, Nr. 435, p. 483. It is worth noting the full sentence: "I look forward to what Kepler has to say concerning the extravagances of Saturn," that is, concerning the old (solved) anagram, not the new one.

p. 146 **"Five out the Sun, is rotating a melancholy, monolithic red smearing."**: *Opere*, v. 11, Nr. 455, p. 15; *KGW*, v. 16, Nr. 604, p. 357. Kepler's guess was *nam Jovem gyrari macula hem rufa testatur* (My my! A red scar doth testify to circle Jupiter). As before, the fact that Jupiter really does have a red spot, which was not discovered for hundreds of years, appears to be one of the more alarming coincidences in the history of science. Caspar, *Kepler*, p. 201, translates the top part of this letter.

p. 146 **"Hello. Carnal mother Venus imitates Moon-God Cynthia in its figure."**: *Opere*, v. 11, Nr. 451, p. 12; *KGW*, v. 4, p. 348. Galileo's actual solution was *Cynthiae figuras aemulatur mater amorum* (the mother of love imitates the forms of Cynthia).

p. 147 **"those who wish me ill."**: *Opere*, v. 11, Nr. 497, p. 71; Shea and Artigas, *Rome*, p. 29.

p. 147 **Galileo had arrived "in good health"**: *Opere*, v. 11, Nr. 505, p. 79; Shea and Artigas, *Rome*, p. 31.

p. 147 **men would come and go, talking.**: This is an inverted poke (again) at Eliot's *The Love Song of J. Alfred Prufrock*. The making of the modern condition is slow, but it is sure!

p. 148 **Kepler's Diopter**: *"Dioptrice seu Demonstratio eorum quae visui & visibilibus propter Conspicilla non ita pridem inventa accidunt. Praemissa Epistolae Galilei di iis, que post editionem Nuncii Siderii ope Perspicilli; Nova & admiranda in caelo deprehensa sunt. Item Examen prefationis Ioannis Penae Galli in Optica Euclidis, de usu Optices in philosophia* (Diopter, or, Proofs concerning visions and sightings with a certain Instrument which happened to be invented not so long ago)." Preceded by *Letters from Galileo* on this, composed after the Sidereal Messenger with the help of the *Perspicillum*, wherein new and admirable things are found in the sky. Also, a consideration of the preface to Euclid's *Optics*, by Jean Pena the Gaul, on the uses of *Optics* in philosophy.

p. 148 **knew exactly what they said.**: From Bucciantini, *Keplero*, p. 200: "It is almost certain that Kepler did not inform Galileo of his hurried project to insert the letters of Giuliano de Medici [their intermediary] into the introduction of *Diopter*."

p. 148 **reinvented for his own purposes.**: Caspar, *Kepler*, p. 198; Neugebauer, *Ancient*, p. 893.

p. 148 **"by way of the inherent force of mathematics."**: *KGW*, v. 4, p. 331.

p. 148 **"this is not an easy book to understand"**: *KGW*, v. 4, p. 334; *Kepler*, Caspar, p. 199.

p. 149 **"but flip it around."**: *KGW*, v. 4, p. 387. *Diopter* has never been translated to English, but the article Malet, Antoni. "Kepler and the telescope." *Annals of science* 60.2 (2003): 107–36, has been enormously helpful. Also Ronchi, Vasco, trans. Edward Rosen. *Optics: the science of vision*. Courier Corporation, 1991, pp. 43–54. I have skirted around Kepler's optical work, but readers interested in the history of optics would do very well to begin with Ronchi's lovely book.

p. 149 **"convex and concave, respectively."**: Galilei, *Sidereus*, p. 37. Simplified. Bucciantini et al., *European*, p. 1, notes that some of Galileo's *perspicilla* had a biconvex objective lens.

p. 149 **"proved from its causes."**: *Opere*, v. 11, Nr. 455, p. 16; *KGW*, v. 16, Nr. 604, p. 358.

p. 150 **"not even the author seemed to understand it."**: This diary entry is from Jean Tarde, *Opere*, v. 19, pp. 589–90.

p. 150 **habit of taking his colleagues out to lunch**: Kepler does this with Martin Hasdale, *Opere*, v. 10, Nr. 375, p. 418. I thought it was such a lovely and timeless petty detail that I had to include it. I expect he was doing the same with Marius.

p. 150 **Io, Europa, Ganymede, Callisto**: Kepler proposed this to Simon Marius, which Marius reported in his *Mundus Iovialis*; Pasachoff, Jay M. "Simon Marius's Mundus Iovialis: 400th Anniversary in Galileo's Shadow." *Journal for the History of Astronomy* 46.2 (2015): p. 229; Owen, Tobias. "Jovian satellite nomenclature." *Icarus* 29.1 (1976): pp. 161–2. Marius had a relatively tame spat with Galileo on the discovery of the four planets, which is beyond the scope of this telling. Rosen has composed a short biography in *Dictionary of Scientific Biography* which is the most serviceable introduction for an interested reader.

p. 151 **"cardinal embraced me warmly"**: *Opere*, v. 11, p. 31; Shea and Artigas, *Rome*, p. 31.

p. 151 **"ornamented rooms, palaces, gardens, and more."**: *Opere*, v. 11, Nr. 517, p. 89.

p. 151 **formally verified . . . two days later.**: In the response to Cardinal Bellarmine; *Opere*, v. 11, Nr. 520, pp. 92–3; English translation in Brodrick, *Bellarmine*, p. 144.

p. 151 **just to have something to laugh at.**: This was the publications of Francesco Sizzi, see *Opere*, v. 11, Nr. 517, p. 91; Drake, Stillman. "Galileo Gleanings III: A Kind Word for Sizzi." *Isis* 49.2 (1958): 155–65.

p. 151 **China and Europe had known about sunspots for millennia**: See the first chapter in Reeves and Van Helden's superlative Galilei, *Sunspots*, pp. 9–24.

p. 151 **There were only new things under the Sun.**: Riffing on Ecclesiastes 1:9, "There is nothing new under the sun."

p. 151 **fearing the Sun would burn their eyes.**: This is from Otto von Maelcote; *Opere*, v. 11, Nr. 810, p. 445; *KGW*, v. 17, Nr. 641, p. 37; Galilei, *Sunspots*, p. 47.

p. 151 **"prefer to gain a friend."**: *Opere*, v. 11, Nr. 517, p. 91.

p. 152 **that his son was gay.**: Bindman, *Pedagogy*, p. 36; Freedberg, David. *The eye of the lynx: Galileo, his friends, and the beginnings of modern natural history*, p. 66. University of Chicago Press, 2003. Gabrieli, *Contributi*, p. 4. From Gabrieli, Cesi's relationship with his mother (she loved him too much) is also of some psychological interest. I have certainly indulged a fair amount of pot-boiling in my biography of Cesi; he is too much fun for me to deny it. I do not, from what I have read, believe the accusations of homosexuality were true, although it is very much possible. A great many things can happen when there is so much repression going on.

p. 152 **a total freak.**: After reading through several commentaries on Cesi, as well as a not trivial amount of the source material, I have come to agree whole–heartedly with Richard Westfall's analysis in "Galileo and the Accademia Dei Lincei," *Galluzzi, Paolo. "Novità celesti e crisi del sapere." Novità celesti e crisi del sapere. P. Galluzzi (Editor). Supplemento agli Annali dell'Istituto e Museo di Storia della Scienza, Monografia N. 7, Florence, Italy. (1983)*, pp. 189–200. My one complaint is that Westfall shows too much restraint.

p. 152 **"Love for woman is profane"**: See Biagioli, Mario. "Knowledge, Freedom, and Brotherly Love: Homosociality and the Accademia dei Lincei." *Configurations* 3.2 (1995), p. 141, for a convenient listing of Cesi's woman-hatred.

p. 152 **"an emasculating repose."**: Carutti, Domenico. *Breve storia della Accademia dei Lincei*. R. Accademia coi tipi del Salviucci, 1883. p. 8.

p. 152 **"Sky-Wanderer . . . The Illuminated One . . . Eclipsoid . . . Mr. Tardy."**: I have had a great deal of fun with these translations. They are *Caelivagus* (Federico Cesi), *Illuminatus* (Johannes Eck), *l'Eclissato* (Anastasio De Filiis), and *Tardigrado* (Francesco Stelluti).

p. 152 **"but what is hidden inside."**: Eamon, William. *Science and the Secrets of Nature: Books of Secrets in Medieval and Early Modern Culture*, p. 231. Princeton, NJ: Princeton University Press, 1996. Slight changes.

p. 152 **"lynxiness" . . . "confraternity."**: Lynxiness, that is, *lincealita*. See Bindman, *Pedagogy*, p. 3, for a convenient listing of "brothership" terms.

p. 152 **One member called it a religion.**: This was Giambattista Della Porta, author of *Natural Magic*, after joining, see Gabrieli, *Contributi*, p. 14.

p. 153 **"I laugh at them madly."**: Bindman, *Pedagogy*, p. 47; Gabrieli, *Contributi*, p. 5. Slight changes.

p. 153 **"Galileo has the eyes of a lynx."**: This was written to Martin Horky on June 7, 1610! See *KGW*, v. 16, Nr. 580, p. 315.

p. 153 **"The Drinkers" . . . "The Lazy Men."**: Drake, *Studies*, p. 80.

p. 153 **"Praise God."**: From the *Praescriptiones* for the academy, which was finally published in 1624. Reprinted in Odescalchi, Baldassare. *Memorie istorico critiche dell'Accademia de'Lincei e del Principe Federico Cesi, Secondo Duca d'Aquasparta, fondatore e principe della Medesima*. Salvioni, 1806, pp. 307–17. Also Drinkwater-Bethune, *Life*, p. 37; Drake, *Studies*, p. 81.

p. 154 **"old discoverer."**: I began collecting names for the telescope a while ago, only to find most of them and more given conveniently in Bucciantini et al., *European*, p. 10. See also Allan-Olney, *Private*, p. 54.

p. 154 **"it was that boy, Cesi."**: Virtually all information on the naming of the telescope comes from Rosen, Edward. "The Naming of the Telescope." *New York, H. Schuman* (1947). Page 31 has the key letter. Page 67 has the key deduction, "the term telescope was originally devised by Demisiani a publicly unveiled by Cesi at the banquet in Galileo's honor on April 14, 1611." Rosen's deductions appear to be universally followed, though, as he states, the historical record is less than exemplary.

p. 154 **"I am a man possessed by the devil." . . . black magic.**: Evans, *Rudolf*, p. 198. This was relayed by Philipp Lang, a member of the court, who the fairly impartial Evans refers to as an "evil genius."

p. 154 **"devoured my time and money."**: *KGW*, v. 17, Nr. 710, p. 137, ln. 27. It is not entirely clear from the line whether or not Kepler was forced to stay, although I don't think it likely.

p. 155 **"while they are jokes."**: This painting is *Vertumnus* by Giuseppe Arcimbaldo, 70.5 cm x 57.5 cm. This portrait was meant to indicate Rudolf as a providential god, but with the hindsight of historical and aesthetic sense, it appears to me to indicate the exact opposite. Kepler must have known about the work, although I cannot find any comment from him. On Galileo, see *Opere*, v. 5, p. 190; Galilei, *Sunspots*, p. 257; Panofsky, Erwin. "Galileo as a Critic of the Arts: Aesthetic Attitude and Scientific Thought." *Isis* 47.1 (1956): p. 7.

p. 155 **a rube, a charlatan, a coward**: Christianson, *Island*, pp. 366–72, is an excellent English-language precis of Grindely, *Zeit*, p. 38, 59, 60, 192.

p. 156 **"affliction of body which I sustained."**: This account is from William Lithgow, a traveling English Protestant who was taken as a spy by the Spanish Inquisition (he did not expect this). This account is very long in the source and has been fiercely restructured and abridged. From G.R. Scott, *A History of Torture* (1959), pp. 172–6; Lithgow, *The Totall Discourse of the Rare Adventures & Painefull Peregrinations of long Nineteene Yeares Travayles from Scotland to the most famous Kingdomes in Europe, Asia and Affrica* (1906), pp. 398–407.

p. 156 **the length of time it takes to say one "Our Father."**: Grindely, *Zeit*, p. 254. Notably, during his torture, Tengnagel repudiated what he said beforehand, replacing what appears to have been the truth with an uncertainty.

p. 157 **commissioned by Salviati**: see Allì Caracciolo, *Salviati* ("Homo Novus"), p. 104; Gabrieli, *Contributi*, p. 986. Ironically, given what I say directly after this line, these paintings do not appear to have survived.

p. 157 **"of good mind, sense, and quality."**: *Opere*, v. 11, Nr. 720, p. 351; Gabrieli, *Contributi*, p. 973.

p. 157 **under Galileo's sway.**: From circumstantial evidence. It is easy to argue the opposite occurred. I suspect there was mutual influence. This is sometimes called the "Galilean turn" (away from Della Porta); see Bindman, *Pedagogy*, p. 74.

p. 157 **a special "Friend of the Lynx" title**: Not exclusively for Castelli; see Gabrieli, *Contributi*, p. 16. "Friend of the Lynx" is a little too official; Gabrieli just says it was a "special category of 'friend' of the Academy."

p. 157 **Cesi wrote Galileo . . . Sagredo was investigating . . . Salviati cajoled**: *Opere*, v. 11: Nr. 732, p. 366, Nr. 687, p. 315, Nr. 668, p. 290, respectively. Salviati wanted to discuss the playwright Ruzzante.

p. 157 **named Maffeo Barberini**: The best source on Maffeo Barberini's life is still his official biography by Andrea Nicoletti, *Della Vita di Papa Urbano VIII, e historia del suo pontificato*, although it not easily accessible. The best English resource is Rietbergen, Peter. *Power and religion in baroque Rome: Barberini cultural policies*. Brill, 2006. V. 28 of Ludwig Pastor's *History of the Popes* is also useful.

p. 157 **"sign of a rare intelligence."**: *Opere*, v. 11, Nr. 690, p. 318. Galileo's letter is Nr. 684, translated in Appendix Two of Galilei, Sunspots, p. 337.

p. 158 **"you who introduced us."**: *Opere*, v. 12, Nr. 999, p. 53; Gabrieli, *Contributi*, p. 977. Gabrieli p. 986 points to Salviati's exact date of death as March 22, 1614.

p. 158 **hour to read; Cesi did so himself**: Westfall, "Galileo and the Accademia dei Lincei" in *Celesti*, p. 197. It is not clear to me that Galileo was present, although I imagine he must have been.

p. 158 **"our esteemed academician Galilei . . ."**: Arrighetti, Niccolò. *Delle lodi del Sig. Filippo Salviati*, Florence, Giunti, 1614, pp. 34–5. Notably, p. 38 contains the phrase "grand machine of the universe." Translated with some freedom.

p. 158 **a history of the Counter-Reformation.**: Wilding, *Idol*, p. 104. The book was the tenth volume of Caesar Baronius's *History of the Council of Trent*—775 pages.

p. 159 **"Ipernicus" . . . "serve you as my master."**: *Opere*, v. 11, Nr. 793, p. 427; Shea and Artigas, *Rome*, p. 52; Drake, *Work*, p. 197.

p. 159 **"The Hounds of God."**: I myself am here punning on "dogmatic." See the colored plates for the Dominicans in Heilbron, *Galileo*. In this author's mind, the only true hound of God was, quite plainly, Johannes Kepler.

p. 159 **"the satisfaction he pursues."**: Ricci-Riccardi, *Caccini*, p. 21. I found the Italian very confusing here. Caccini was preaching in a major cathedral at about the age of thirty-five, which is very young indeed.

p. 159 **"Friars, at times, have an excessive temper."**: Ricco-Riccardi, *Caccini*, p. 31. "Excessive temper" could also be "excess of humors."

p. 159 **"so I laugh at them."**: *Opere*, v. 11, Nr. 827, p. 461; Shea and Artigas, *Rome*, p. 53. Slightly abridged.

p. 160 **"because it had a broad and flat shape."**: Drake, *Studies*, p. 165. Edited for ease of reading. Drake's essay here is the clearest on this conflict in English. There is a more thorough analysis in Biagioli, *Courtier*, chapter 3. I agree very much with Biagioli that there was a "lack of closure" in this instance.

p. 160 **"even if it is thinner than paper."**: Drake, *Work*, p. 173. Italics added. The man with the ebony chips was Ludovico delle Colombe. *Columba* is Italian for "dove" or "pigeon." This conflict resulted in Galileo writing a small treatise called *Discourse on Floating Bodies*, but I do not think this work and dispute have much to contribute to my story, so I have not mentioned them. Drake has published a translation in his *Cause, Experiment, and Science* (1981) and was also responsible for the republication of a much older translation by Thomas Salusbury.

p. 160 **"anti-Galileists, out of respect for Aristotle."**: Shea and Artigas, *Rome*, p. 51.

p. 160 **"joking, mockery, and ridicule."**: This line is ripped with slight changes from Milan Kundera's *Book of Laughter and Forgetting*. I first read the book a very long time ago, and have only recently realized how much effect it has had on this book—especially the section "The Angels." From the section "Petrarch Condemns Boccaccio's Laughter": "'Laughter, on the other hand,' Petrarch went on, 'is an explosion that tears us away from the world and throws us back into our own cold solitude. Joking is a barrier between man and the world.'" Compare this laughter to the laughter I assign to Kepler, or to Galileo himself while in Venice.

p. 161 **"certain friend of mine" . . . "behind the painting."**: Galilei, *Sunspots*, pp. 60, 331. The interlocutor was Marc Welser. I have chosen to largely skip over the fairly complex conflicts of Galileo's 1611–5 period, using them exclusively as setup to Galileo's trial. *On Sunspots* is a very rich, thorough, and well-constructed account of the sunspot disputes on their own terms, which I highly recommend.

p. 161 **"the insults of spots"**: Galilei, *Sunspots*, p. 67.

p. 161 **"the heaven of the Sun."**: Ibid., p. 69.

p. 161 **"fancies once imposed upon him."**: Ibid., p. 95, edited for ease of reading.

p. 161 **"Some arise and others disappear."**: Ibid., *Sunspots*, p. 109.

p. 161 **"this matter will be considered closed."**: Ibid., *Sunspots*, p. 103, edited for ease of reading.

p. 161 **a technique written in Kepler's *Optics***: I think we may safely deduce from this that Galileo either never received, read, or understood the copy of Kepler's *Optics* he requested earlier in the text, or in *Opere*, v. 10, Nr. 402, p. 441. He quotes from it much later in *The Assayer*, however.

p. 161 **social cues from etiquette books**: Biagioli, *Courtier*, p. 114, has a small list of etiquette books which Galileo owned or read. To allow a bit of character analysis, Galileo is very often accused of being intentionally mean. It seems to me more accurate that he suffered from an inability to understand the indirect social cues society is filled with.

p. 161 **"scorches worse than sour words"**: *Opere*, v. 11, Nr. 795, p. 429, Galilei, *Sunspots*, p. 239.

p. 162 **"you must face the storm."**: *Opere*, v. 11, Nr. 849, p. 485; Galilei, *Sunspots*, p. 248. Sent by Ludovico Cigoli.

p. 162 ***saufen***: Rublack, *Witch*, p. 50.

p. 162 **"through the Alps"**: *KGW*, v. 9, p. 12.

p. 162 **specifically for her well-being.**: *KGW.* v. 5, p. 225: "I hope for a rectified future; at this point in time I have established what will have to be done, and chiefly taken up the cause of renewal for the good of my wife and work."

p. 163 **"stinky breath" . . . "torn between so many parties?"**: *KGW*, v. 17, Nr. 669, p. 80. There is a longer account in Koestler, *Sleepwalkers*, pp. 399–405, which is tainted by Koestler's sexual predilections. Koestler attempts to show that Kepler's actions here are like Koestler's understanding of his scientific work, in that he is "sleepwalking" into the "correct choice" "without a will of his own." But Kepler, even within Koestler's own translation, considers his actions here to be a moral (and therefore conscious) failing. The correct metaphor, which I have here pursued, is that of being drunk.

p. 163 **"without reason or computation."**: *KGW*, v. 9, p. 9.

p. 163 *The New Measurement of Wine Barrels*: *Nova Stereometria Dolorium Vinariorum, in Primis Austriaci, figurae omnium aptissimae; et Usus in eo Virgae Cubicae compendiosissimus et plane singularis. Accessit Stereometriae Archimediae Supplementum* (The New Measurement of Wine Barrels, Principally Austrian, Suiting All Shapes; and the Appropriate Usage of the Measuring Rod for Cubics and Planes. Adjoined by a Supplement on Archimedean Measurement). Commonly known as *Stereometria Doliorum*, this abbreviation misses the absolutely hilarious "new" in the title. A small amount of this work has been translated to English in Struik, *A Source Book for Mathematics* (1969), pp. 192–7, which points to a full German translation "in Ostwald's *Klassiker*, No. 165, ed. R. Klug (Engelmann, Leipzig, 1908)."

p. 164 **magical codex**: I am dissenting here with current scholarly opinion, mostly from Voelkel, *Composition*, and Stephenson, *Physical*, which considers the work as Kepler's effort to be persuasive. I believe that if you eat, sleep, and breathe Kepler long enough, this is how the work will begin to appear, but such a reading could not be reasonably held by his contemporaries or any common reader. Kepler recognized his own obscurity: "Only a few copies of the book were printed, as it had so to speak hidden the teaching about celestial causes in thickets of calculations."

p. 164 **Archimedes was being reforged**: Archimedes was, of course, not dead in the Middle Ages, but he was not celebrated alongside Euclid and Aristotle, as he is today. The standard reference for this is Marshall Clagett's epic sourcebook *Archimedes in the Middle Ages*, see especially the very first essay, "The Impact of Archimedes on Medieval Science": "I think we must conclude that he played a modest but nevertheless important part in that thin tradition of Greek mathematics and physics that trickled down through the Middle Ages into modern times."

p. 165 **"Until now" . . . "true figure of a barrel"**: *KGW*, v. 9, p. 38, 43. Translated with a bit of freedom.

p. 165 **"Aetna" . . . "belly button" . . . "with a foreskin."**: *KGW*, v. 9, pp. 42, 43, 46. Mathematicians and scientists today rarely, if ever, inform their audience that they are uncircumsized.

p. 166 **"there's Comedy in Susanna."**: *KGW*, v. 17, Nr. 672, p. 89, lns. 1–20, 35, 43. This friend was, I kid you not, one "Dr. Seuss" (Johannes Seussius). I have had a lot of fun with this poem, and it should be understood as an abridged, groovy reinterpretation.

p. 166 **Kepler's stepdaughter scolded him.**: *KGW*, v. 17, Nr. 634, p. 24.

p. 166 **she enjoyed sex**: Rublack, *Witch*, p. 147; Kepler, *Harmony*, p. 360, "a joyful wife, that is, one who perceives what is happening to her with pleasure, and helps her husband with suitable motions." See also Kepler, *Harmony* pp. 309–10.

p. 166 **and being cared for by.**: Kepler's trip up the Danube is in *KGW*, v. 17, Nr. 783, p. 254; Kepler, *Somnium*, p. 184. This is the same trip he makes following his mother, where he reads Vincenzo Galilei. His stepdaughter was named Regina. His other daughter, confusingly, was named Susanna. Kepler is protective concerning his sister Margaretha in *KGW*, v. 15, Nr. 376, p. 315. Rublack's *Witch* is chock full of info on the Kepler women (though a little free in its sourcing, it remains a very helpful and scholarly work); see pp. 42, 125 (deafness), 203, 292, especially.

p. 166 **the same small town**: Kepler's mother moved around a bit, from Eltingen to Weil der Stadt to Leonberg, but these were all basically the same place. I am reminded of William Cobbett: "It is a great error to suppose that people are rendered stupid by remaining always in the same place."

p. 167 **"wise-woman"**: From the helpful comments in *KGW*, v. 11, pt. 2, p. 477. This is a euphemism. Kepler thought she was hard-working and cared about education, but was fundamentally ignorant. But he thought almost everyone was ignorant. One of the defenses he offered in her trial was that she was senile.

p. 167 **"the mother who gives birth to science."**: *KGW*, v. 11, pt. 2, p. 33; Kepler, *Somnium*, p. 36. Kepler really did write this thinking of his own mother; it concerns a character in his *Somnium* named Fiolxhilde, who was based on her. Rublack, *Witch*, p. 283, attempts to show Fiolxhilde was not based on Kepler's mother (I disagree.).

p. 167 **"take back her spell, before I die!"**: Rublack, *Witch*, p. 76. The "town governor" was Lukas Einhorn.

p. 167 **"another cursed spell"**: There is a very convenient listing of twenty preliminary accusations in *OO*, v. 8, pt. 1, p. 443. They grow increasingly irreverent.

p. 168 **"my heart burst from my body."**: Caspar, *Kepler*, p. 240; *KGW*, v. 17, Nr. 725, p. 155; *OO*, v. 8, pt. 1, p. 363.

p. 168 **"Johannes Kepler. Alas."**: Caspar, *Kepler*, p. 254; Koestler, *Sleepwalkers*, p. 386.

p. 168 **"nothing to confess."**: Caspar, *Kepler*, p. 255.

p. 169 **She had been caring for his little son**: Favaro, *Maria*, p. 97; Heilbron, *Galileo*, p. 164. Many texts draw on an irrelevant marriage certificate to suggest Marina Gamba lived and remarried. This story has enough names that I do not refer to Virginia and Livia by their new convent titles after taking the cloth, which were Sister Maria Celeste and Sister Arcangela, respectively.

p. 169 **"waiting room."**: Galilei, *Father*, p. 17.

p. 169 **Lesbianism was uncommon but not unheard of**: Most notably Benedetta Carlini, detailed enjoyably in Judith C. Brown's book *Immodest Acts: The Life of a Lesbian Nun in Renaissance Italy*. Much more common than this were affairs with men from outside the convent.

p. 169 **steal away a knife . . . and end themselves.**: Virginia reports a very similar event to Galileo; see Galilei, *Father*, p. 93; Allan-Olney, *Private*, p. 161. I do not mean to construct a world of hysterical suicidal gay nuns and have been shocked to find such fantasies repeatedly exampled. This does no damage to the happiness of other nuns, the potentiality of celibacy and female-driven communities, or the spiritual and economical value of a life spent in service to God. It does damage to the kind of society which forces women into it. Religion, like therapy, works wonders on the willing, at best.

p. 169 **eating disorder in her twenties.**: Galilei, *Father*, p. 53. I am inducing a great deal from scanty historical evidence, but one has to wonder after this whether her routine "purges" were entirely medical.

p. 170 **He put a baby in her**: Federico Cesi married Artemesia Colonna in 1614, born 1600, died 1616. Her youth makes it unsurprising she died in childbirth. The article "Dispersed Collections of Scientific Books: The Case of the Private Library of Federico Cesi (1585–1630)" by Maria Teresa Biagetti in *Lost Books: Reconstructing the Print World of Pre–Industrial Europe*, p. 389, states it was "in the aftermath of the premature birth of twins," although I have seen not seen this cause of death repeated elsewhere.

p. 170 **cousin of the late Filippo Salviati**: This was Isabella Salviati, married in 1617. I could not figure out their actual relation, but it is not brother and sister, as some secondary sources report. His first bride was also related to a Lynx, Fabio Colonna.

p. 170 **"philosophical militia."**: Bindman, *Pedagogy*, p. 111.

p. 170 **his mentor's post at university.**: The University of Pisa, not Padua.

p. 171 **"Only Her Ladyship contradicted me . . ."**: *Opere*, v. 11, Nr. 956, p. 605; Drake, *Work*, p. 222; Galilei, *Affair*, p. 47. I have censored this quote of all insignificant names and byzantine syntax. The Duchess's "friend" is Cosimo Boscaglia, professor of philosophy. The "other Medici", both times, was Don Antonio. The additional people in the Duchess's chambers were the Archduchess and Don Paolo Giordano.

On the Castelli affair, Heilbron, *Galileo*, p. 205 notes "The speculations of mathematicians are a sort of drunkenness."

p. 172 **"a great, perhaps the greatest, advantage."**: *Opere*, v. 5, p. 285; Galilei, *Affair*, p. 53; Drake, *Work*, p. 227. I have been unable to resist the neologism "speak truth to power" (there is no "to power"). I have added "as the Bible" for clarity.

p. 173 **quoted from the pulpit**: It is a legendary piece of Galileo historiography that Caccini quoted from the Book of Acts, but there is no primary source confirmation of this, and the earliest mention is from the 1770s; see Galilei, *Affair*, p. 330.

p. 173 **it is said, all of mathematics.**: This was said by Galileo; see Galilei, *Affair*, p. 55.

p. 173 **a happy accident.**: Tommaso claimed in his deposition that Lorini showed him Galileo's letter to Castelli only after his sermon, see Galilei, *Affair*, p. 138.

p. 173 **men of letters traded epistles**: Galileo does this several times, sending Kepler's letters to his friends, and Kepler published Galileo's letters as well. Tycho Brahe collected letters from many friends for publication. It was expected behavior, although this is not to say that Galileo expected it; from Galilei, *Affair*, p. 62 (abridged): "It is, you see, a private letter, to be read only by him. He let it be copied without my knowledge."

p. 174 **"I wish you the best."**: Ricci-Riccardi, *Caccini*, p. 69; Santillana, *Crime*, p. 104. Strongly abridged. Sent by Matteo. Ricci calls these "the dreadful words of January 2." I have had a lot of fun translating them, and they are not especially faithful except in tone. Though Tommaso continued to inveigh against Galileo, it does appear that he actually did stop preaching.

p. 174 **"let himself go so far."**: *Opere*, v. 12, Nr. 1065, p. 123; Drake, *Work*, p. 239. Here meaning Pisa.

p. 175 **"a secret, as I am sure you will."**: *Opere*, v. 19, p. 297; Galilei, *Affair*, p. 134. Strongly abridged.

p. 175 **"has hit me again!"**: *Opere*, v. 5, p., 291, Galilei, *Affair*, p. 55; Drake, *Work*, p. 240.

p. 175 **"I should pluck out my eyes before I sin."**: Galilei, *Affair*, p. 58; Drake, *Work*, p. 243. Abridged. Written to Piero Dini.

p. 175 **"I do not wish to quarrel with anyone"**: I am jumping ahead here slightly, to the *Letter to the Grand Duchess*. Galilei, *Affair*, p. 91.

p. 175 **Benedetto furnished him with a huge base of theological scholarship.**: Westfall, *Essays*, "Bellarmino and Galileo," p. 19, says that Galileo received the help of Castelli. I think this is very likely and have followed it, but have not found any irrefutable evidence for it.

p. 175 **Some hundred odd pages of notes**: These were published posthumously as "Considerations on the Copernican Opinion," beginning on *Opere*, v. 5, p. 349. They are translated in the second chapter of Galilei's *Affair*. I do not know exactly how many pages of notes there were; my off-the-cuff guess stems from scans of Galileo's other letters I have seen.

p. 175 **"All cannot be written."**: Ricci–Riccardi, *Caccini*, p. 111.

p. 176 **a little man at the front desk**: This would have been "Reverend Father Michelangelo Segizzi, O.P., Master of Sacred Theology and Commissary General of the Holy Roman and Universal Inquisition, etc." See Galilei, *Affair*, p. 136. It is not entirely clear to me, but I believe this was taking place in the Palace of the Holy Office, which is what I'm describing. I thought the details necessary for narrative, but this paragraph is a bit of imagination.

p. 176 **the confessor to a local nunnery.**: And this he did; see Ricci-Riccardi, *Caccini*, p. 195. It seems to me Santillana, *Crime*, and Ricci both exaggerate Caccini's involvement. I believe he had no further influence on Galileo's life, excepting some extremely minor intrigues which stopped about 1620. Many selections from the letters of the Caccini brothers also appear in v. 18 of *Opere*.

p. 176 **"I do not even know what he looks like."**: Galilei, *Affair*, p. 139. Other quotes relating to Caccini's brief deposition are therein.

p. 176 **"first author cannot recognize them."**: *Opere*, v. 12, Nr. 1085, p. 146; Shea and Artigas, *Rome*, p. 64. From Giovanni Ciampoli. Abridged.

p. 177 **"He is now preaching in Rome."**: *Opere*, v. 12, Nr. 1089, p. 150; Shea and Artigas, *Rome*, p. 67. Slightly abridged.

p. 177 **"For you will please no one and fight quite a few."**: "*Iudicium populi numquam contempseris unus / Ne nulli placeas, dum vis contemnere multos.*" From *Distichs of Cato*.

p. 177 **"Hell is the center of the Earth, not the World."**: This is from Paolo Antonio Foscarini's *Sopra l'opinione de' Pittagorici, e del Copernico. Della mobilita' della terra*. It is has been archaically Englished by Thomas Salusbury. Both editions can be found online. This narrative mostly focuses on Galileo's perspective, but from a theological view, it is difficult to express what a bombshell this book was.

p. 177 **"the salvation we stress about."**: *Opere*, v. 5, p. 340; Galilei, *Affair*, p. 112.

p. 178 **"the number of intellectuals."**: *Opere*, v. 5, p. 322, Galilei, *Affair*, p. 98.

p. 178 **"the tiny brains of common people."**: *Opere*, v. 5, p. 333, Galilei, *Affair*, p. 106. One thinks of Edmund Burke's famous, "Along with its natural protectors and guardians, learning will be cast into the mire, and trodden down under the hoofs of a swinish multitude."

p. 178 **"than help them up."**: *Opere*, v. 5, p. 311, Galilei, *Affair*, p. 89.

I will offer some brief commentary on Galileo's *Letter to the Grand Duchess* here. I have seen it praised widely. It is garbage. What Galileo needed to do here, assuming there was no compromise, was demonstrate the predominance and mutual exclusion of physical reasoning over the literal interpretation of scripture in philosophizing about nature. What he does, instead, is assume this, then proceed to spend thirty-five pages telling theologians what they ought to do in light of this apparent fact. I would be shocked if it convinced a single soul. The kindest reading I could offer is that Galileo had no desire to convince anyone and simply felt a moral imperative to bear witness to the ludicrous gaff he was increasingly certain the Catholic Church was about to make. I have tried not to come down on it as hard as I feel it ought to be, critically speaking; I agree with Rosen's analysis (in "Galileo's Misstatements about Copernicus") that its enormous failures are mostly due to "the intensity of his struggle," as well as some disabilities in Galileo's character. It contradicts itself in multiple locations (I have only provided my favorite in the text.) It is poorly organized and shoddily crafted. We know from *The Assayer* that Galileo was a talented polemicist and writer; that is not on display here. It is not even particularly good sophistry. I hardly know what to say. I consider all praise of the work to be presentist backreading. It really is a piece a trash. Inasmuch as his own safety, the opinions of others, or literary merit were Galileo's concern, I consider him to have entirely botched this absolutely pivotal moment of his biography.

p. 178 **"without the restoration of my reputation."**: *Opere*, v. 12, Nr. 1164, p. 223. To Curzio Picchena.

p. 179 **"to make life sober and secluded."**: *Opere*, v. 12, Nr. 1142, p. 203.

p. 179 **a halo of false light.**: See the useful Watson, Peter G. "The enigma of Galileo's eyesight: Some novel observations on Galileo Galilei's vision and his progression to blindness." *Survey of ophthalmology* 54.5 (2009): 633. This was reported about a year later, but I suppose it was beginning about now.

p. 179 **Janus, the two-faced god.**: Janus, the Roman god of duality and doorways. The doors to his temple were closed during peacetime and open during wartime. I am describing a door to the Villa Medici. It may have been constructed after 1616.

p. 179 **turn it back on its course.**: Stolen almost wholesale from the beginning of Proust's *Swann's Way*. For the pact between young Robert and his sister, see Brodrick, *Blessed*, v. 1, pp. 12–3. Brodrick also professes young Roberto disliked getting up in the mornings, and in later life he would need an alarm to do so (Brodrick, *Saint*, p. 161).

p. 179 **twelve children in twice as many years**: I have not been able to locate the birthdates of most of Robert's family (the names of each sister are not even present in any major biographies; the names of each brother only one.) I have extrapolated from his own birthdate. His mother, Cinzia Cervini, was married at twelve and gave birth to Robert, her third son, at sixteen (astounding!).

Submission is an unusual characterization of Bellarmino, but he always was: not even to the pope, but to the Church and the Bible. Though Roberto's nickname was "The Hammer of the Heretics," a hammer is an instrument.

p. 180 **He was loath to come back down.**: There is some added drama here; Roberto says he turned up an old footstool and was wearing "a string." This is from his autobiography, reproduced in Brodrick, *Blessed*, v. 1, pp. 460–81. Ryan Grant from the independent publisher Mediatrix Press has embarked on an esteemable project to translate Bellarmine's full *Omnia*, and has published the autobiography (with much else) in fine English under the title "The Autobiography of St. Robert Bellarmine."

p. 180 **"Thus, we have words."**: From Bellarmino, Roberto Francesco Romolo. *Disputationum Roberti Bellarmini: E. Societate Jesus, R.E. Cardinalis: De Controversiis Christianae Fidei* . . . Apud Josephum Giuliano, 1856, v. 5, *Concio III*, p. 22, with an interesting quote for theme: "There will be signs in the sun, moon and stars. On the earth, nations will be in anguish and perplexity at the roaring and tossing of the sea." Slightly abridged.

p. 180 **"Water is the foe."**: Brodrick, *Blessed*, v. 1, pp. 93–4.

p. 181 **"different reckoning of things than we have now."**: Selections from *Hell and Its Torments*, TAN Books, 1990.

p. 181 **a middle–aged canvasser**: This canvasser was Paschase Bröet, one of the first Jesuits.

p. 181 **"I cannot stop this aching in my heart."**: Brodrick, *Blessed*, v. 1, p. 31.

p. 181 **diverting corrupt Church funds**: I have tried to detail the Jesuits as best I can given a few paragraphs, but their literature is massive and not at all the subject of this volume. In general, the question of "Jesuit science" is a strange one. No one, for example, asks whether or not science has produced good Jesuits.

A very readable work to begin with is J. W. O'Malley's *The Jesuits: A History from Ignatius to the Present*. A strong collection of essays is *Jesuit Science and the Republic of Letters*, ed. Mordechai Feingold. Agustin Udias's *Jesuit Contribution to Science: A History* is a painfully specific overview, but very worthy reference.

p. 181 **Johannes Kepler . . . Jesuit persuasion.**: The obvious book on this topic being Burke–Gaffney's *Kepler and The Jesuits.*

p. 181 **Roman College might dare pass for Copernicans.**: Thinking specifically of Christopher Grienberger; see Sharratt, Michael. *Galileo: Decisive Innovator*, pp. 105–6. Cambridge: Cambridge University Press, 1999. After the condemnation of Copernicanism, the Jesuits entirely abandoned the Copernican position. (Yet so, technically, did Galileo).

p. 182 **"would that not advance a strange absurdity?"**: Bachelet, Xavier–Marie Le. *Bellarmin Et La Bible Sixto–Cleméntine: Études Et Documents Inédits*, p. 29. Gabriel Beauchesne & Cie, Éditeurs Ancienne Librairie Delhomme & Briguet Rue De Rennes, 1911. This was actually sent back when Roberto was in Louvain, to Cardinal Sirleto. Similar opinions are advanced on p. 161, and (as Bachelet points out) in early chapters of *The Controversies* focusing on the Septuagint and Tanakh.

p. 182 **"There's no escaping a book"**: freely translating Brodrick, *Blessed*, v. 1, p. 132: *Non te fugit quid sit libros edere.*

p. 182 **"equal for learning in all of Christendom"**: Brodrick, *Blessed*, v. 1, p. 400.

p. 183 **"How many obstacles surround the truth!"**: *KGW*, v. 19, p. 333; Burke-Gaffney, *Kepler & Jesuits*, p. 15.

p. 183 **Federico Cesi . . . Bellarmino's . . . private exchange of letters.**: Gabrieli, *Contributi*, p. 415. This was around 1604, and is the earliest mention of their interaction I could find.

p. 183 **"the light which leads to eternal life."**: from Bruno, Giordano, and Christian Bartholmess. *La cena de le ceneri: descritta in cinque dialoghi per quattro interlocutori con tre considerazioni circa doi suggetti.* v. 36. G. Daelli, 1864; p. 35, hereafter cited as *Cena*; Yates, *Hermetic*, p. 253. This quote, more than any other, summarizes my general feelings about the Galileo affair.

p. 183 **"even he could not entirely remove it."**: Bruno, *Cena*, p. 21; Yates, *Hermetic*, p. 236.

p. 183 **"Dear God, save me from the Papacy."**: From his autobiography; *A Papatu liber me Domine.* The episode is recounted in depth in Brodrick, *Blessed*, v. 2, chapter XXI.

p. 184 **"definite knowledge"**: *Opere*, v. 11, Nr. 515, p. 87; best English translation in Brodrick, *Blessed*, v. 2, p. 343.

p. 184 **By his own admission, Roberto did not hear of the matter again for five years.**: According to Piero Dini; Galilei, *Affair*, p. 58. This is the only word of Galileo and Bellarmino meeting in Galileo's second trip to Rome. Drake, *Work*, p. 47, expresses the fact that they met, along with Cesare Baronius, in Padua around 1600, but it is unsourced and has not been repeated elsewhere. In *Opere*, Baronius is mentioned only in Galileo's *Letter to the Grand Duchess.*

p. 184 **"undoubtedly against scripture."**: *Opere*, v. 12, Nr. 1071, p. 129; Shea and Artigas, *Rome*, p. 60.

p. 184 **"I have always believed Copernicus spoke."**: *Opere*, v. 12, Nr. 1110, p. 171; Brodrick, *Blessed*, v. 2, p. 358.

p. 185 **"this proposition receives the same judgment."**: Galilei, *Affair*, p. 146; *Opere*, v. 19, p. 321. Slightly abridged. The men on the council were: Petrus Lombardus, Hyacintus Petronius, Raphael Riphoz, Michael Angelus, Hieronimus de Casalimaiori, Thomas de Lemos, Gregorius Nunnius Coronel, Benedictus Justinianus, Raphael Rastellius, Michael a Neapoli, and Jacobus Tintus.

p. 185 **They were friends.**: All these examples come from Brodrick's *Saint*, pp. 7, 15, 23, 87, 122, 129, 168, 184, 211. Both their fathers had the same name as well. Both had their books forbidden by the Church. The similarities are entertaining.

p. 186 **his headache would go away.**: Brodrick reports repeatedly that Bellarmino suffered from severe headaches all his life. I have been unable to resist here a quotation from Bulgakov's *Master and Margarita*, when Ha-Notsri is dragged before the martyr Pontius Pilate; Bellarmino here is the saintly Pontius, while Bruno is the simpleton Christ.

p. 186 **"you should be so degraded."**: Rowland, *Giordano Bruno*, p. 274. Rowland has done a hilarious job translating the legalese of a state-mandated murder, but I have still simplified it just a bit. The ruling makes it sound like Bellarmino was physically present when it was read to Bruno, but the matter is not entirely clear to me.

p. 187 **"He has failed."**: Sent to Curzio Picchena, Secretary of state. The "friar" is assumedly Tommaso Caccini. *Opere*, v. 12, Nr. 1187, p. 243; Finocchiaro, Maurice A. *The Trial of Galileo: Essential Documents, p. 108* Hackett Publishing, 2014.

p. 187 **"I am not mentioned."**: *Opere*, v. 12, Nr. 1187, p. 243. Slightly abridged.

p. 187 **"and declared a heretic."**: *Ibid.,* Nr. 1198, p. 257. Slightly abridged.

p. 187 **"at the hands of Cardinal Bellarmino."**: Ibid., Nr. 1195, p. 254.

p. 188 **"put my mind at ease."**: Ibid., Nr. 1189, p. 248; Shea and Artigas, *Rome*, pp. 89–90. Abridged.

p. 188 **"Poke not again this slumbering giant."**: *Opere*, v. 12, Nr. 1202, p. 261. From Curzio Picchena. Actually says "sleeping dog."

p. 188 **"twenty-sixth day of May, 1616."**: Brodrick, *Blessed*, v. 2, p. 372; Allan-Olney, *Private*, p. 100. Abridged.

p. 189 **Christian fundamentalist**: In my uppity vulgar-speak, I have referred to Bellarmino as a fundamentalist, but this is entirely anachronistic, and many of the associations it brings to mind, though valuable, are inaccurate.

p. 189 **"heat will be milder than here."**: *Opere*, v. 12, Nr. 1209, p. 265; Nr. 1215, p. 271. "In Rome" is added.

p. 189 **"a horde of cardinals."**: *KGW*, v. 17, Nr. 827, p. 328. To Vincenzo Bianchi.

p. 189 **"to castrate the friars"**: *Opere*, v. 12, Nr. 1119; said by Piero Guiccardini.

p. 190 **"The ungrateful brute."**: Favaro's *Opere* and the Drake and O'Malley translation have unfortunately not included marginalia for the comet controversy. These are assorted comments from Santillana, *Crime*, p. 152.

p. 190 **one of his more ambitious students**: This was Mario Guiducci. He was taken from law to science by Galileo, and went on the become a Lynx. He is the subject of book 37 of Favaro's *Amici*.

p. 190 **"you absolute idiot!"**: These annotations are helpfully given in Besomi, *Ottavio.* "Galileo Reader and Annotator.": *The Inspiration of Astronomical Phenomena VI*, v. 441. 2011. Galileo's annotations are vital; they are a shortcut to his intention.

p. 190 **true name of his enemy**: It was Horatio Grassi. There is some further mention of him in Heilbron, *Galileo*, and Pietro Redondi, *Galileo Heretic*, but his only importance to the Galileo narrative is that he was a religious.

p. 190 **The word "annihilate"**: *Opere*, v. 13, Nr. 1429, p. 499. The friend was Giovanni Ciampoli.

p. 190 **"nothing offensive to morality."**: Galilei, *Comets [Assayer]*, p. 152. From Niccolo Riccardi. Riccardi would have similar involvement in Galileo's *Dialogue*; Drake, *Work*, pp. 313, 337–9.

p. 191 **"but that is not how things stand."**: *Opere*, v. 6, p. 232; Galilei, *Comets [Assayer]*, pp. 183–4. This sentence is followed by the most oft-quoted line Galileo has ever written.

"Philosophy is written in this grand book—I mean the universe—which stands completely open to our gaze, but it cannot be understood unless one first learns to comprehend the language and interpret the characters in which it is written. It is written in the language of mathematics, and its characters are triangles, circles, and other geometrical figures, without which it is humanly impossible to understand a single word of it; without these, one is wandering about in a dark labyrinth."

This quote is regularly cited as evidence of a revolutionary new way of thinking about the world, but such a claim is lazy thinking and masturbatory self-congratulation. It is actually routine mathematical posturing of no significance whatever. Grassi makes similar (though less severe) comments in his work, which *The Assayer* is a response to. I quoted a very similar statement earlier in the eulogy of Filippo Salviati. Martin Horky says something similar in *Peregrination*. The Pythagoreans said similar things. Here, for example, is Robert Grossteste from about the year 1200:

"The consideration of lines, angles and figures is of the greatest utility since it is impossible for natural philosophy to be known without them . . . All causes of natural effects have to be given through lines, angles and figures, for otherwise it is impossible for the reason why to be known in them"

Alternately, here is Chuang Tzu, from about 300 B.C.E.:

"These names and processes can be examined, and however minute, can be recorded. The principles determining the order in which they follow one another, their mutual influences, now acting directly, now revolving; how, when they are exhausted, they revive; and how they come to an end only to begin all over again—these are the properties belonging to things. Words can describe them, and knowledge can reach them."

p. 191 **"the mirror two or three times."**: *Opere*, v. 6, p. 277; Galilei, *Comets [Assayer]*, p. 232.

p. 191 **"smaller its number of followers."**: *Ibid.*, p. 237; Galilei, *Comets [Assayer]*, p. 189.

p. 191 **"befouling the earth beneath them."**: Ibid.

p. 192 **"might compare thoughts with us poets."**: *Opere*, v. 12, Nr. 1349, p. 415. Abridged, modernized. This is from Virginio Cesarini; "my friends" and "us poets" is Giovanni Ciampoli. The "as I do him" refers to Cesarini stating he will add Galileo to the dedication of his book; this is an exchange of goodwill.

p. 192 **many months and even years**: *Opere*, v. 18, Nr. 4008. The notes of this study are in *Opere* v. 9, p. 59; the time of its writing is unclear, but I place it around 1620. The most important English analysis of this is Panofsky, Erwin. "Galileo as a critic of the arts: Aesthetic attitude and scientific thought." *Isis* 47.1 (1956). Heilbron's *Galileo*, pp. 16–23, contains a useful though at times incoherent discussion (e.g., "Galileo: 'I've never seen sweat go white except around a horse's testicles.' He objected to explicit descriptions.")

p. 193 **"a bad sort of dense."**: *Opere*, v. 9, p. 66.

p. 193 **"which tries hard but appears not to."**: This is my own translation of Canto 16, Stanza 9 of Tasso's *Jerusalem Delivered*. The last two lines are a reference to the infamous Italian *sprezzatura*.

p. 193 **"yet this island is very small."**: *Opere*, v. 9, p. 138. Abridged.

p. 193 **"unspeakable disorder for five days."**: *Opere*, v. 13, Nr. 1452, p. 27. Sagredo died on March 5, 1620.

p. 193 **They traded portraits just months ago . . . living room.**: Wilding, *Idol*, pp. 9–10.

p. 193 **"pasty white teen."**: *Opere*, v. 12, Nr. 1138, p. 200; Heilbron, Galileo, p. 82.

p. 193 **plates of truffle and peach.**: Ibid., Nr. 1350, p. 416.

p. 193 **"appear much more beautiful."**: Ibid., Nr. 1339, p. 404.

pp. 193–4 **birds, martens, marmots, and wolves.**: Favaro, *Amici*, Bk. 8, p. 26.

p. 194 **"My idol"**: *Opere*, v. 16, Nr. 3283, p. 414.

p. 194 **stolen his fingers.**: This account from Brodrick, *Blessed*, v. 2, chapter 23–4, especially pp. 450, 456. Roberto Bellarmino died on September 17, 1621.

p. 194 **"my concern for your sake."**: Galilei, *Father*, p. 1. I have mostly followed Sobel's translations, which are not accurate but flow very smoothly to the modern ear. They are printed side-by-side with the Italian.

p. 194 **chasing down silk cloth**: *Opere*, v. 10, Nr. 34, p. 46.

p. 194 **"carried a particular affection."**: *Opere*, v. 13, Nr. 1885, p. 429.

p. 194 **"easy remedy unless they want it."**: Bindman, *Pedagogy*, p. 80. The full quote is excellent, but sad. From Johann-Baptist Winther.

p. 194 **"material for a composition."**: *Opere*, v. 13, Nr. 1479, p. 48.

p. 194 **an ocular cyst.**: This is from the 1624 portrait by Ottavio Leoni; see the color plates in Heilbron, *Galileo*.

p. 195 **Death came for the Pope.**: For simplicity, I let the reader imagine this is Pope Paul V, But it was actually Pope Gregory XV (Alessandro Ludovisi), who had a brief two year reign after Paul's death.

p. 195 **shut up about their brilliance**: "As to Latin composition, the Pope has none who equal him," wrote Fulvio Testi. See Rietbergen, Peter. *Power and religion in Baroque Rome: Barberini cultural policies*, p. 95. Brill, 2006. This whole chapter is very informative. Fulvio's comment was probably true if only because Latin poetry was clearly a dying art.

p. 195 **"by your art, your labor—Galileo."**: Surprisingly, this poem is not in Favaro's *Opere*. Favaro indicates Maphaei S. R. E. Card. Barberini, nunc Urbani PP. VIII, *Poemata*. Antwerp, office of Balthasaris Moreti, 1634, pp. 278–2. It is not clear to me that this is the poem's earliest publication date. Santillana, *Crime*, p. 156, prints some of the Latin and mentions it is not a large part of Galileo mythos but should be. This remains the case. The only partial translation I have found is in Reston, James. *Galileo: A life*, p. 189. Beard Books, 2000. Here are all nineteen stanzas with my loose translation.

Cum Luna caelo fulget, et auream	When the moon shines in heaven
Pompam Sereno pandit in ambitu	With calm unraveling its golden parade
Ignes coruscantes, voluptas	Surrounded by glittering fire
Mira trahit, retinetque visus.	What strange pleasures draw us in.
Hic emicantem suspicit Hesperum,	Suddenly the evening is on me,
Dirumque Martis sidus, et orbitam	The dread Martian star, the circle
Lactis coloratum nitore;	Which glistens like milk
Ille tuam Cynosura lucem.	And the star which lights the way.
Seu Scorpii cor, sive Canis facem	The heart of the Scorpion, torch of the Dog,
Miratur alter, vel Iovis asseclas,	Their glories, or the servants to Jupiter,
Patrisque Saturni, repertos	Father to Saturn, discovered by
Docte tuo Galilaee vitro.	You, learned Galileo, with your glass.
At prima Solis cum reserat diem	But first the light of the Sun springs forth
Lux orta, puro Gangis ab aequore	Which unlocks the day, diffuse
Se sola diffundit, micansque	Across the placid Ganges, trembling,
Intuitus radiis moratur.	As seen in its lingering rays.
Non una vitae sic ratio genus	A singular focus invites its specific death:
Mortale ducens pellicit: horrida	Such a life happily grabs the sword
Hic bella per flammas et enses	And plunges through the flames

Laetus init, meditans triumphos.	Into battle, thinking only of triumph.
Est, pacis ambit qui bonus artibus	The good of peace is tangent to its arts
Ad clara rerum munia provehi.	Beneath the duty of its own advancement.
Illum Peruanas ad oras	The Peruvians by the coast
Egit amor malesuadus auri.	Imagine love of gold to be ill–advised.
Hunc sumptuosus dum Siculae iuvat	There it is lavish, while Sicilians delight
Mensae paratus, spes alit aleae	In its measure, their hopes nourished by
Mendacis, ac fundis avitis	Gambling deceits, losing the gains of
Exuit, et laribus paternis.	Ancestors, and homes of their fathers.
Nil esse regum sorte beatius,	Nothing could be a kinglier fate,
Mens et cor aeque concipit omnium,	Mind and heart acting as one
Quos larva rerum, quos inani	In their evil object, their vane and empty
Blanda rapit specie cupido.	Desire consuming their vision.
Non semper extra quod radiat iubar,	Not always is there splendor within
Splendescit intra: respicimus nigras	That radiant outburst: consider the black
In sole (quis credat?) retectas	In the sun uncovered (who would believe?)
Arte tua Galilaee labes.	By your art, your labor—Galileo.
Sceptri coruscat gloria regii	Glory is brandished like a scepter
Ornata gemmis; turba satellitum	Adorned by a jewel; the mob of followers
Hinc inde praecedit, colentes	Expands in all directions, servants
Officiis comites sequuntur.	Nourished by the party line.
Luxu renidet splendida, personat	These luxuries radiate splendor,
Cantu, superbit deliciis domus:	Sing their song, proud of their homey role:
Sunt arma, sunt arces, et aurum:	They are soldiers, castles, golden all:
Iussa libens populus capessit.	The people seize upon it eagerly.
At si recludas intima, videris	But if you look within, you will see
Ut saepe curis gaudia suspicax	That often joy is plagued by trouble
Mens icta perturbet. Promethei	Slapping the mind. Otherwise an eagle
Haud aliter laniat cor ales.	Would not have eaten Prometheus's heart.
Cui sensa mentis providus abdita	Does a prudent king believe what is
Rex credat? Aut quos caverit? Omnium	Foreign to the mind? Or be wary of it?
Sincera, seu fallax, eodem	From truths and falsehoods all alike
Obsequio tegitur voluntas.	Some will prefer to be protected.
Fugit potentum limina Veritas:	Truth flees the threshold of the powerful:
Quamquam salutis nuntia nauseam	Though the messenger is safe, prejudice
Invisa proritat, vel iram:	Provokes nausea, or anger:
Saepe magis iuvat hostis hostem.	They often prefer the enemy's enemy.
Ictus sagitta Rex Macedo videt	The Macedonian king hit by an arrow
Non esse prolem se Iovus. Irrita	Could not see the children of Jupiter.

Xerxem tumentem spe trecentis	Three hundred at Thermopylae with sarissae
Thermopylae cohibent sarissis;	Held back Xerxes, swollen with hope.
Docentque fractum clade, quid aulici	They taught that he could break, as princely words
Sint verba plausus. Ut nocet, ut placet	May be struck. As it may hurt, so it may please
Stillans adulatrix latenti	Speech dripping with hidden flattering
Lingua favos madidos veneno!	Like the juice of wet honeycomb!
Haec in theatri pulvere barbarum	It stained the savage Commodus
Infecit atro sanguine Commodum,	With dark blood in the amphitheater.
Probrisque foedavit Neronem.	By it did shame find Nero.
Perdidit illecebris utrumque.	So were both lured to their ruin.
Artes nocendi mille tegit dolis	Countless are those to be damaged,
Imbuta: Quis tam Lynceus aspicit	Who protect against the grief–giving arts:
Quod vitet? Intentus canentis	How could Lynceus see what he avoids?
Mercurii numeris, sopore	Intent upon the song of Mercury,
Centena claudens lumina, sensibus	Limping asleep with a hundred eyes,
Abreptus, aures dum vacuas melos	Robbed of his senses, ears soothed
Demulcet, exemplum peremptus	By empty melody, such was Argus Panoptes
Exitii grave praebet Argus.	Whose destruction leaves a grave example.

As a poet, I consider Barberini to be decently talented but much held back by his didacticism. This poem has an almost obscene prophetic aspect to it, but it seems (?) to be wholly positive towards Galileo. Barberini is alluding to the downfall of Aristotelianism throughout. Galileo indicates to Barberini "the depths of his talent, that he should choose to depict ignorance."

p. 196 **"has been revived."**: *Opere*, v. 13, Nr. 1576, p. 130.

p. 196 **A sun was to be painted . . . interior of his carriage.**: Heilbron, *Galileo*, p. 254; Shea and Artigas, *Rome*, p. 122.

p. 196 **"When is he coming?"**: Santillana, *Crime*, p. 157.

p. 196 ***"il microscopio."***: *Opere*, v. 13, Nr. 1719, p. 264. This was Johannes Faber.

p. 196 **"one must be young."**: *Opere*, v. 13, Nr. 1628, p. 175; Shea and Artigas, *Rome*, p. 110.

p. 196 **"through which you proceed."**: *Opere*, v. 13, Nr. 1749, p. 295. Said by Giovanni Ciampoli. Abridged.

p. 196 ***"Discourse on the Tides."***: Galilei, *Affair*, p. 197. Abridged. This is from the oft–referenced "Letter to Ingoli."

There is an overabundance of work on Galileo's tidal theory. It does not deserve the controversy it has received. The most notable contributions are Aiton, Eric J. "Galileo's Theory of the Tides." *Annals of Science* 10.1 (1954): 44–57; Burstyn, Harold L. "Galileo's Attempt to Prove that the Earth Moves." *Isis* 53.2 (1962): 161–85; Aiton, Eric J., and Harold L. Burstyn. "Galileo and the Theory of the Tides." *Isis* 56.1 (1965): 56–63; and Palmieri, Paolo. "Re–examining Galileo's Theory of Tides." *Archive for History of Exact Sciences* 53.3–4 (1998): 223–375.

I believe Aiton's analysis is entirely definitive, but the theory is not such an embarrassment as they make out. It explained a good deal and was of such a piece with Galileo's way of looking at the world (especially before his dialogues) that it does not seem to me surprising that he came up with it. Galileo's unusual exclamations of uncertainty here are also quite winsome; he did not consider his explanation vital to the Copernican cause because it was true, but because it was more relevant for nonspecialist readers than any astronomical argument. It seems to me many historians try to "save the phenomenon" here, where the phenomenon is Galileo's mythic genius. Other historians view false theory as inherently blameworthy, so it is difficult to get an explanation with any emotional maturity.

p. 196 **"the motion of the seawater basins."**: Galilei, *Affair*, p. 122.

Concerning tides, it is worth noting Kepler's brief statements on the matter from *Kepler, Nova*, p. 26: "Gravity is a mutual disposition among kindred bodies to attract and unite." "The reason for the ebb and flow of the sea is that the attractive power of the Moon extends all the way to the Earth, and calls its waters forth." "Should the Earth quit attracting its waters to itself, all the seas would be lifted up, and flow onto the Moon."

Galileo (In Galilei, *Dialogue*, p. 462) responded, "I am more astonished at Kepler than any other, for lending his ear to occult properties and other puerilities." Kepler's hypothesis, though clearly a guess, was in broad strokes entirely correct.

p. 197 **"twice at intermediate speeds."**: Galilei, *Affair*, p. 124.

p. 197 **"to our minds" . . . "despairing over ever understanding it."**: Galilei, *Dialogue*, pp. 448, 447.

p. 197 **"It is empty."**: Galilei, *Affair*, p. 133.

p. 198 **adulation of the crowd.**: Mascardi, Agostino. *Le Pompe del Campidoglio per la Sta. di NS Vrbano VIII quando piglio il posesso. appresso l'herede di Bartolomeo Zannetti*, p. 22; Reston, James. *Galileo: A life*, pp. 189–90. Beard Books, 2000. Mascardi's account is wonderfully ebullient and much worth a read.

p. 198 **"decree would never have been made."**: *Opere*, v. 14, Nr. 1993, p. 88. From Benedetto Castelli, as heard from Federico Cesi, as heard from Tommaso Campanella, who was the Dominican friar in question.

p. 198 **"The seasons turn like a wheel"**: *Opere*, v. 18, p. 75.

p. 198 **"astronomical tyro."**: Kepler, *Secret*, p. 39.

p. 199 **"this is just some astrological fun" . . . "It does not matter."**: Kepler, *Secret*, pp. 119, 125.

p. 199 **footnotes as original text.**: This is an exaggeration; 44 pages to 77; Field, *Cosmology*, p. 74. Field's discussion of these notes (pp. 73–95) is the best available, and I think they pretty much hit the nail on the head. It is not clear how quickly Kepler wrote his notes, but it was fast. A week is my guess.

p. 199 **"stems from this little book."**: Kepler, *Secret*, p. 39. Slightly abridged.

p. 199 **"my friends would like to hear about it."**: *KGW*, v. 18, Nr. 893, p. 43. This was Matthias Bernegger.

p. 200 **"This is really beyond expectation."**: *KGW*, v. 17, Nr. 748, p. 193.

p. 200 **"that primer by Maestlin."**: *KGW*, v. 16, Nr. 619, p. 389.

p. 200 **"making changes and editions."**: *KGW*, v. 7, p. 359. Charles Glenn Wallis translated books 4 and 5 of the *Epitome* for the Great Books of the Western World series, v. 16. These translations are notoriously poor. Wallis, however, was the first to translate substantial sections of works by Copernicus and Kepler into English, in his early twenties, before history of science was an established profession. His contribution is appreciated.

Kepler's *Epitome* certainly deserves the difficult full translation. Under a pedagogical view of the history of science, which this author is highly amenable to, it is his most important work.

p. 200 **"with the tedium of proofs eliminated"**: *KGW*, v. 7, p. 251.

p. 200 **"parroting back the doctrine of the spheres"**: *KGW*, v. 7, p. 7. There is not so much information on the creation of *Epitome* (as compared to *New Astronomy*), so most of the quotations here come from the prefaces for its different books.

p. 200 **"the essence of nautical things, involve astronomy . . ."**: *KGW*, v. 7, p. 23.

p. 200 **"But all of me is Copernican!"**: *KGW*, v. 17, Nr. 846, p. 364. Kepler meant his work. He was worried censorship would come to Austria: "I may need to renounce astronomy as my livelihood, after I give my opinion of this dogmatism, if none may ever contradict; and finally, renounce Austria, my Austria, if it will not be home to free philosophy."

The discussion of the Epitome in Rothman, *Strife*, chapter 4 has been very helpful.

p. 200 **"wherever I find it . . . spiritual mutt."**: Caspar, *Kepler*, pp. 217, 259.

p. 201 **"so subject to human frailties, I cling."**: *KGW*, v. 18, Nr. 1072, pp. 331–3, lns. 39–41, 58–60, 114–6.

p. 201 **"new hypotheses on lunar motion."**: *KGW*, v. 7, p. 360. I have not touched on the mathematics in Kepler's *Epitome*, but interested readers should refer to Stephenson's peerless study in *Physical*, pp. 138–202.

p. 201 **"is a qualitative relation."**: Kepler, *Harmony*, p. 296. Italics added. Kepler defines this relation as "of unity" on p. 490. The basis of this unity is, obscurely, the circle, but I have tried to avoid the thickets of specificity within number-mysticism.

p. 201 **"brute beasts rather than human nature."**: *Harmony*, p. 217.

p. 202 **"similar to the five regular solids."**: *KGW*, v. 6, p. 299; *Harmony*, p. 407. This figure is today known as a Kepler-Poinsot polyhedron. It is truly remarkable that such a figure had never been mathematical analyzed before (that I can tell). In modern times it is especially important as forming the archetypal example of "monster-barring" in Imre Lakatos's wonderful *Proofs and Refutations*.

p. 204 **"The structure is very intricate."**: Kepler, *Harmony*, pp. 106, 108. This pattern has been mentioned especially because of its inspiration on Roger Penrose's studies of aperiodic tilings. This, the most complex tiling Kepler provided, has fivefold pentagonal symmetry. Penrose, in open acknowledgement of Kepler, reduced this tiling to a smaller set, and went on to produce fully aperiodic tilings. In 1982, Dan Schectman discovered quasicrystals, physical structures which are modeled by aperiodic and nonperiodic tilings. If we today suspect Kepler's *Harmony* is no longer relevant, the mathematical ideas, at least, beg to differ.

p. 204 **"without respect to person."**: Kepler, *Harmony*, p. 264, from a truly perplexing appendix "On the Three Means."

p. 204 **"subjected to men."**: Kepler, *Harmony*, p. 397; Stephenson, *Heavens*, p. 131.

p. 204 **"benumbed while it does not."**: *KGW*, v. 6, p. 105; Kepler, *Harmony*, p. 147.

p. 204 **notate a bar of music after one listen.**: He does so, mostly successfully, in *Harmony*, p. 217.

p. 205 **"sounds do not exist in heaven"**: Kepler, *Harmony*, p. 449.

p. 205 **"miserable worm that I am"**: From Kepler, *Harmony*, p. 491. This is, noticeably, the end of the book.

p. 205 **distances from the Sun.**: Stephenson, *Heavens*, p. 145, contains a full list of Kepler's possibilities: "its periodic time; its daily eccentric arcs; its daily delays (morae) in the eccentric arcs; its apparent daily arc, or angular motion, seen as if by an observer in the Sun; and finally, its true daily path, measured along the path, rather than an arc of a circle."

p. 206 **"from out which every epoch flows."**: *KGW*, v. 6, p. 324; *Harmony*, p. 443.

p. 206 **unreasonable for his hysteria**: I have rubbed up against language trying to express myself here, and I am unsure who has come away the victor. If we take hysteria in its common meaning of "ungovernable emotional excess," then Kepler clearly was not;

he was governed by reason. But the governance of reason does not limit emotion, only belief. With reason emotion has no excess. The term melts away; Kepler makes it melt. The term becomes "experiencing life with maximal intensity." I have tried to force the reader into this redefinition.

p. 207 **a day or two's work at most.**: Kepler ends his *Harmony*, p. 498, with the date of completion, 17/27 May 1618. The perfectly terse Gingerich, Owen, "10.9. The origins of Kepler's Third Law," *Vistas in astronomy* 18 (1975): 595–601, is worth a look for those interested in more details.

p. 207 **"a most wonderful contemplation."**: *KGW*, v. 6, p. 290; Stephenson, *Heavens*, p. 129; *Harmony*, p. 391.

p. 207 **"Tengnagel has given up" . . . "birthing pains torment me."**: *KGW*, v. 15, Nr. 431, p. 491; v. 16, Nr. 560, p. 295; v. 17, Nr. 734, p. 173; Nr. 783, p. 254; v. 18, Nr. 965, p. 145; Nr. 983, p. 181, respectively.

p. 208 **"the famous and generous". . . . "how to read, the ass."**: *KGW*, v. 8, pp. 267, 278, 401. I have changed Kepler's speech from second to third person. The postface, pp. 470–5, has been very helpful.

p. 208 **"wandering off the course of reason."**: Galilei, *Comets* [Appendix to Hyperaspistes], p. 344; *KGW*, v. 18, p. 417.

p. 208 **"Tycho is a nullity."**: Galilei, *Comets*, [*Assayer*], p. 185.

p. 208 **"too soft to accept the motion of the Earth"**: Kepler, *Harmony*, p. 404; *KGW*, v. 6, p. 297. Though speaking generally, this obviously means the Jesuits.

p. 208 **The Tychonic universe was not even up for discussion.**: There is only one paper I have found directly addressing Galileo's rejection of the Tychonic system; Margolis, Howard. "Tycho's system and Galileo's Dialogue." *Studies in History and Philosophy of Science* 22 (1991): 259–75. I find it unconvincing. The rejection forms a through-thread in Heilbron's biography, but it is not coherently organized or justified. The truth seems to be that Galileo never published very much on it at all. We can guess to why as we like. I like to imagine his rejection derived from his research in physics. No sensible physics (at least, certainly not Galilean) could map onto the Tychonic system; to accept Tychonic (or Ptolemaic) astronomy was to deny the planets as physical entities.

p. 209 **"maledictions put upon my person."**: From the first page of the preface to Scipione Chiaramonti's *Apologia Pro Antitychone*. Slightly abridged.

p. 209 **"will bring back morality."**: *KGW*, v. 18, Nr. 1045, p. 295. Such an inquiry no longer exists, if it was sent.

p. 209 **survive it, and turned back.**: A dramatic reading of *KGW*, v. 18, Nr. 1037, p. 278.

p. 209 **"bitch being deprived of her pups!"**: *KGW*, v. 18, Nr. 1024, p. 258.

p. 211 **"the whole blessed family."**: This translation is mostly a very, very extreme abridgment of the full translation offered in Gingerich, Owen, "Johannes Kepler and the new astronomy," *Quarterly Journal of the Royal Astronomical Society* 13 (1972): 360–73; however, it also refers to the original Latin in *KGW*, v. 10, pp. 36–44. I have not touched at all upon the content of the *Rudolphine Tables*; those interested should consult Bialas, Volker. *Die Rudolphinischen Tafeln von Johannes Kepler*. 1968; and Gingerich, Owen. "A Study of Kepler's Rudolphine Tables." *Actes du XIe Congrès International d'Histoire des Sciences*. 1965.

p. 211 **of those beings made up of chords which vibrate at the slightest zephyr**: This phrase was learned from the diary of Alice James, a profoundest woman—the only thing more extreme than her illness was her ability to say "yes" to life. "Permanent smile" is a metaphor developed by Bill Callahan. All the worser phrases are, I'm afraid, my own.

At the end of Kepler's life, he had outstanding plans to publish Tycho's observations, as well as a book on eclipses called *Hipparchus*, but these plans were not important lifelong commitments in the vein of *Harmony* or *Rudolphine Tables*. His eldest son, Ludwig, is detailed in Appendix B of Ed Rosen's translation of *Somnium*. His eldest daughter, Susanna, married Jacob Bartsch just before Kepler's death. Jacob and Susanna are detailed in Appendix A of *Somnium*.

p. 212 **"I am stunned."**: *Opere*, v. 13, Nr. 1604, p. 155.

p. 212 **"any real debauchery."**: *Opere*, v. 13, Nr. 1682, p. 228.

p. 212 **"withholding nothing in their presence."**: *Opere*, v. 13, Nr. 1880, pp. 422–3. Abridged. This "overseer" is Francesco Crivelli, an agent of the Duke of Bavaria. Galileo's nephew is, confusingly, also named Vincenzo.

p. 212 **"he bursts into laughter."**: *Opere*, v. 13, Nr. 1892, p. 437.

p. 212 **"hardships and horrors."**: *Opere*, v. 13, Nr. 1855, p. 393.

p. 212 **"what really spooks me . . . poisonous hatred of the clergy."**: *Opere*, v. 13, Nr. 1892, p. 437; Nr. 1818, p. 358.

p. 212 **"as I had pennies."**: Allan-Olney, *Private*, p. 138; *Opere*, Nr. 1747, p. 294; Favaro, *Maria*, p. 149.

p. 212 **"not contradict you" . . . "weak-minded"**: *Opere*, v. 13, Nr. 1805, p. 347; Nr. 1876, p. 416; Heilbron, *Galileo*, p. 266.

p. 213 **"She is still so terrible."**: *Opere*, v. 12, Nr. 1422, p. 494; Favaro, *Maria*, p. 127. In the line above his brother Michelangelo states that Galileo actively desired "*haver nota della mia famiglia.*" On p. 106, Favaro suggests Galileo's mother was cut off from her granddaughters after they entered the convent. Galileo's mother was Giulia Ammannati. She died in August 1620.

p. 213 **"more collars" . . . "such transgressions."**: Galilei, *Father*, pp. 15, 21. It is unclear what this refers to, but I expect it was either matters of finance or religion and have framed it as such.

p. 213 **"Send me" . . . "I enjoyed it last time."**: Galilei, *Father*, pp. 53, 135, 163.

p. 213 **"pin our every hope on you."**: Very free translation of Galilei, *Father*, p. 53.

p. 213 **the brief moments in which she had children.**: Allan-Olney, *Private*, p. 114 has the key insight: "What Sister Maria Celeste wanted was home life."

p. 214 **"live to a ripe old age."**: Virginia mentions Galileo's garden negatively three times; see Galilei, *Father*, pp. 55–6, 69–71. This quote is a mishmash of the latter two. It is an overt juxtaposition, but if that is not enough, Virginia also writes about Galileo's scholarly work directly in the same manner on p. 111.

p. 214 **"I am alone in this world."**: Dark. Galilei, *Father*, p. 137.

p. 214 **"tumble into darkness."**: *Opere*, v. 14, Nr. 1971, p. 60. Slightly abridged.

p. 214 **The Dialogue**: I have foregone complete footnotes for this section. All quotes are from Drake's translation, except where noted.

p. 214 **"(Nor shall a good Peripatetic lack a place.)"**: In *Opere*, v. 7, p. 31, this is not literally parenthetical, though it reads like it should be. It is so in Drake's translation, p. 7.

p. 215 **"limitations of the human mind."**: Strongly abridged.

p. 215 **"broken with our faith."**: Drake, *Dialogue*, p. 279. This line is important. The preface was forced by Catholic censors, while this was not.

p. 215 **a vague set of guiding principles.**: The best demonstration of this is Finocchiaro's abridged translation, which divides the *Dialogue* into themed sections, none of which have much to do with any another.

p. 216 **"they never knew they knew."**: *Opere*, v. 7, p. 276. The word I translate as "undermining," *scalzare*, is worthy of note.

p. 216 **"form of violence."**: *Opere*, v. 7, 171. *violenza*.

p. 216 **"thrown an unlimited distance."**: From Aristotle's *Complete Works*, Jonathan Barnes (ed.), 1984, *On The Heavens* §14 p. 51. I feel this line is important to cite to indicate that Aristotle never suggested the ship experiment. There is a separate ship metaphor used earlier in the text that may have served as Galileo's inspiration.

p. 217 **This invariance principle**: I do not mean to seem fancy by throwing in this term, but it is a very important and timeless one I believe the reader can follow. If forced to define science, I would say it is the discovery and application of invariance principles. If forced to define art, I would say it is the handling of variation.

p. 217 **"extrusion by terrestrial whirling."**: Drake, *Dialogue*, p. 198. These demonstrations are ponderously irrelevant to the issue at hand. This is especially interesting because of the obvious experiments that would indicate they were so. In a footnote Drake correctly points out that they are approaching methods from calculus, but there is little else redemptive about them.

p. 218 **"I am lost at sea."**: Gabrieli, *Contributi*, "Cesi e Caetani," p. 135. To Giovanni Faber. Basically a paraphrase, actually: "In a tough conflict such as this, neither Seneca nor Epictetus are good enough. By good God's favor, I hope to free myself from the sea in which I find myself, thwarted by foreign powers and my own impotence."

p. 218 **"With trembly hands" . . . "stopping up the piss."**: Lovely. *Opere*, v. 14, Nr. 2042, p. 127. From Francesco Stelluti. Federico Cesi died on April Fools, 1630.

p. 218 **Master Censor**: This was Father Niccolo Riccardi. I have opted to skip over the considerable issues concerning Galileo's imprimatur, because I find the issue to be inextricably boring (especially in the context of my narrative). See Shea and Artigas, *Rome*, pp. 142–55, for the most coherent rendition. The fact was that Galileo got one through technically legitimate channels.

p. 218 **"My life is wasted."**: Galilei, *Affair*, p. 208. Westfall has rightly pointed out the question is not "Why was *Dialogue* delayed so long?" but rather "How in hell's name was it ever allowed to be printed in the first place?" (Westfall, *Essays*, "Patronage and the Publication of the Dialogue").

p. 219 **"the outcome of my hard work."**: Galilei, *Affair*, p. 208.

p. 219 **purblinded so badly**: *Opere*, v. 14, Nr. 2250, 2257, and especially 2256.

p. 219 **"almost entirely from you."**: *Opere*, v. 14, Nr. 2295, p. 378. The disciple was Buonaventura Cavalieri. Slightly abridged.

p. 220 **"which I crave."**: *Opere*, v. 14, Nr. 2300, p. 386. "Crave" is *tanto avidamente desideravo.*

p. 220 **"the will of he who caused it."**: *Opere*, v. 14, Nr. 2325, p. 411.

p. 220 **drainage systems for the city.**: *Opere*, v. 13, Nr. 1818, p. 358; Favaro, *Amici*, Bk. 21, p. 44.

p. 220 **"he could reply asap."**: *Opere*, v. 14, Nr. 2277, p. 360; Geymonat, Ludovico. "Galileo Galilei: A biography and inquiry into his philosophy of science." (1965), p. 137; Shea and Artigas, *Rome*, p. 162.

p. 220 **"Scheiner has great influence"**: Galilei, *Affair*, p. 206. From Mario Guiducci.

p. 220 **byzantine in its construction.**: *Opere*, v. 14, Nr. 2301, p. 387. This is basically my opinion as well.

p. 221 **no questions asked.**: Brodrick, *Blessed*, v. 1, chapter 8; p. 269 on. Also Santillana, *Crime*, p. 90. The mere act of prohibition was actually relatively insignificant; the outright ban (not "pending corrections") was much more damning.

p. 221 **"despite malicious envy."**: *Opere*, v. 14, Nr. 2286, p. 372. From Fulgenzio Micanzio.

p. 221 **"Never hope about anything on Earth."**: *Opere*, v. 14, Nr. 2269, p. 351. Abridged. Galileo claims this is a quote from Petrach; he might be trying to remember Canzone 294: truly, hope deceives.

p. 221 **"I will leave next Sunday."**: *Opere*, v. 14, Nr. 2318, p. 402. Abridged. Sent on October 6, 1632, a Saturday; "next Sunday" was thus October 14.

p. 221 **"the fact that I aspired."**: *Opere*, v. 14, Nr. 2324, p. 407. Freely translated.

p. 221 **and hypochondria.**: *Opere*, v. 19, p. 334.

p. 221 **"subterfuge" . . . tie you up like a hog**: Slight exaggeration, *Opere*, v. 19, p. 335.

p. 222 **absence of Church dogma**: Specific theological issues are explained at length in Wisan, Winifred Lovell. "Galileo and God's creation." *Isis* 77.3 (1986): 473–86. This article is the best I have found at expatiating the reasoning going on in the mind of Pope Urban, but I do not feel it is satisfactory. The most viable answer is that old men with undeserved power have arbitrary whims and anger easily. Historians will not generally find this truism satisfactory. I do.

p. 222 **They asked him many questions.**: Quotations from the trial are taken from Finocchiaro's *Affair* except where noted.

p. 224 **"promised to obey."**: Galilei, *Affair*, p. 147; Santillana, *Crime*, p. 126. Abridged. There has been a great deal of argument concerning this injunction's authenticity (most notably in Santillana's *Crime*; also by Emil Wohlwill). It is now widely accepted as legitimate. I have no opinion on the matter, and its authenticity is irrelevant to my purposes.

p. 224 **"He teaches *ad nauseam*."**: Galilei, *Affair*, p. 264. This is all from the important report by Melchior Inchofer, a Jesuit. I have mostly refused to enter the horrifying labyrinth of Church politics circa 1600, but many further details are to be found in Blackwell, Richard J. *Behind the Scenes at Galileo's Trial: Including the First English Translation of Melchior Inchofer's Tractatus Syllepticus*. University of Notre Dame Press, 2008.

p. 225 **never shown the instruments of torture.**: The "torture debate" is one of the great points of Galileo historiography. Current scholarship seems to have resolved that Galileo gave testimony under the verbal threat of torture, but was neither tortured nor shown the instruments of torture. I agree. It seems clear to me, however, that the use of torture would have been unusual but legally justified, which is an issue still under dispute. See chapter 11 of Finocchiaro's lovely *Retrying Galileo*; also p. 11; also Santillana, *Crime*, p. 297.

p. 225 **would never renounce Catholicism.**: *Opere*, v. 19, p. 411.

p. 226 **"bird of prey."**: Galilei, *Father*, p. 253.

p. 226 **convent in a fit.**: Galilei, *Father*, p. 267. These two "students" were Niccolo Aggiunti and Geri Bocchineri.

p. 227 **"for the next three years."**: Galilei, *Affair*, p. 291.

p. 227 **"and lasts forever."**: Opere, v. 16, Nr. 2970, p. 116; Santillana, *Crime*, p. 223; Favaro, *Maria*, p. 203. Abridged.

p. 228 **"I would lost the pleasure."**: An animated retelling from Drake, Stillman. "Galileo Gleanings I: Some Unpublished Anecdotes of Galileo." *Isis* 48.4 (1957): 393. This anecdote also begins Drake's *Work*. I very much agree with Drake that this story is entirely true.

p. 228 **all new material whatsoever.**: *Opere*, v. 16, Nr. 3075, p. p. 209.

p. 228 **"popular"**: *Opere*, v. 17, Nr. 3780, p. 370. Galileo would have preferred the title *Discourse on Motion*.

p. 228 **"Archimedes of our age" . . . "read with admiration" . . . "illuminated this work."**: Galilei, *Two*, pp. 148, 41, 126, respectively. Referring to Luca Valerio, Buonaventura Cavalieri, and Bishop di Guevara, respectively. The book by Cavalieri which is "read with admiration" is the same that Galileo had been insulted by right before his trial.

p. 229 **"will be rigidly maintained."**: Galilei, *Two*, p. 215. Most historians of Galileo, most notably Alexandre Koyré in his *Galileo Studies*, but also Drake, do not credit Galileo with the discovery of inertia. I am more lax in my constraints, and would give him the title for comments such as this.

p. 230 **In another demonstration**: The basis of this demonstration is a famous paradox from Aristotle's *Mechanics* known as "Aristotle's Wheel." Galileo discusses it at great length.

p. 231 **right eye had ceased all functionality.**: *Opere*, v. 17, Nr. 3513, p. 126.

p. 231 **"my own person."**: *Opere*, v. 17, Nr. 3635, p. 247. Slightly abridged, freely translated.

p. 231 **His son Vincenzo.**: Vincenzo is the subject of book 13 in Favaro's *Amici*.

p. 231 **"Most Reverend Father."**: *Opere*, v. 18, Nr. 3992, p. 179. Freely translated.

p. 232 **"novelties things in the heavens."**: *Opere*, v. 16, Nr. 3259, p. 391. Sharply abridged, slightly edited; I'm just aiming to capture the bitterness here. A proper translation is in Galilei, *Sunspots*, p. 330.

p. 232 **a month's salary in the black market**: Heilbron, *Galileo*, p. 345. It sold for six scudi; *Two New Sciences*, by comparison, sold for two, and did not sell as well; see Raphael, Renée, and Renée Jennifer Raphael. *Reading Galileo: Scribal Technologies and the Two New Sciences*. JHU Press, 2017. p. 15. *Two New Sciences* was wisely ignored by the Church and as a result had mediocre sales. It was a slow burn, however, and seems to me correctly regarded since the eighteenth century as the most impactful book of Galileo's scientific career. Notably, most copies wound up in England, the obvious heir to Galilean science.

p. 232 **be called industry**: This term is really not yet appropriate, but I am trying to allude what is to come, and why more than any other reason I find these subjects of continuing social relevance. I am steadfast in my conviction that, if not for the aftereffects of the industrial revolution, the public would (rightly) not care about science at all.

p. 232 **bust of his ambivalent mug.**: An obvious reference to Vincenzo Viviani, who asked to serve under Galileo at the age of seventeen in 1639. He revered Galileo as a god for the rest of his life, writing a ludicrous hagiography, erecting a statue of Galileo's face above the door to his house, and compiling the first collected edition of Galileo's works. Though Viviani was the most egregious, other students came to Galileo, most notably Evangelista Torricelli.

p. 232 **John Milton**: This meeting has acquired a greater body of scholarly discussion than the entirety of *Two New Sciences*. Its veracity is based on a glib but unobscure comment Milton made in his *Areopatigica*, which appears plausible but has been called into question. *Paradise Lost* is littered with allusions to Galileo's discoveries, but the only direct references to Galileo I found as reading through are in bk. 1 ln. 287, bk. 3 ln. 589, and bk. 5 ln. 262.

I admit to having been extremely confused by the nature of Milton's allusions to Galileo, which seemed totally arbitrary until I read the illuminating paper by Neil Harris, "Galileo as Symbol: The Tuscan Artist in Paradise Lost." *Annali dell'Istituto e Museo di storia della scienza di Firenze* 10.2 (1985): 3–29. I have followed Harris's interpretation, which is obviously derived from William Empson's *Seven Types of Ambiguity* and *Milton's God*.

p. 233 **"the Tuscan artist views."**: *Paradise Lost*, bk. 1, ln. 288.

p. 233 **"regions in the moon."**: *Paradise Lost*, bk. 5, ln. 263.

p. 233 **Thomas Hobbes.**: Drake, *Work*, p. 370; Jesseph, Douglas M. "Galileo, Hobbes, and the Book of Nature 1." *Perspectives on science* 12.2 (2004): p. 196. This visit is much more questionable than Milton's and has attracted less attention, although Hobbes was far more influenced by Galileo and seems to me rather a Protestant Galileo incarnate. Also useful has been Martinich, Aloysius P. *The two gods of Leviathan: Thomas Hobbes on religion and politics*. Cambridge University Press, 2003.

p. 233 **"restraint over themselves."**: Hobbes, *Leviathan*, Pt. 2, chpt. 17, sect. 1.

p. 234 **"their lawful sovereigns."**: Hobbes, *Leviathan*, Pt. 3, chpt. 33, sect. 1. Deemphasized.

p. 234 **"language of truth and reason"**: This is from Voltaire. Finocchiaro, *Retrying Galileo*, p. 117.

p. 234 **"cheat them with skill"**: Finocchiaro, *Retrying Galileo*, pp. 168, 170. Slight changes.

p. 234 **the pacifist whose theoretical contributions allowed for the nuclear bomb**: Surely a mean-spirited way to refer to Albert Einstein, but not a libelous one. It is like handing a gun to children. From "On the method of theoretical physics." *Philosophy of science* 1.2 (1934): p. 164.

p. 234 **"Hell is other people."**: This is from Jean-Paul Sartre's *No Exit*. I very much view, but have not stressed, Galileo's story as the first properly existential drama available in history (so, I think did Bertolt Brecht in his play *Galileo*). This line correctly read is absolutely a poke at the entire field of Galileo studies.

p. 234 **"I have ended up in Hell on Earth."**: *Opere*, v. 18, Nr. 3972, p. 154; Heilbron, Galileo, p. 356.

p. 234 **he soiled himself.**: A malicious reading of *Opere*, v. 17, Nr. 3780, p. 370. I am trying to demonstrate abjection; the opening paragraph of my entire Galileo story was a misdirect, at least in terms of the body.

p. 234 **"calling me, calling me . . ."**: *Opere*, v. 16, Nr. 2927, pp. 84–5. I apologize for breaking chronology so much here, but my primary concern in this section is to establish tragic effect.

p. 235 **standards, and discipline.**: Another reference to the industrial revolution through E.P. Thompson's important concept of time-discipline.

p. 235 **field, waiting for it to begin.**: I have in mind the classic line by Rumi: "Beyond rightdoing and wrongdoing there is a field. I will meet you there." I am not a spiritualist like Rumi, though.

p. 235 **"lapse of so many years."**: *Opere*, v. 8, p. 349. Translated in full in Drake, *Work*, p. 422 on.

p. 237 **"I am ready to depart."**: This is Walter Savage Landor's famous "Dying Speech of an Old Philosopher." Goodbye!

p. 239 **according to the opening of the *Optics*.**: Kepler, *Optics*, p. 14.

p. 239 **made clear that he sided entirely with Copernican cosmology.**: See, for example, Kepler, *Optics*, pp. 336, 342.

p. 239 **more or less uncensored.**: Cf. Voelkel, James R. "Publish or perish: Legal contingencies and the publication of Kepler's Astronomia *Nova*." *Science in context* 12.01 (1999): 33–59. Voelkel spins a very different narrative concerning the legal embattlement surrounding *New Astronomy*, something I have largely hashed over. They would place censorship as paramount in its making. I value Voelkel's work very highly, but believe they push their perspective too far. The political is not the starting point from which *New Astronomy* departs; it is only another actor of the many involved in its creation.

p. 239 **"absolutely, perfectly, geometrically equivalent."**: Kepler, *Nova*, p. 106.

p. 240 **up and down the street eight times.**: Immanuel Kant is, of course, not the subject of this book. This is possibly, but not at all necessarily, untrue; see Stuckenberg, *John Henry Wilbrandt. The Life of Immanuel Kant* (Stuckenburg reports "half past three," which seems to be incorrect). Macmillan and Company, 1882, p. 163. For a picture of his house, see the centerfold between pp. 234–5 in Kuehn, Manfred. *Kant: A biography*. Cambridge University Press, 2001.

p. 241 **150 years after Johannes Kepler.**: This entire passage was written before I encountered any written mention relating Kepler to Kant (although many vice versa, as Kant mentioned Kepler by name several times in his writing). I was, therefore, extremely pleased to see my intuition repeated in E.J. Aiton's article, "Johannes Kepler in Light of Recent Research," p. 83. It lies at the very heart of the earliest Kepler scholarship, which was philosophical in nature, see Ernst Cassirer, *The Individual and the Cosmos* *in Renaissance Philosophy*, Dover, p. 165. The nature of Kant's 'Copernican revolution' metaphor is very strange; he really did quite the opposite of Copernicus.

p. 241 **assumptions in astronomy may yield true conclusions.**: Kepler, *Nova*, p. 217.

p. 241 **first inequality . . . second inequality**: To be a bit more didactic, the first and second inequalities (sometimes called anomalies) simply refer to deviations from uniform motion around the ecliptic. They are distinguished by their periodicity. Generally speaking, the first inequality has a period the length of each trip around the ecliptic, and the second inequality has a period respective to the time for each syzygy. However, this need not necessarily be the case; these expressions are purely mathematical.

p. 242 **attempted to explain the second inequality**: Copernicus was not successfully in this, and wisely acknowledged as much, see Swerdlow, *Mathematical*, pp. 157–61.

p. 242 **"the planet is truly stripped of its second inequality."**: Kepler, *Nova*, p. 110.

p. 242 **of this type, but they were rare.**: Particularly in lunar theory, which was far more accurate at opposition than quadrature. Copernicus had a complicated schema for reconciling true and mean opposition; Swerdlow, *Mathematical*, p. 39, 194, 274–5.

p. 243 **has Earth draw closer to Mars than the Sun.**: To be a little more specific, this sort of recalculation would lead Kepler to be among the first to attempt to correct the wrong solar parallax measurements which had plagued Brahe earlier when testing for a parallax in Mars. See Galilei, *Comets*, p. 371, ft. 3, or Kepler, *Nova*, pp. 149–50.

p. 243 **"'prelude'—Tycho's favorite word."**: Kepler, *Nova*, p. 99, edited to fit my translations. Kepler switches to Greek to write "progymnasmata." This calculation directly relates to the one he did under the section "Two Families"; I do not think the word choice was a coincidence.

p. 243 **"occur where Mars is the greatest distance from the Sun"**: Kepler, *Nova*, p. 189. Replacing "the star" with "Mars" for clarity.

p. 243 **the shape of astronomy to come**: I follow Koyré's thoughts on the matter of Kepler's equant, *Astronomical Revolution*, p. 185.

p. 244 **"algebra too forsakes us here!"**: Kepler, *Nova*, p. 187.

p. 244 **"I am insufficiently skilled."**: This was sent to Herwart von Hohenburg, who was meant to send it to François Viète. Koyré, *Astronomical Revolution*, p. 399.

p. 244 **"weaken the sinews of the problem."**: Kepler, *Nova*, p. 186.

p. 245 **"fifth year has now gone by since I took up Mars."**: Kepler, *Nova*, p. 190.

p. 245 **"agreement with the observations during sunset, is nonetheless false . . ."**: Kepler, *Nova*, p. 208. The technical term for sunset is "acronychal."

p. 246 **an accuracy unheard of**: After Y. Maeyama, *Kepler's Hypothesis Vicaria*, *Archive for history of exact sciences* 41.1 (1990), p. 63.

p. 246 **"have become the material for a great part of the present work."**: Kepler, *Nova*, p. 211.

p. 246 **"the very nature of things."**: Kepler, *Nova*, p. 274.

p. 246 **"the more convenient it was for metaphysics in particular."**: Kepler, *Nova*, p. 309

p. 246 **"An oval is thus substituted for a circular path."**: Kepler, *Nova*, p. 269

p. 247 **round noncircular shapes have no geometrical definition.**: Ellipses were well-known from Apollonius's *Conics*. There can be many Euclidean ovals, but they are of increasing complexity and bare no obvious relation to one another. See *Mathographics*, Robert A. Dixon, pp. 5–10 for some enjoyable visual examples. The golden egg is especially beautiful.

p. 247 **bit by bit, in awkward steps.**: I have skipped over why this was so hard for Kepler, which is a valid question, given that his oval is plausibly generated by standard Ptolemaic mechanisms. But his planets do not travel at Ptolemy's traditional uniform speed (with respect to the equant), as far as I can tell, so the answer had to do with Kepler's physical hypothesis; see Koyré, *Astronomical Revolution*, pg 229. I have skipped over this in my efforts to summarize the maze that is *New Astronomy* into a linear narrative, whereby I sketch Kepler's physical hypothesis at the end.

p. 247 **continued to draw the problem as a circle.**: Kepler does this because he is here considering the distance from the Earth and Sun, which approaches a circle very closely. But the fact that he continues to model Mars's orbit with a circle afterward really shows the pull circularity had upon the medieval mind.

p. 247 **"This procedure is mechanical and tedious."**: Kepler, *Nova,* p. 309.

p. 247 **"how great the time increment is for any one of them."**: Kepler, *Nova,* p. 309.

p. 247 **between the infinite and finite there is no proportion.**: I do not know the origin of this phrase. It is a famous part of Pascal's wager. I first saw it in Koyré, *From the Closed World to the Infinite Universe*, p. 34. It appears in the works of Albertus Magnus and, in a changed form, the works of Thomas Aquinas.

p. 248 **called the "optical equation," because**: I assume this is the reason. I don't believe Kepler ever actually explains the names, but merely says he has "grown accustomed to calling them" so. At exactly this point, in the context of Kepler's area law, the "triangle of the optical equation" becomes synonymous with "the physical equatio." I have saved the discussion of Kepler's physics, however, for the last vignette.

p. 248 **"whole sine."**: The whole sine referred, to be exact, to the sine of the right angle about the center, that is, simply the radius, but in the illustration I have let it refer to the triangle itself for the sake of brevity. While this abuse of terminology may enrage some historians of mathematics, I feel confident it will make things clearer to the average reader.

p. 249 **first ever graph of a function.**: Kepler called it a *conchoid.* Kepler also added a slight decrease at the base of the triangles (Kepler, *Nova,* p. 314) though he does not reference it, and I think it just represents the same decrease taken from the "opposite side" of the function, the sine wave, rather than the bottom axis. I have followed Stephenson (Kepler's *Physical Astronomy*, p. 82) in removing it. Kepler said that "as if by miracle" this method worked because the error canceled out the other error caused by using a circle rather than an oval.

p. 249 **"thought it was "not important at this time."**: *KGW,* v. 3, p. 267; Kepler, *Nova*, p. 313. Kepler suggested from this that the distance and area law were not the same, see Stephenson, Physical, p. 83, or *KGW,* v. 15, Nr. 281.

p. 250 **"find some plane figure equal to the sum of distances."**: Kepler, *Nova,* p. 314.

p. 250 **"'bears blind pups' happened to me."**: *KGW,* v. 3, p. 288, Kepler, *Nova,* p. 339. As much as I can tell, this is not a proverb. Kepler made it up!

p. 250 **"ignorant of his own riches."**: Kepler, *Nova,* p. 172.

p. 250 **"did not know how to make proper use of his own riches."**: Caspar, *Kepler,* p. 87. Tense edited.

p. 250 **He had mixed distances with areas**: Davis, E. L. "Kepler's "Distance Law"-Myth, not Reality." *Centaurus* 35.2 (1992): 103–20, contests that this was ever the case. This is not a popular opinion. While I certainly do not know as much about *New Astronomy* as Davis, I am afraid that I must, nonetheless, disagree.

p. 250 **"I entered into new labyrinths."**: *KGW,* v. 3, pg 288, Kepler, *Nova,* p. 339. As I feel the remaining sections are quite important, I have begun to check Donahue's translation against Kepler's Latin. The translation is tremendously fine. I have removed chapter names and made small amendments (as I do) but stuck with the translation throughout.

p. 250 **vicarious hypothesis, modified into an oval.**: At this point, Kepler had moved from oval paths to oval first inequalities, abandoning spherical orbits. I combine this with his discovery of the ellipse. As in general discussions of books, I must begin to dismiss so many specifics as to be laughable. Kepler finds this new oval by modulating between the vicarious hypothesis and a version which shifts the mean anomaly to the point of bisection between the equant and Earth/Sun. It is in Kepler, *Nova,* pp. 346–7, and Whiteside offers a good rundown on p. 9 of his report.

p. 250 **"where is the geometer who will show us how to do this?"**: Kepler, *Nova,* p. 353

pp. 250–1 **"as though Mars's path was a perfect ellipse . . ."**: From Whiteside, Derek T. "Keplerian planetary eggs, laid and unlaid, 1600–1605." *Journal for the History of Astronomy* 5.1 (1974): 11. Every one of Kepler's essential devices in *New Astronomy* seems to have exactly one wonderful mathematical article on it. Whiteside's is exemplary.

p. 251 **"Let it be a perfect ellipse"**: Kepler, *Nova*, p. 354. "It" is the symbols representing the orbit, which I have edited out. It has been commonly said that Kepler discovered the ellipse by way of a triangulation procedure, which only occurs in the later chapters of *New Astronomy*; I consider that triangulation so insignificant that my account does not mention it. Cf. Wilson, Curtis. "Kepler's derivation of the elliptical path." *Isis* 59.1 (1968): 4–25. A similar triangulation procedure was used at the beginning of Kepler's researches on Mars, but this only gave Kepler justification for his equant; see Wilson again or Voelkel, *Composition*, p. 106. I take it for granted that the final bridge, which Kepler crossed from oval to ellipse was, in essence, an aesthetical one.

p. 251 **"from the belly towards the two ends."**: Kepler, *Nova,* p. 338. Kepler had written this regarding his discovery of a non-circular first inequality, but I have merged this fact with his discovery of the elliptical first inequality a few chapters later, because the distinction is simply too subtle for this brief summary.

p. 252 **"And I began to reason . . ."**: Kepler, *Nova*, p. 407. The "greatest optical equation" is, I believe, formed by the point on the orbit whose perpendicular to the apsidal line goes through the Sun. On Kepler's model (and any model with a low eccentricity) it will form a triangle close to that of the whole sine, but if it is equal to the whole sine, the orbit is a perfect circle.

p. 252 **it were a true *ovum*, a fat egg.**: Why did he do this? Koestler engages in his most obscene speculation of *The Sleepwalkers* when he postulates "some unconscious biological bias," (p. 330), which is literally incoherent. Whiteside merely supposes it "seemed most natural" for Kepler to angle the second anomaly so as to produce an egg, rather than an ellipse. Stephenson, in *Physical*, p. 123, suggests why this was Kepler's most natural next move mathematically—I am not wholly convinced, and still think that it was a fundamentally unnecessary step. Regardless, my form of storytelling is nothing without my character's words, and so I simply follow Kepler's own telling of his mistake, which was lively and self-critical.

p. 253 **A picture of the Sun.**: This account comes from Donahue's translation of Kepler's *Optics*, especially p. 423, 347–9 and Donahue's introduction to that work. I found the account somewhat confusing (as did Donahue Kepler, apparently) but hope I have understood it.

p. 253 **"in that I use my own observations."**: Kepler, *Nova,* p. 153. Clown show, literally, "spectaculum ridiculum," *KGW*, v. 3, p. 124.

p. 253 **"He also bestowed a moving soul."**: Kepler, *Nova,* p. 83.

p. 253 **"walking sticks for finding their appointed road."**: Kepler, *Nova,* p. 84. Kepler professes a multitude of absurdities, which I have largely hashed over.

p. 253 **Kepler's strange elliptical orbit**: Very easily, I believe, with a reasonable approximation from a single epicycle and a solar equant. Whitehead has shown in "Planetary Eggs" that any number of approximating ovals would have been accepting of the data, had Kepler found them. But I suspect he would have rejected them anyway; they were not elegant.

p. 254 **optical equation was the "physical equation."**: The angle formed from the center of the orbit to the end of the arc to the equant. Named as much, I assume, because the work initially began by assuming it was equant distance which provided a physical explanation to planetary movement. See Koyré, *The Astronomical Revolution*, Appendix I, p. 365. Kepler's replacement of the physical equation of the equant with his area law (Kepler, *Nova*, p. 311) is the first statement of his "second law" (the physical equation then became the triangle of the optical equation).

p. 254 **Kepler invented astrophysics.**: a claim I assumed early, but did not see stated so often. It appears in Gingerich, *The Eye of Heaven*, p. 321. In any case, Kepler surely invented an astrophysic.

p. 254 **foison Sun**: Ezra Pound, *Villonaud for This Yule*: "The ghosts of dead loves everyone / That make the stark winds reek with fear / Lest love return with the foison sun / And slay the memories that me cheer / (Such as I drink to mine fashion) / Wineing the ghosts of yester–year. / Where are the joys my heart had won? / (Saturn and Mars to Zeus drawn near!)"

p. 254 **"other reflections will soon be provided."**: Kepler, *Nova*, p. 285.

p. 254 **"I think I would have deserved an equal hearing."**: Kepler, *Nova*, p. 280. "Machine" has been changed to "world–machine" for clarity.

p. 255 **"since it also makes its nest in the Sun."**: Kepler, *Nova*, p. 283.

p. 255 **"the motive power from the Sun coincides with light."**: Kepler, *Nova*, p. 281.

p. 255 **"William Gilbert has proved it."**: Kepler, *Nova*, p. 412.

p. 255 **"seeking the Sun or fleeing it."**: Kepler, *Nova*, p. 418.

p. 255 **"concerning its details, however, I have doubts."**: Kepler, *Nova*, p. 418.

p. 256 **"to work through such questions thoroughly."**: Kepler, *Nova*, p. 480. *KGW*, v. 3, p. 408. It is interesting to note that this rather anticlimactic ending is also how Ptolemy concluded the *Almagest*.

Index